T0419648

MATHEMATICS RESEARCH DEVELOPMENTS

STATISTICS

VOLUME 2

MULTIPLE VARIABLE ANALYSIS

MATHEMATICS RESEARCH DEVELOPMENTS

Additional books and e-books in this series can be found
on Nova's website under the Series tab.

MATHEMATICS RESEARCH DEVELOPMENTS

STATISTICS

VOLUME 2

MULTIPLE VARIABLE ANALYSIS

KUNIHIRO SUZUKI

Library of Congress Cataloging-in-Publication Data

ISBN: 978-1-53615-122-0

Published by Nova Science Publishers, Inc. † New York

CONTENTS

Preface		**vii**
Chapter 1	A Correlation Factor	**1**
Chapter 2	A Probability Distribution for a Correlation Factor	**17**
Chapter 3	Regression Analysis	**69**
Chapter 4	Multiple Regression Analysis	**117**
Chapter 5	Structural Equation Modeling	**169**
Chapter 6	Fundamentals of a Bayes' Theorem	**179**
Chapter 7	A Bayes' Theorem for Predicting Population Parameters	**223**
Chapter 8	Analysis of Variances	**277**
Chapter 9	Analysis of Variances Using an Orthogonal Table	**305**
Chapter 10	Principal Component Analysis	**339**
Chapter 11	Factor Analysis	**361**
Chapter 12	Cluster Analysis	**431**
Chapter 13	Discriminant Analysis	**453**
References		**461**
About the Author		**463**
Index		**465**
Nova Related Publications		**469**

PREFACE

We utilize statistics when we evaluate TV program rating, predict a result of voting, prepare stock, predict the amount of sales, and evaluate the effectiveness of medical treatments. We want to predict the results not on the base of personal experience or images, but on the base of the corresponding data. The accuracy of the prediction depends on the data and related theories. It is easy to show input and output data associated with a model without understanding it. However, the models themselves are not perfect, because they contain assumptions and approximations in general. Therefore, the application of the model to the data should be careful. We should know what model we should apply to the data, what are assumed in the model, and what we can state based on the results of the models.

Let us consider a coin toss, for example. When we perform a coin toss, we obtain a head or a tail. If we try the coin toss three times, we may obtain the results of two heads and one tail. Therefore, the probability that we obtain for heads is 2/3, and the one that we obtain for tails is 1/3. This is a fact and we need not to discuss this any further. It is important to notice that the probability (2/3) of getting a head is limited to this trial. Therefore, we can never say that the probability that we obtain for heads with this coin is 2/3, in which we state general characteristics of the coin. If we perform the coin toss trial 400 times and obtain heads 300 times, we may be able to state that the probability of obtaining a head is 2/3 as the characteristics of the coin. What we can state based on the obtained data depends on the sample number. Statistics gives us a clear guideline under which we can state something is based on the data with corresponding error ranges.

Mathematics used in statistics is not so easy. It may be a tough work to acquire the related techniques. Fortunately, the software development makes it easy to obtain results. Therefore, many members who are not specialists in mathematics can do statistical analysis with these softwares. However, it is important to understand the meaning of the model, that is, why some certain variables are introduced and what they express, and what we can state based on the results. Therefore, understanding mathematics related to the models is invoked to appreciate the results.

In this book, we treat models from fundamental ones to advanced ones without skipping their derivation processes as possible as I can. We can then clearly understand the assumptions and approximations used in the models, and hence understand the limitation of the models.

We also cover almost all the subjects in statistics since they are all related to each other, and the mathematical treatments used in a model are frequently used in the other ones.

We have many good practical and theoretical books on statistics [1]-[10]. However, these books are oriented to special direction: fundamental, mathematical, or special subjects. I want to add one more, which treats fundamental and advanced models from the beginnings to the advanced ones with a self contained style. I also aim to connect theories to practical subjects.

This book consists of three volumes:

- The first volume treats the fundamentals of statistics.
- The second volume treats multiple variable analysis.
- The third volume treats categorical and time dependent data analysis

This volume 2 treats multiple variable analysis.

We frequently treat many variables in the real world. We need to study relationship between these many variables.

First of all, we treat correlation and regression for two variables, and it is extended to multiple variables.

We also treat various techniques to treat many variables in this volume.

The readers can understand various treatments of multiple variables.

We treat the following subjects.

(Chapter 1 to 4)

Up to here, we treat statistics of one variable. We then study correlation and regression for two variables. The probability function for the correlation function is also discussed, and it is extended to multiple regression analysis.

(Chapter 5)

We treat data which consist of only explanatory variables. We assume an objective variable based on the explanatory variables. The analysist can assume the image variables freely, and he can discuss the introduced variables quantitatively based on the data. This process is possible with the structural equation modeling. We briefly explain the outline of the model.

(Chapter 6 and 7)

We treat a Bayes' theorem which is the extension of conditional probability, and is vital to analyze problems in real world. The Bayes' theorem enables us to specify the probability of the cause of the event, and we can increase the accuracy of the probability by updating it using additional data. It also predict the two subject which are parts of vast network. Further, we can predict population prametervalues or the related distribution of a average, a variance, and a ratio using obtained data. We further treat vey complex data using a hierarchical Bayesian theory.

(Chapter 8 and 9)

We treat a variance analysis and also treat it using an orthogonal table which enables us to evaluate the relationship between one variable and many parameter variables. Performing the analysis, we can evaluate the effectiveness of treatment considering the data scattering.

(Chapter 10 to 13)

We then treat standard multi variable analysis of principal component analysis, factor analysis, clustering analysis, and discriminant analysis.

In principal component analysis, we have no objective variable, and we set one kind of an objective variable that clarify the difference between each members. The other important role of the analysis is that it can be used to lap the variables which have high correlation factors. A high correlation factor induces unstable results in multi variable analysis, which can be overcome by the proposed method associated with the principal component analysis.

In factor analysis, we also have no objective variable. We generate variables that explain the obtained results.

Cluster analysis gives us to make groups that resemble each other.

In discriminal analysis, we have two groups. We decide that a new sample belongs to which group.

We do hope that the readers can understand the meaning of the models in statistics and techniques to reach the final results. I think it is not easy to do so. However, I also believe that this book helps one to accomplish it with time and efforts.

I tried to derive any model from the beginning for all subjects although many of them are not complete. It would be very appreciated if you point out any comments and suggestions to my analysis.

Kunihiro Suzuki

Chapter 1

A CORRELATION FACTOR

ABSTRACT

Analysis of a relationship between two variables is an important subject in statistics. The relationship is expressed with a covariance, and it is extended to a non-scale parameter of correlation factor. The correlation factor has a value range between -1 and 1, and it expresses the strength of the relationship between two variables. The ranking data have a special data form, and the corresponding correlation factor has its special form and is called as a Spearman's correlation factor. We also discuss a pseudo correlation factor where the other variable makes as if two variables have a relationship each other.

Keywords: covariance, correlation factor, ranking data, Spearman's correlation factor, pseudo correlation

1. INTRODUCTION

We guess that smoking influences the lifespans based on our feeling. We want to clarify the relationship quantitatively. We obtain a data set of smoking or no-smoking and their lifespans. Using the data, we should obtain a parameter that expresses the significance of a relationship between smoking and the lifespan, and want to decide whether the two items have a relationship or not. It is important to study the relationship between the two variables quantitatively. We focus on the two quantitative data and study to quantify the relationship in this chapter.

2. A COVARIANCE

We consider the case where we obtain various kinds of data from one sample. Table 1 shows concrete data of scores of mathematics, physics, and language for 20 members.

Table 1. Data for scores of mathematics, physics, and language

Member ID	Mathematics	Physics	Language
1	52	61	100
2	63	52	66
3	34	22	64
4	43	21	98
5	56	60	86
6	67	70	35
7	35	29	43
8	53	50	47
9	46	39	65
10	79	78	43
11	75	70	72
12	65	59	55
13	51	56	75
14	72	79	82
15	49	48	81
16	69	73	99
17	55	57	57
18	61	68	81
19	41	42	75
20	34	36	63

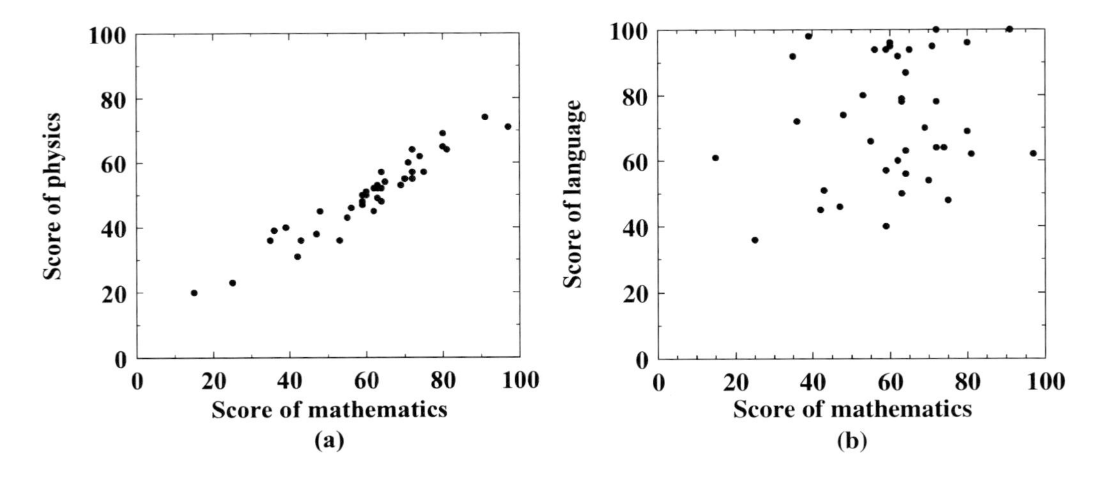

Figure 1. Scatter plots. (a) Sore of mathematics vs score of physics. (b) Score of mathematics vs score of language.

We can see the relationship between scores of mathematics and physics or between scores of mathematics and language by obtaining scatter plots as shown in Figure 1 (a) and (b).

We can qualitatively say that the member who gets a high score of mathematics is apt to get a high score of physics, and the member who gets a low score of mathematics is apt to get a low score of physics. We can also say that the score of mathematics and that of language have no clear relationship.

We want to express the above discussions quantitatively.

We consider the relationship between scores of mathematics and physics, or between scores of mathematics and language.

The average of mathematics is given by

$$\mu_x = E[X] = \frac{\sum_{i=1}^{N} x_i}{N} \tag{1}$$

where we express the score of mathematics for each member as x_i, and N is the number of members, and N is 20 in this case.

The average of physics or language is given by

$$\mu_y = E[Y] = \frac{\sum_{i=1}^{N} y_i}{N} \tag{2}$$

where we express the score of them for each member as y_i.

The variance of mathematics is given by

$$\sigma_{xx}^{(2)} = E\left[(X - \mu_x)^2\right] = \frac{\sum_{i=1}^{N} (x - \mu_x)^2}{N} \tag{3}$$

The variance of physics or language is given by

$$\sigma_{yy}^{(2)} = E\left[(Y - \mu_y)^2\right] = \frac{\sum_{i=1}^{N} (y - \mu_y)^2}{N} \tag{4}$$

Note that we express the variances as $\sigma_{xx}{}^{(2)}$ and $\sigma_{yy}{}^{(2)}$ instead of $\sigma_x{}^{(2)}$ and $\sigma_y{}^{(2)}$ to extend them to a covariance which is defined soon after.

In the two variables, we newly introduce a covariance, which is given by

$$\begin{aligned}\sigma_{xy}{}^{(2)} &= E\left[\left(X-\mu_x\right)\left(Y-\mu_y\right)\right] \\ &= \frac{\sum_{i=1}^{N}\left(x-\mu_x\right)\left(y-\mu_y\right)}{N}\end{aligned} \tag{5}$$

In the above discussion, we implicitly assume that the set is a population. We assume that the sample size is n.

A sample average of a variable x is denoted as $\bar{x}$ and is given by

$$\bar{x} = \frac{\sum_{i=1}^{n} x_i}{n} \tag{6}$$

A sample average of y is given by

$$\bar{y} = \frac{\sum_{i=1}^{n} y_i}{n} \tag{7}$$

An unbiased variance of x is given by

$$s_{xx}{}^{(2)} = \frac{\sum_{i=1}^{n}\left(x-\bar{x}\right)^2}{n-1} \tag{8}$$

A sample variance of x is given by

$$S_{xx}{}^{(2)} = \frac{\sum_{i=1}^{n}\left(x-\bar{x}\right)^2}{n} \tag{9}$$

An unbiased variance of y is given by

$$s_{yy}^{(2)} = \frac{\sum_{i=1}^{n}(y-\bar{y})^2}{n-1} \tag{10}$$

A sample variance of y is given by

$$S_{yy}^{(2)} = \frac{\sum_{i=1}^{n}(y-\bar{y})^2}{n} \tag{11}$$

We can also evaluate the variances given by

$$s_{xy}^{(2)} = \frac{\sum_{i=1}^{n}(x-\bar{x})(y-\bar{y})}{n-1} \tag{12}$$

$$S_{xy}^{(2)} = \frac{\sum_{i=1}^{n}(x-\bar{x})(y-\bar{y})}{n} \tag{13}$$

We study characteristics of the covariance graphically as shown in Figure 2.

The covariance is the product of deviations of two variables with respect to each average. The deviation is positive in the first and third quadrant planes, and negative in the second and fourth quadrant planes. Therefore, if the data are only on the first and third quadrants, the product is always positive, and hence sum becomes large. If the data is scattered on all quadrant planes, the product sign is not constant, and some terms vanish each other, and hence the sum becomes small. The covariance between scores of mathematics and physics is 132.45, and that between scores of mathematics and language is 40.88. Therefore, the magnitude of the covariance expresses the strength of the relationship qualitatively.

Let us consider the 100 point examination, which corresponds to Table 1. We can convert these scores to 50 point one simply by divided by 2. The corresponding scatter plots are shown in Figure 3. The relationship must be the same as that of Figure 1(a). However, the corresponding covariance is 40.88, which is much smaller than that of 100 points. Since we only change the scale in both figures, we expect the same significance of the relationship. Therefore, we want to obtain a parameter that expresses the strength of relationship independent of the data scaling, which is not the covariance.

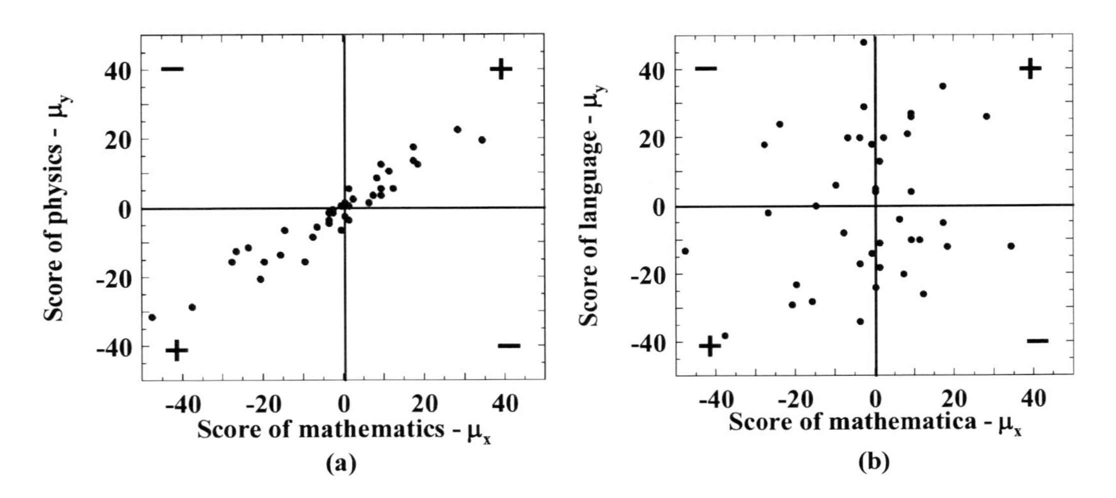

Figure 2. Scattered plots of deviations with respect to each variable's average. (a) Sore of mathematics vs score of physics. (b) Score of mathematics vs score of language.

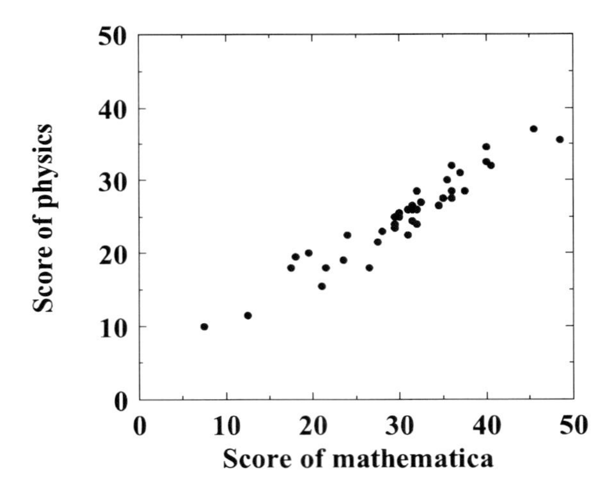

Figure 3. Scatter plots under 50 points.

3. A Correlation Factor

A correlation factor ρ_{xy} is introduced as

$$\rho_{xy} = \frac{\sigma_{xy}^{(2)}}{\sqrt{\sigma_{xx}^{(2)}}\sqrt{\sigma_{yy}^{(2)}}} \tag{14}$$

We can also define a sample correlation factor given by

$$r_{xy} = \frac{s_{xy}{}^{(2)}}{\sqrt{s_{xx}{}^{(2)}}\sqrt{s_{yy}{}^{(2)}}} = \frac{S_{xy}{}^{(2)}}{\sqrt{S_{xx}{}^{(2)}}\sqrt{S_{yy}{}^{(2)}}} \tag{15}$$

We obtain the same correlation factor of $\rho_{xy} = 0.94$ for Figure 1(a) and Figure 3 as is expected independent of the scale.

We evaluate a following expectation given by

$$\begin{aligned}
&E\left[\left\{\lambda\left(X-\mu_x\right)+\left(Y-\mu_y\right)\right\}^2\right] \\
&= E\left[\lambda^2\left(X-\mu_x\right)^2 + 2\lambda\left(X-\mu_x\right)\left(Y-\mu_y\right)+\left(Y-\mu_y\right)^2\right] \\
&= \lambda^2\sigma_{xx}^{(2)} + 2\lambda\sigma_{xy}^{(2)} + \sigma_{yy}^{(2)} \\
&= \sigma_{xx}^{(2)}\left(\lambda^2 + 2\frac{\sigma_{xy}^{(2)}}{\sigma_{xx}^{(2)}}\lambda\right) + \sigma_{yy}^{(2)} \\
&= \sigma_{xx}^{(2)}\left[\left(\lambda + \frac{\sigma_{xy}^{(2)}}{\sigma_{xx}^{(2)}}\right)^2 - \left(\frac{\sigma_{xy}^{(2)}}{\sigma_{xx}^{(2)}}\right)^2\right] + \sigma_{yy}^{(2)} \\
&= \sigma_{xx}^{(2)}\left(\lambda + \frac{\sigma_{xy}^{(2)}}{\sigma_{xx}^{(2)}}\right)^2 - \frac{\left(\sigma_{xy}^{(2)}\right)^2}{\sigma_{xx}^{(2)}} + \sigma_{yy}^{(2)} \geq 0
\end{aligned} \tag{16}$$

where λ is an arbitrary constant. Since the formula has a quadratic form, we insist on

$$-\frac{\left(\sigma_{xy}^{(2)}\right)^2}{\sigma_{xx}^{(2)}} + \sigma_{yy}^{(2)} \geq 0 \tag{17}$$

Therefore, we obtain

$$\frac{\left(\sigma_{xy}^{(2)}\right)^2}{\sigma_{xx}^{(2)}\sigma_{yy}^{(2)}} = \left[\frac{\sigma_{xy}^{(2)}}{\sqrt{\sigma_{xx}^{(2)}\sigma_{yy}^{(2)}}}\right]^2 = \rho_{xy}{}^2 \leq 1 \tag{18}$$

Finally, the range of the correlation factor is given by

$$-1 \le \rho_{xy} \le 1 \tag{19}$$

If there is no correlation, that is, two variables are independent of each other, ρ_{xy} is 0. If the correlation is significant, the absolute value of ρ_{xy} approaches to 1.

The above theory is for a population set. The sample correlation factor r_{xy} is simply given by

$$r_{xy} = \frac{s_{xy}^{(2)}}{\sqrt{s_{xx}^{(2)} s_{yy}^{(2)}}} \tag{20}$$

This is also expressed with sample variances given by

$$r_{xy} = \frac{s_{xy}^{(2)}}{\sqrt{s_{xx}^{(2)} s_{yy}^{(2)}}} = \frac{S_{xy}^{(2)}}{\sqrt{S_{xx}^{(2)} S_{yy}^{(2)}}} \tag{21}$$

4. A Rank Correlation

There is a special data form where we rank the team. If we evaluate m teams, the corresponding data is from 1 to m . We want to obtain the relationship between each member's evaluation. The theory is exactly the same for the correlation factor. However, the data form is special, and we can derive a special form of a correlation factor, which is called as a Spearman's correlation factor.

Let us derive the correlation factor.

Table 2 shows the ranking of each member to the football teams.

We want to know the resemblance of each member, which is evaluated by the correlation factor.

The correlation factor is simply evaluated by the procedure shown in this chapter, which is shown in Table 3. It is clear that member 1 resembles with member 3.

Table 2. Ranking of each member to teams

Member ID	Team					
	A	B	C	D	E	F
m1	2	3	5	1	6	4
m2	4	5	1	6	3	2
m3	1	3	4	2	6	5

Table 3. Correlation factor of each member

	m1	m2	m3
m1		-0.771429	0.8857143
m2			-0.6
m3			

We denote the data of the member 1 as $a_1, a_2, \cdots, a_n$, and the data of the member 2 as $b_1, b_2, \cdots, b_n$. The correlation factor is evaluated as

$$r = \frac{\sum_{i=1}^{n} (a_i - \bar{a})(b_i - \bar{b})}{\sqrt{\sum_{i=1}^{n} (a_i - \bar{a})^2 \sum_{i=1}^{n} (b_i - \bar{b})^2}} \tag{22}$$

The averages for a and b are the same and is decided by the team number, and is given by

$$\bar{a} = \bar{b} = \frac{n+1}{2} \tag{23}$$

A variance is also decided only by the team number, and is given by

$$\begin{aligned}
\sum_{i=1}^{n} (a_i - \bar{a})^2 &= \sum_{i=1}^{n} (b_i - \bar{b})^2 \\
&= \left(1 - \frac{n+1}{2}\right)^2 + \left(2 - \frac{n+1}{2}\right)^2 + \left(3 - \frac{n+1}{2}\right)^1 + \cdots + \left(n - \frac{n+1}{2}\right)^2 \\
&= \sum_{i=1}^{n} \left(i - \frac{n+1}{2}\right)^2 \\
&= \sum_{i=1}^{n} i^2 - (n+1) \sum_{i=1}^{n} i + n \left(\frac{n+1}{2}\right)^2 \\
&= \frac{(2n+1)n(n+1)}{6} - (n+1)\frac{n(n+1)}{2} + n\frac{(n+1)^2}{4} \\
&= \frac{n(n+1)(4n+2-6n-6+3n+3)}{12} \\
&= \frac{n(n+1)(n-1)}{12}
\end{aligned} \tag{24}$$

We then evaluate a covariance.
Further, we obtain

$$\begin{aligned}\left(a_i - b_i\right)^2 &= \left[\left(a_i - \bar{a}\right) - \left(b_i - \bar{b}\right)\right]^2 \\ &= \left(a_i - \bar{a}\right)^2 + \left(b_i - \bar{b}\right)^2 - 2\left(a_i - \bar{a}\right)\left(b_i - \bar{b}\right) \\ &= 2\left(a_i - \bar{a}\right)^2 - 2\left(a_i - \bar{a}\right)\left(b_i - \bar{b}\right)\end{aligned} \tag{25}$$

Therefore, we obtain

$$\left(a_i - \bar{a}\right)\left(b_i - \bar{b}\right) = \left(a_i - \bar{a}\right)^2 - \frac{1}{2}\left(a_i - b_i\right)^2 \tag{26}$$

Finally, we obtain the Spearman's correlation factor form as

$$\begin{aligned}r &= \frac{\sum_{i=1}^{n}\left(a_i - \bar{a}\right)^2 - \frac{1}{2}\sum_{i=1}^{n}\left(a_i - b_i\right)^2}{\sum_{i=1}^{n}\left(a_i - \bar{a}\right)^2} \\ &= 1 - \frac{6\sum_{i=1}^{n}\left(a_i - b_i\right)^2}{n\left(n-1\right)\left(n+1\right)}\end{aligned} \tag{27}$$

This gives the same results in Table 3.
The maximum value for correlation factor is 1, and hence we set

$$1 = 1 - \frac{6\sum_{i=1}^{n}\left(a_i - b_i\right)^2}{n\left(n-1\right)\left(n+1\right)} \tag{28}$$

This leads to

$$a_i = b_i \tag{29}$$

The minimum value of the correlation factor is -1, and hence we set

$$-1=1-\frac{6\sum_{i=1}^{n}(a_i-b_i)^2}{n(n-1)(n+1)} \tag{30}$$

This leads to

$$\sum_{i=1}^{n}(a_i-b_i)^2=\frac{1}{6}n(n-1)(n+1) \tag{31}$$

We can easily guess this is the case when one data is $1,2,\cdots,n$, and corresponding the other data is $n,n-1,\cdots,1$, respectively. The corresponding term is

$$\begin{aligned}
&\sum_{i=1}^{n}i(n-i+1)\\
&=n\sum_{i=1}^{n}i-\sum_{i=1}^{n}i^2+\sum_{i=1}^{n}i\\
&=n\frac{1}{2}n(n+1)-\frac{1}{6}n(n+1)(2n+1)+\frac{1}{2}n(n+1)\\
&=\frac{1}{2}n(n+1)\left[n-\frac{1}{3}(2n+1)+1\right]\\
&=\frac{1}{6}n(n+1)(n+2)
\end{aligned} \tag{32}$$

This agrees with Eq. (31)。

Random can be expressed by $r=0$, this expressed by

$$0=1-\frac{6\sum_{i=1}^{n}(a_i-b_i)^2}{n(n-1)(n+1)} \tag{33}$$

This leads to

$$\sum_{i=1}^{n}(a_i-b_i)^2=\frac{1}{6}n(n-1)(n+1) \tag{34}$$

5. A Pseudo Correlation and a Partial Correlation Factor

Even when the correlation factor r_{xy} for two variables of x and y is large, we cannot definitely say that there is a causal relationship.

Let us consider the case of the number of sea accident and amount of ice cream sales. The back ground of the relationship is the temperature. If the temperature increases, the number of people who go to the sea increases, and the number of the accident increases, and the amount of ice cream sales may increase. The relationship between the accident number and the ice cream sales is related to the same one parameter of the temperature.

The second example is the relationship between annual income and running time for 50 m. The income increases with age, and the time increase with the age. We then obtain the relationship between the income and the time thorough the parameter of the age.

The above feature can be evaluated as the followings.

When there is a third variable u, and x and y both have correlation relationships, we then obtain a strong correlation factor between x and y even if there is no correlation between them. The correlation under this condition is called as pseudo correlation. We want to obtain the correlation eliminating the influence of the variable u.

The regression relationship between x and u is given by

$$X_i = \bar{x} + \frac{\sqrt{S_{xx}^{(2)}}}{\sqrt{S_{uu}^{(2)}}} r_{xu} \left(u_i - \bar{u} \right) \tag{35}$$

Similarly, the regression relationship between y and u is given by

$$Y_i = \bar{y} + \frac{\sqrt{S_{yy}^{(2)}}}{\sqrt{S_{uu}^{(2)}}} r_{yu} \left(u_i - \bar{u} \right) \tag{36}$$

The residuals for x and y associated with the u are given by

$$\begin{aligned} e_{x_i} &= x_i - X_i \\ &= x_i - \bar{x} - \frac{\sqrt{S_{xx}^{(2)}}}{\sqrt{S_{uu}^{(2)}}} r_{xu} \left(u_i - \bar{u} \right) \end{aligned} \tag{37}$$

$$
\begin{aligned}
e_{y_i} &= y_i - Y_i \\
&= y_i - \bar{y} - \frac{\sqrt{S_{yy}^{(2)}}}{\sqrt{S_{uu}^{(2)}}} r_{yu} \left(u_i - \bar{u}\right)
\end{aligned}
\tag{38}
$$

These are the variables eliminating the influence of u. The correlation factor between these variables is the one eliminating the influence of u and is called as a causal correlation factor.

The causal correlation factor is denoted as $r_{xy,u}$, and is given by

$$
r_{xy,u} = \frac{S_{e_x e_y}^{(2)}}{\sqrt{S_{e_x e_x}^{(2)} S_{e_y e_y}^{(2)}}}
\tag{39}
$$

where

$$
\begin{aligned}
S_{e_{xi} e_{xi}}^{(2)} &= \frac{1}{n}\sum_{i=1}^{n}\left[x_i - \bar{x} - \frac{\sqrt{S_{xx}^{(2)}}}{\sqrt{S_{uu}^{(2)}}} r_{xu}\left(u_i - \bar{u}\right)\right]\left[x_i - \bar{x} - \frac{\sqrt{S_{xx}^{(2)}}}{\sqrt{S_{uu}^{(2)}}} r_{xu}\left(u_i - \bar{u}\right)\right] \\
&= \frac{1}{n}\sum_{i=1}^{n}\left(x_i - \bar{x}\right)^2 + \frac{1}{n}\frac{S_{xx}^{(2)}}{S_{uu}^{(2)}} r_{xu}^2 \sum_{i=1}^{n}\left(u_i - \bar{u}\right)^2 - 2\frac{\sqrt{S_{xx}^{(2)}}}{\sqrt{S_{uu}^{(2)}}} r_{xu} \frac{1}{n}\sum_{i=1}^{n}\left(x_i - \bar{x}\right)\left(u_i - \bar{u}\right) \\
&= S_{xx}^{(2)} + \frac{S_{xx}^{(2)}}{S_{uu}^{(2)}} r_{xz}^2 S_{uu}^{(2)} - 2S_{xx}^{(2)} r_{xu}^2 \\
&= S_{xx}^{(2)}\left(1 - r_{xz}^2\right)
\end{aligned}
\tag{40}
$$

Similarly, we obtain

$$
S_{e_{yi} e_{yi}}^{(2)} = S_{yy}^{(2)}\left(1 - r_{yu}^2\right)
\tag{41}
$$

On the other hand, we obtain

$$
\begin{aligned}
S^{(2)}_{e_{xi}e_{yi}} &= \frac{1}{n}\sum_{i=1}^{n}\left[x_i - \bar{x} - \frac{\sqrt{S^{(2)}_{xx}}}{\sqrt{S^{(2)}_{uu}}} r_{xu}\left(u_i - \bar{u}\right)\right]\left[y_i - \bar{y} - \frac{\sqrt{S^{(2)}_{yy}}}{\sqrt{S^{(2)}_{uu}}} r_{yu}\left(u_i - \bar{u}\right)\right] \\
&= \frac{1}{n}\sum_{i=1}^{n}\left(x_i - \bar{x}\right)\left(y_i - \bar{y}\right) - \frac{1}{n}\sum_{i=1}^{n}\left(x_i - \bar{x}\right)\frac{\sqrt{S^{(2)}_{yy}}}{\sqrt{S^{(2)}_{uu}}} r_{yu}\left(u_i - \bar{u}\right) \\
&\quad - \frac{1}{n}\sum_{i=1}^{n}\left(y_i - \bar{y}\right)\frac{\sqrt{S^{(2)}_{xx}}}{\sqrt{S^{(2)}_{uu}}} r_{xu}\left(x_i - \bar{x}\right) + \frac{1}{n}\sum_{i=1}^{n}\left[\frac{\sqrt{S^{(2)}_{xx}}}{\sqrt{S^{(2)}_{uu}}} r_{xu}\left(u_i - \bar{u}\right)\right]\left[\frac{\sqrt{S^{(2)}_{yy}}}{\sqrt{S^{(2)}_{uu}}} r_{yu}\left(u_i - \bar{u}\right)\right] \\
&= S^{(2)}_{xy} - r_{yz}\frac{\sqrt{S^{(2)}_{yy}}}{\sqrt{S^{(2)}_{zz}}} S^{(2)}_{xz} - r_{xz}\frac{\sqrt{S^{(2)}_{xx}}}{\sqrt{S^{(2)}_{zz}}} S^{(2)}_{yz} + r_{xz} r_{yz}\frac{S^{(2)}_{xx} S^{(2)}_{yy}}{S^{(2)}_{zz}} S^{(2)}_{zz} \\
&= S^{(2)}_{xy} - \sqrt{S^{(2)}_{yy}}\sqrt{S^{(2)}_{xx}} r_{xu} r_{yu} - \sqrt{S^{(2)}_{yy}}\sqrt{S^{(2)}_{xx}} r_{yu} r_{xu} + \sqrt{S^{(2)}_{xx}}\sqrt{S^{(2)}_{yy}} r_{xu} r_{yu} \\
&= S^{(2)}_{xy} - \sqrt{S^{(2)}_{yy}}\sqrt{S^{(2)}_{xx}} r_{xu} r_{yu}
\end{aligned}
\tag{42}
$$

Finally, we obtain

$$
\begin{aligned}
r_{xy,u} &= \frac{S^{(2)}_{xy} - \sqrt{S^{(2)}_{yy}}\sqrt{S^{(2)}_{xx}} r_{xu} r_{yu}}{\sqrt{\left(1 - r^2_{xu}\right) S^{(2)}_{xx}\left(1 - r^2_{yu}\right) S^{(2)}_{yy}}} \\
&= \frac{r_{xy} - r_{xu} r_{yu}}{\sqrt{\left(1 - r^2_{xu}\right)\left(1 - r^2_{yu}\right)}}
\end{aligned}
\tag{43}
$$

The other try is to take data considering the stratification. For example, if we take data considering the age, we take correlation selecting the limited age range with no relation between the income and running time.

The above means the followings.

If we have some doubt about the correlation and have some image of the third parameter that should be the cause of the correlation, we should take data for the limited range of the third parameter. This means that we should take data of two variables of x and y with the limited range of the third parameter of Δu where we can approximate that the value of u is constant. We can then evaluate the real relationship between x and y.

6. CORRELATION AND CAUSATION

We should be careful about that a high correlation factor does not directly mean causation. On the other hand, if two variables have strong causation relationship, we can always expect a high correlation factor. Therefore, a high correlation is a requirement, but is not a sufficient condition.

For example, there is implicitly the other variable, and two variables relate strongly to this implicit variable. We then observe a strong correlation. However, this strong correlation is related to the implicit variable.

The relationship between two variables is summarized in Figure 4. The key process is whether there is an implicit variable.

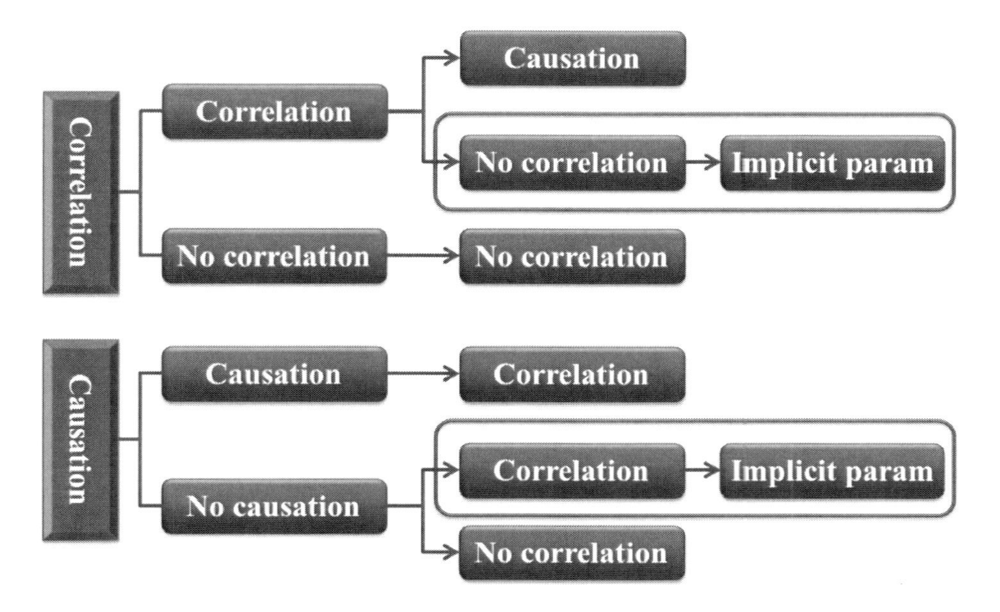

Figure 4. Summary of the relationship between correlation and causation.

SUMMARY

To summarize the results in this chapter, the covariance between variable x and y is given by

$$\sigma_{xy}{}^{(2)} = \frac{1}{N}\sum_{i=1}^{N}(x-\mu_x)(y-\mu_y)$$

This expresses the relationship between two variables. This can be extended to the correlation factor given by

$$\rho_{xy} = \frac{\sigma_{xy}{}^{(2)}}{\sqrt{\sigma_{xx}{}^{(2)}}\sqrt{\sigma_{yy}{}^{(2)}}}$$

where $\sigma_{xx}^{(2)}$ and $\sigma_{yy}^{(2)}$ are variances of variables x and y given by

$$\sigma_{xx}^{(2)} = \frac{1}{N}\sum_{i=1}^{N}\left(x-\mu_x\right)^2$$

$$\sigma_{yy}^{(2)} = \frac{1}{N}\sum_{i=1}^{N}\left(y-\mu_y\right)^2$$

The correlation factor is given by

$$-1 \le \rho_{xy} \le 1$$

When the data are order numbers, the related correlation factor is called as a Spearman's factor and is given by

$$\begin{aligned} r &= \frac{\sum_{i=1}^{n}\left(a_i-\bar{a}\right)^2 - \frac{1}{2}\sum_{i=1}^{n}\left(a_i-b_i\right)^2}{\sum_{i=1}^{n}\left(a_i-\bar{a}\right)^2} \\ &= 1-\frac{6\sum_{i=1}^{n}\left(a_i-b_i\right)^2}{n\left(n-1\right)\left(n+1\right)} \end{aligned}$$

If the correlation factor is influenced by the third variable u, we want to know the correlation factor eliminating the influence of u and is called as a causal correlation factor, and is given by.

$$r_{xy,u} = \frac{r_{xy}-r_{xu}r_{yu}}{\sqrt{\left(1-r_{xu}^2\right)\left(1-r_{yu}^2\right)}}$$

Chapter 2

A PROBABILITY DISTRIBUTION FOR A CORRELATION FACTOR

ABSTRACT

We derive a distribution function for the correlation factor, and the distribution function with a zero correlation factor is obtained as the special case of it. Using the distribution function, we evaluate the prediction range of the correlation factor, and the validity of the relationship.

Keywords: covariance, correlation factor, prediction, testing

1. INTRODUCTION

We can easily obtain a correlation factor with one set of data. However, it should have an error range, which we have no idea up to here. The accuracy of the correlation factor should depend on the sample number. We want to evaluate the error range for the correlation factor. We therefore study a probability distribution function for a correlation factor. We can then predict the error range of the correlation factor.

We perform matrix operation in the analysis here. The basics of matrix operation are described in Chapter 15 of volume 3.

2. PROBABILITY DISTRIBUTION FOR TWO VARIABLES

We consider a probability distribution for two variables which have relation to each other.

We first consider two independent normalized variables z_1 and z_2, which follow standard normal distributions. The corresponding joint distribution is given by

$$f(z_1, z_2)dz_1dz_2 = f(z_1)f(z_2)dz_1dz_2$$
$$= \frac{1}{\sqrt{2\pi}}e^{-\frac{z_1^2}{2}} \frac{1}{\sqrt{2\pi}}e^{-\frac{z_2^2}{2}} dz_1dz_2$$
$$= \frac{1}{2\pi}\exp\left[-\frac{1}{2}\left(z_1^2 + z_2^2\right)\right]dz_1dz_2 \tag{1}$$

We then consider two variables with a linear connection of z_1 and z_2, which is given by

$$\begin{cases} x_1 = az_1 + bz_2 \\ x_2 = cz_1 + dz_2 \end{cases} \tag{2}$$

where

$$ad - bc \neq 0 \tag{3}$$

Eq. (2) can be expressed with a matrix form as

$$\begin{pmatrix} x_1 \\ x_2 \end{pmatrix} = \begin{pmatrix} a & b \\ c & d \end{pmatrix}\begin{pmatrix} z_1 \\ z_2 \end{pmatrix} \tag{4}$$

We then obtain

$$\begin{pmatrix} z_1 \\ z_2 \end{pmatrix} = \begin{pmatrix} a & b \\ c & d \end{pmatrix}^{-1}\begin{pmatrix} x_1 \\ x_2 \end{pmatrix}$$
$$= D\begin{pmatrix} d & -b \\ -c & a \end{pmatrix}\begin{pmatrix} x_1 \\ x_2 \end{pmatrix} \tag{5}$$

where D is a determinant of the matrix, given by

$$D \equiv \frac{1}{ad - bc} \tag{6}$$

We then have

$$\begin{cases} z_1 = dDx_1 - bDx_2 \\ z_2 = -cDx_1 + aDx_2 \end{cases} \tag{7}$$

The corresponding Jacobian (see Appendix 1-5 of volume 3) is given by

$$\begin{aligned} |J| &= \begin{vmatrix} \dfrac{\partial z_1}{\partial x_1} & \dfrac{\partial z_1}{\partial x_2} \\ \dfrac{\partial z_2}{\partial x_1} & \dfrac{\partial z_2}{\partial x_2} \end{vmatrix} \\ &= \begin{vmatrix} dD & -bD \\ -cD & aD \end{vmatrix} \\ &= \left| adD^2 - bcD^2 \right| \\ &= |D| \end{aligned} \tag{8}$$

Therefore, we obtain

$$\begin{aligned} f(z_1, z_2)dz_1dz_2 &= f(x_1, x_2)dx_1dx_2 \\ &= \frac{|D|}{2\pi}\exp\left[-\frac{(dx_1 - bx_2)^2 + (-cx_1 + ax_2)^2}{2}D^2\right]dx_1dx_2 \end{aligned} \tag{9}$$

We can evaluate averages and variances of x_1 and x_2 as

$$E[x_1] = E[az_1 + bz_2] = aE[z_1] + bE[z_2] = 0 \tag{10}$$

$$E[x_2] = E[cz_1 + dz_2] = cE[z_1] + dE[z_2] = 0 \tag{11}$$

$$\begin{aligned} E\left[x_1^2\right] &= E\left[(az_1 + bz_2)^2\right] \\ &= a^2 + b^2 \equiv \sigma_1^{(2)} \end{aligned} \tag{12}$$

$$\begin{aligned} E\left[x_2^2\right] &= E\left[(cz_1 + dz_2)^2\right] \\ &= c^2 + d^2 \equiv \sigma_2^{(2)} \end{aligned} \tag{13}$$

$$\begin{aligned} Cov[x_1, x_2] &= E[x_1 x_2] - E[x_1]E[x_1] \\ &= E[(az_1 + bz_2)(cz_1 + dz_2)] \\ &= acE[z_1^2] + bdE[z_2^2] + (ad + bc)E[z_1 z_2] \\ &= ac + bd \equiv \sigma_{12}^{(2)} \end{aligned} \tag{14}$$

The correlation factor is given by

$$\begin{aligned} \rho &= \frac{\sigma_{12}^{(2)}}{\sqrt{\sigma_1^{(2)}}\sqrt{\sigma_2^{(2)}}} \\ &= \frac{ac + bd}{\sqrt{a^2 + b^2}\sqrt{c^2 + d^2}} \end{aligned} \tag{15}$$

We then obtain

$$\begin{aligned} 1 - \rho^2 &= \frac{\sigma_1^{(2)}\sigma_2^{(2)} - \sigma_{12}^{(2)}\sigma_{21}^{(2)}}{\sigma_1^{(2)}\sigma_2^{(2)}} \\ &= \frac{\sigma_1^{(2)}\sigma_2^{(2)} - \left[\sigma_{12}^{(2)}\right]^2}{\sigma_1^{(2)}\sigma_2^{(2)}} \\ &= \frac{(a^2 + b^2)(c^2 + d^2) - (ac + bd)^2}{\sigma_1^2 \sigma_2^2} \\ &= \frac{(ad - bc)^2}{\sigma_1^{(2)}\sigma_2^{(2)}} \end{aligned} \tag{16}$$

The determinant is then expressed by

$$|D| = \frac{1}{|ad - bc|} = \frac{1}{\sqrt{(1 - \rho^2)\sigma_1^{(2)}\sigma_2^{(2)}}} \tag{17}$$

We preform calculation in the exponential term of $f(x_1, x_2)$ as

$$
\begin{aligned}
&-\frac{\left(dx_1 - bx_2\right)^2 + \left(-cx_1 + ax_2\right)^2}{2} D^2 \\
&= -\frac{1}{2}\left[\left(d^2 + c^2\right)x_1^{\ 2} - 2\left(bd + ca\right)x_1 x_2 + \left(b^2 + a^2\right)x_2^{\ 2}\right] D^2 \\
&= -\frac{1}{2}\left[\sigma_2^{(2)} x_1^{\ 2} - 2\rho\sqrt{\sigma_1^{(2)}\sigma_2^{(2)}}\, x_1 x_2 + \sigma_1^{(2)} x_2^{\ 2}\right] \frac{1}{\sigma_1^{(2)}\sigma_2^{(2)}\left(1-\rho^2\right)} \\
&= -\frac{1}{2\left(1-\rho^2\right)}\left[\sigma_2^{(2)} \frac{x_1^{\ 2}}{\sigma_1^{(2)}} - 2\rho \frac{x_1 x_2}{\sqrt{\sigma_1^{(2)}\sigma_2^{(2)}}} + \frac{x_2^{\ 2}}{\sigma_2^{(2)}}\right]
\end{aligned}
\tag{18}
$$

We then have $f\left(x_1, x_2\right)$ as

$$
f\left(x_1, x_2\right) = \frac{1}{2\pi\sqrt{\left(1-\rho^2\right)\sigma_1^{(2)}\sigma_2^{(2)}}} \exp\left[-\frac{1}{2\left(1-\rho^2\right)}\left(\frac{x_1^{\ 2}}{\sigma_1^{(2)}} - 2\rho\frac{x_1 x_2}{\sqrt{\sigma_1^{(2)}\sigma_2^{(2)}}} + \frac{x_2^{\ 2}}{\sigma_2^{(2)}}\right)\right] \tag{19}
$$

This is the probability distribution for two variables that follow a normal distribution with $\left(0, \sigma_1^{(2)}\right)$ and $\left(0, \sigma_2^{(2)}\right)$, and the correlation factor between the two variables is ρ.

We need to generalize this for a certain averages instead of 0. Therefore, we introduce two variables as

$$
\begin{cases}
u_1 = x_1 + \mu_1 \\
u_2 = x_2 + \mu_2
\end{cases}
\tag{20}
$$

These two variables follow standard distributions with $\left(\mu_1, \sigma_1^{(2)}\right)$ and $\left(\mu_2, \sigma_2^{(2)}\right)$. The Jacobian is not changed, and hence

$$
f\left(x_1, x_2\right) dx_1 dx_2 = f\left(u_1, u_2\right) du_1 du_2 \tag{21}
$$

We then obtain a probability distribution as

$$
f\left(u_1, u_2\right) = \frac{1}{2\pi\sqrt{\left(1-\rho^2\right)\sigma_1^{(2)}\sigma_2^{(2)}}} \exp\left[-\frac{1}{2\left(1-\rho^2\right)}\left(\frac{\left(u_1 - \mu_1\right)^2}{\sigma_1^{(2)}} - 2\rho\frac{\left(u_1 - \mu_1\right)\left(u_2 - \mu_2\right)}{\sqrt{\sigma_1^{(2)}\sigma_2^{(2)}}} + \frac{\left(u_1 - \mu_2\right)^2}{\sigma_2^{(2)}}\right)\right] \tag{22}
$$

u_1 and u_2 are the dummy variables, and we can change them to any notation, and hence we use common variables x and y, and obtain

$$f(x,y)dxdy = \frac{1}{2\pi\sqrt{(1-\rho^2)\sigma_{xx}^{(2)}\sigma_{yy}^{(2)}}} \exp\left[-\frac{1}{2(1-\rho^2)}\left(\frac{(x-\mu_x)^2}{\sigma_{xx}^{(2)}} - 2\rho\frac{(x-\mu_x)(y-\mu_y)}{\sqrt{\sigma_{xx}^{(2)}\sigma_{yy}^{(2)}}} + \frac{(y-\mu_y)^2}{\sigma_{yy}^{(2)}}\right)\right]dxdy \tag{23}$$

This is the probability distribution for two variables that follow normal distributions with $\left(\mu_1, \sigma_1^{(2)}\right)$ and $\left(\mu_2, \sigma_2^{(2)}\right)$. The correlation factor for the two variables is ρ.

The probability associated with $x - y$ plane region is given by

$$P = \iint_A f(x,y)dxdy \tag{24}$$

where A is the region in the $x - y$ plane.

3. The Probability Distribution for Multi Variables

We can extend the probability distribution for multi variables.

We consider m variables. The variables can be expressed as a form of vector as

$$\mathbf{x} = \begin{pmatrix} x_1 \\ x_2 \\ \vdots \\ x_m \end{pmatrix} \tag{25}$$

The corresponding probability distribution can be expressed by

$$f(x_1, x_2, \cdots, x_m) = \frac{1}{(2\pi)^{\frac{m}{2}}\sqrt{|\Sigma|}} \exp\left(-\frac{D^2}{2}\right) \tag{26}$$

where

$$\Sigma = \begin{pmatrix} \sigma_{11}^{(2)} & \sigma_{12}^{(2)} & \cdots & \sigma_{1m}^{(2)} \\ \sigma_{21}^{(2)} & \sigma_{22}^{(2)} & \cdots & \sigma_{2m}^{(2)} \\ \vdots & \vdots & \ddots & \vdots \\ \sigma_{m1}^{(2)} & \sigma_{m2}^{(2)} & \cdots & \sigma_{mm}^{(2)} \end{pmatrix} \tag{27}$$

and D is given by

$$\begin{aligned} D^2 &= \left(x_1-\mu_1, x_2-\mu_2, \cdots, x_m-\mu_m\right) \begin{pmatrix} \sigma^{11(2)} & \sigma^{12(2)} & \cdots & \sigma^{1m(2)} \\ \sigma^{21(2)} & \sigma^{22(2)} & \cdots & \sigma^{2m(2)} \\ \vdots & \vdots & \ddots & \vdots \\ \sigma^{m1(2)} & \sigma^{m2(2)} & \cdots & \sigma^{mm(2)} \end{pmatrix} \begin{pmatrix} x_1-\mu_1 \\ x_2-\mu_2 \\ \vdots \\ x_m-\mu_m \end{pmatrix} \\ &= \sum_{i=1}^{m}\sum_{j=1}^{m}\left(x_i-\mu_i\right)\left(x_j-\mu_j\right)\sigma^{ij(2)} \end{aligned} \tag{28}$$

where

$$\mu^{[k]} = \begin{pmatrix} \mu_1^{[k]} \\ \mu_2^{[k]} \\ \vdots \\ \mu_m^{[k]} \end{pmatrix} \tag{29}$$

$$\begin{pmatrix} \sigma^{11(2)} & \sigma^{12(2)} & \cdots & \sigma^{1m(2)} \\ \sigma^{21(2)} & \sigma^{22(2)} & \cdots & \sigma^{2m(2)} \\ \vdots & \vdots & \ddots & \vdots \\ \sigma^{m1(2)} & \sigma^{m2(2)} & \cdots & \sigma^{mm(2)} \end{pmatrix} = \begin{pmatrix} \sigma_{11}^{(1)} & \sigma_{12}^{(2)} & \cdots & \sigma_{1m}^{(2)} \\ \sigma_{21}^{(2)} & \sigma_{22}^{(2)} & \cdots & \sigma_{2m}^{(2)} \\ \vdots & \vdots & \ddots & \vdots \\ \sigma_{m1}^{(2)} & \sigma_{m2}^{(2)} & \cdots & \sigma_{mm}^{(2)} \end{pmatrix}^{-1} \tag{30}$$

We do not treat this distribution here, but use it in Chapter 4 where multi-regression is discussed.

4. Probability Distribution Function for Sample Correlation Factor

We pick up n pair data from a population set. The obtained data are expressed by $\left\{\left(x_1, y_1\right), \left(x_2, y_2\right), \cdots, \left(x_n, y_n\right)\right\}$, and the sample correlation factor r is given by

$$
\begin{aligned}
r &= \frac{1}{\sqrt{S_{xx}^{(2)} S_{yy}^{(2)}}} \frac{1}{n} \sum_{i=1}^{n} (x_i - \bar{x})(y_i - \bar{y}) \\
&= \frac{1}{\sqrt{S_{xx}^{(2)} S_{yy}^{(2)}}} \left(\frac{1}{n} \sum_{i=1}^{n} x_i y_i - \bar{x}\,\bar{y} \right)
\end{aligned}
\tag{31}
$$

where

$$
S_{xx}^{(2)} = \frac{1}{n} \sum_{i=1}^{n} (x_i - \bar{x})^2 \tag{32}
$$

$$
S_{yy}^{(2)} = \frac{1}{n} \sum_{i=1}^{n} (y_i - \bar{y})^2 \tag{33}
$$

We start with the probability distribution for (x, y), and modify it for the sample correlation factor.

The simultaneous probability distribution for (x, y) is given by

$$
f(x, y) = \frac{1}{2\pi \sqrt{\sigma_{xx}^{(2)} \sigma_{yy}^{(2)}} \sqrt{1-\rho^2}} \exp\left\{ -\frac{1}{2(1-\rho^2)} \left[\frac{(x-\mu_x)^2}{\sigma_{xx}^{(2)}} - \frac{2\rho (x-\mu_x)(y-\mu_y)}{\sqrt{\sigma_{xx}^{(2)} \sigma_{yy}^{(2)}}} + \frac{(x-\mu_y)^2}{\sigma_{yy}^{(2)}} \right] \right\} \tag{34}
$$

where ρ is the correlation factor of the population, $\sigma_{xx}^{(2)}$ and $\sigma_{yy}^{(2)}$ are the variance of x and y, respectively.

We perform variable conversions as

$$
t = \frac{x - \mu_x}{\sqrt{\sigma_{xx}^{(2)}}} \tag{35}
$$

$$
u = \frac{y - \mu_y}{\sqrt{\sigma_{yy}^{(2)}}} \tag{36}
$$

The corresponding Jacobian is evaluated as follows.

$$\frac{\partial x}{\partial t} = \sqrt{\sigma_{xx}^{(2)}} \tag{37}$$

$$\frac{\partial x}{\partial u} = 0 \tag{38}$$

$$\frac{\partial y}{\partial t} = 0 \tag{39}$$

$$\frac{\partial y}{\partial u} = \sqrt{\sigma_{yy}^{(2)}} \tag{40}$$

We then have

$$\begin{aligned}\frac{\partial(x,y)}{\partial(t,u)} &= \begin{vmatrix} \sqrt{\sigma_{xx}^{(2)}} & 0 \\ 0 & \sqrt{\sigma_{yy}^{(2)}} \end{vmatrix} \\ &= \sqrt{\sigma_{xx}^{(2)}\sigma_{yy}^{(2)}}\end{aligned} \tag{41}$$

The probability distribution for simultaneous variable pair (t,u) is then given by

$$\begin{aligned}f(x,y)dxdy &= f(t,u)dtdu \\ &= \frac{1}{2\pi\sqrt{\sigma_{xx}^{(2)}\sigma_{yy}^{(2)}}\sqrt{1-\rho^2}} \exp\left[-\frac{1}{2(1-\rho^2)}\left(t^2 - 2\rho tu + u^2\right)\right]\sqrt{\sigma_{xx}^{(2)}\sigma_{yy}^{(2)}}dtdu \\ &= \frac{1}{2\pi\sqrt{1-\rho^2}} \exp\left[-\frac{1}{2(1-\rho^2)}\left(t^2 - 2\rho tu + u^2\right)\right]dtdu\end{aligned} \tag{42}$$

We further introduce a variable given by

$$v = \frac{u - \rho t}{\sqrt{1-\rho^2}} \tag{43}$$

and convert the variable pair from (t,u) to (t,v).

The corresponding Jacobian is given by

$$\frac{\partial t}{\partial t} = 1 \tag{44}$$

$$\begin{aligned}\frac{\partial t}{\partial v} &= \frac{1}{\frac{\partial v}{\partial t}} \\ &= \frac{1}{-\frac{\rho}{\sqrt{1-\rho^2}}} \\ &= -\frac{\sqrt{1-\rho^2}}{\rho}\end{aligned} \tag{45}$$

$$\frac{\partial u}{\partial t} = 0 \tag{46}$$

$$\frac{\partial u}{\partial v} = \sqrt{1-\rho^2} \tag{47}$$

and we obtain

$$\begin{aligned}\frac{\partial(t,u)}{\partial(t,v)} &= \begin{vmatrix} 1 & -\frac{\sqrt{1-\rho^2}}{\rho} \\ 0 & \sqrt{1-\rho^2} \end{vmatrix} \\ &= \sqrt{1-\rho^2}\end{aligned} \tag{48}$$

The term in the exponential in Eq. (42) is evaluated as

$$\begin{aligned}t^2 - 2\rho tu + u^2 &= (u - \rho t)^2 - \rho^2 t^2 + t^2 \\ &= (1-\rho^2)v^2 + (1-\rho^2)t^2\end{aligned} \tag{49}$$

We then obtain a probability distribution for variable pair of (t, v) as

$$\begin{aligned}f(t,u)dtdu &= g(t,v)dtdv \\ &= \frac{1}{2\pi\sqrt{1-\rho^2}}\exp\left[-\frac{v^2}{2} - \frac{t^2}{2}\right]\sqrt{1-\rho^2}\,dtdv \\ &= \frac{1}{\sqrt{2\pi}}\exp\left(-\frac{v^2}{2}\right)\frac{1}{\sqrt{2\pi}}\exp\left(-\frac{t^2}{2}\right)dtdv\end{aligned} \tag{50}$$

Consequently, v and t follow a standard normal distributions independently.

We convert each sample data as

$$t_i = \frac{x_i - \mu_x}{\sqrt{\sigma_{xx}^{(2)}}} \tag{51}$$

$$u_i = \frac{y_i - \mu_y}{\sqrt{\sigma_{yy}^{(2)}}} \tag{52}$$

We further convert u_i to v_i as

$$v_i = \frac{u_i - \rho t_i}{\sqrt{1-\rho^2}} \tag{53}$$

We know that (t_i, v_i) follows a standard normal distributions independently.

We need to express r with (t_i, v_i).

r is a sample correlation factor for the pair x_i, y_i. It is also equal to the sample correlation factor for the pair t_i, u_i.

$$r = \frac{\sum_{i=1}^{n}(t_i - \bar{t})(u_i - \bar{u})}{\sqrt{\sum_{i=1}^{n}(t_i - \bar{t})^2}\sqrt{\sum_{i=1}^{n}(u_i - \bar{u})^2}} \tag{54}$$

We only know that t and v follow standard distributions independently, but do not know for u. Therefore, we need to convert u to v in Eq. (53).

We do not start with Eq. (54) directly, but start with a variable of q given by

$$q = \sum_{i=1}^{n} v_i^2 \tag{55}$$

Since v_i follows a standard normal distribution, q follows a χ^2 distribution with a freedom of n. Substituting Eq. (53) into Eq. (55), we obtain

$$q = \sum_{i=1}^{n} \left(\frac{u_i - \rho t_i}{\sqrt{1-\rho^2}} \right)^2 = \frac{1}{1-\rho^2} \sum_{i=1}^{n} \left(u_i^2 - 2\rho u_i t_i + \rho^2 t_i^2 \right) \tag{56}$$

We want to relate q to r. Therefore, we investigate last terms in Eq. (56) with respect to the corresponding deviations as

$$\begin{aligned} \sum_{i=1}^{n} u_i^2 &= \sum_{i=1}^{n} \left(u_i - \bar{u} + \bar{u} \right)^2 \\ &= \sum_{i=1}^{n} \left[\left(u_i - \bar{u} \right)^2 + 2\left(u_i - \bar{u} \right)\bar{u} + \bar{u}^2 \right] \\ &= \sum_{i=1}^{n} \left(u_i - \bar{u} \right)^2 + 2\bar{u} \sum_{i=1}^{n} \left(u_i - \bar{u} \right) + \sum_{i=1}^{n} \bar{u}^2 \\ &= \sum_{i=1}^{n} \left(u_i - \bar{u} \right)^2 + n\bar{u}^2 \end{aligned} \tag{57}$$

$$\begin{aligned} \sum_{i=1}^{n} t_i u_i &= \sum_{i=1}^{n} \left(t_i - \bar{t} + \bar{t} \right)\left(u_i - \bar{u} + \bar{u} \right) \\ &= \sum_{i=1}^{n} \left[\left(t_i - \bar{t} \right)\left(u_i - \bar{u} \right) + \bar{t}\left(u_i - \bar{u} \right) + \bar{u}\left(t_i - \bar{t} \right) + \bar{t}\,\bar{u} \right] \\ &= \sum_{i=1}^{n} \left(t_i - \bar{t} \right)\left(u_i - \bar{u} \right) + n\bar{t}\,\bar{u} \end{aligned} \tag{58}$$

$$\begin{aligned} \sum_{i=1}^{n} t_i^2 &= \sum_{i=1}^{n} \left(t_i - \bar{t} + \bar{t} \right)^2 \\ &= \sum_{i=1}^{n} \left[\left(t_i - \bar{t} \right)^2 + 2\left(t_i - \bar{t} \right)\bar{t} + \bar{t}^2 \right] \\ &= \sum_{i=1}^{n} \left(t_i - \bar{t} \right)^2 + 2\bar{t} \sum_{i=1}^{n} \left(t_i - \bar{t} \right) + \sum_{i=1}^{n} \bar{t}^2 \\ &= \sum_{i=1}^{n} \left(t_i - \bar{t} \right)^2 + n\bar{u}^2 \end{aligned} \tag{59}$$

Substituting Eqs. (57), (58), and (59) into Eq. (56), we obtain

$$
\begin{aligned}
q &= \sum_{i=1}^{n}\left(\frac{u_i-\rho t_i}{\sqrt{1-\rho^2}}\right)^2 \\
&= \frac{1}{1-\rho^2}\sum_{i=1}^{n}\left(u_i^2-2\rho u_i t_i+\rho^2 t_i^2\right) \\
&= \frac{1}{1-\rho^2}\left\{\left[\sum_{i=1}^{n}\left(u_i-\bar{u}\right)^2+n\bar{u}^2\right]-2\rho\left[\sum_{i=1}^{n}\left(t_i-\bar{t}\right)\left(u_i-\bar{u}\right)+n\bar{t}\,\bar{u}\right]+\rho^2\left[\sum_{i=1}^{n}\left(t_i-\bar{t}\right)^2+n\bar{t}^2\right]\right\} \\
&= \frac{1}{1-\rho^2}\left\{\sum_{i=1}^{n}\left(u_i-\bar{u}\right)^2-2\rho\sum_{i=1}^{n}\left(t_i-\bar{t}\right)\left(u_i-\bar{u}\right)+\rho^2\sum_{i=1}^{n}\left(t_i-\bar{t}\right)^2\right\} \\
&+\frac{1}{1-\rho^2}\left\{n\bar{u}^2-2\rho n\bar{t}\,\bar{u}+\rho^2 n\bar{t}^2\right\} \\
&= q_0+q_1
\end{aligned}
\tag{60}
$$

$$
\begin{aligned}
q &= \frac{1}{1-\rho^2}\sum_{i=1}^{n}\left(u_i^2-2\rho u_i t_i+\rho^2 t_i^2\right) \\
&= \frac{1}{1-\rho^2}\left\{\left[\sum_{i=1}^{n}\left(u_i-\bar{u}\right)^2+n\bar{u}^2\right]-2\rho\left[\sum_{i=1}^{n}\left(t_i-\bar{t}\right)\left(u_i-\bar{u}\right)+n\bar{t}\,\bar{u}\right]+\rho^2\left[\sum_{i=1}^{n}\left(t_i-\bar{t}\right)^2+n\bar{t}^2\right]\right\} \\
&= \frac{1}{1-\rho^2}\left\{\sum_{i=1}^{n}\left(u_i-\bar{u}\right)^2-2\rho\sum_{i=1}^{n}\left(t_i-\bar{t}\right)\left(u_i-\bar{u}\right)+\rho^2\sum_{i=1}^{n}\left(t_i-\bar{t}\right)^2\right\} \\
&+\frac{1}{1-\rho^2}\left\{n\bar{u}^2-2\rho n\bar{t}\,\bar{u}+\rho^2 n\bar{t}^2\right\} \\
&= q_0+q_1
\end{aligned}
\tag{61}
$$

where

$$
\begin{aligned}
q_0 &\equiv \frac{1}{1-\rho^2}\left\{\sum_{i=1}^{n}\left(u_i-\bar{u}\right)^2-2\rho\sum_{i=1}^{n}\left(t_i-\bar{t}\right)\left(u_i-\bar{u}\right)+\rho^2\sum_{i=1}^{n}\left(t_i-\bar{t}\right)^2\right\} \\
&= \frac{1}{1-\rho^2}\left\{U^2-2\rho TU\frac{\sum_{i=1}^{n}\left(t_i-\bar{t}\right)\left(u_i-\bar{u}\right)}{TU}+\rho^2T^2\right\} \\
&= \frac{1}{1-\rho^2}\left(U^2-2\rho rTU+\rho^2T^2\right)
\end{aligned}
\tag{62}
$$

$$
\begin{aligned}
q_1 &\equiv \frac{n}{1-\rho^2}\left\{\bar{u}^2-2\rho\bar{t}\,\bar{u}+\rho^2\bar{t}^2\right\} \\
&= \frac{n}{1-\rho^2}\left(\bar{u}-\rho\bar{t}\right)^2
\end{aligned}
\tag{63}
$$

We define U and T as

$$U = \sqrt{\sum_{i=1}^{n} \left(u_i - \bar{u} \right)^2} \tag{64}$$

$$T = \sqrt{\sum_{i=1}^{n} \left(t_i - \bar{t} \right)^2} \tag{65}$$

q_1 is also expressed as

$$\begin{aligned} q_1 &= \frac{n}{1-\rho^2} \left(\bar{u} - \rho \bar{t} \right)^2 \\ &= \frac{n}{1-\rho^2} \left(\frac{1}{n} \sum \left(u_i - \rho t_i \right) \right)^2 \\ &= \frac{1}{n} \left(\sum \left(\frac{u_i - \rho t_i}{\sqrt{1-\rho^2}} \right) \right)^2 \\ &= \frac{1}{n} \left(\sum v_i \right)^2 \end{aligned} \tag{66}$$

Since v_i follows a standard normal distribution, $\frac{1}{\sqrt{n}} \sum_{i=1}^{n} v_i$ follows a standard normal distribution. Therefore, q_1 follows a χ^2 distribution.

We analyze q_0 next.

We introduce a variable

$$\varsigma = \frac{\sum_{i=1}^{n} \left(t_i - \bar{t} \right) v_i}{T} \tag{67}$$

and analyze it before doing analysis of q_0. Substituting Eq. (53) into Eq. (67), we obtain

$$\begin{aligned}
\varsigma &= \frac{\sum_{i=1}^{n}(t_i - \bar{t})\frac{(u_i - \rho t_i)}{\sqrt{1-\rho^2}}}{T} \\
&= \frac{1}{\sqrt{1-\rho^2}} \frac{\sum_{i=1}^{n}(t_i - \bar{t})\left[(u_i - \bar{u}) - \rho(t_i - \bar{t})\right]}{T} \\
&= \frac{1}{\sqrt{1-\rho^2}} \left(\frac{\sum_{i=1}^{n}(t_i - \bar{t})(u_i - \bar{u})}{T} - \rho \frac{\sum_{i=1}^{n}(t_i - \bar{t})^2}{T} \right) \\
&= \frac{1}{\sqrt{1-\rho^2}} \left(U \frac{\sum_{i=1}^{n}(t_i - \bar{t})(u_i - \bar{u})}{UT} - \rho \frac{T^2}{T} \right) \\
&= \frac{1}{\sqrt{1-\rho^2}} (rU - \rho T)
\end{aligned} \tag{68}$$

We define q_2 as

$$\begin{aligned}
q_2 &= \varsigma^2 \\
&= \frac{1}{1-\rho^2}\left(r^2U^2 - 2r\rho UT + \rho^2T^2\right)
\end{aligned} \tag{69}$$

Finally, q_0 is related to q_2 as

$$\begin{aligned}
q_0 &= \frac{1}{1-\rho^2}\left(U^2 - 2\rho rTU + \rho^2T^2\right) - \frac{1}{1-\rho^2}\left(r^2U^2 - 2r\rho UT + \rho^2T^2\right) + q_2 \\
&= \frac{1-r^2}{1-\rho^2}U^2 + q_2 \\
&= q_2 + q_3
\end{aligned} \tag{70}$$

where q_3 is defined as

$$q_3 \equiv \frac{1-r^2}{1-\rho^2}U^2 \tag{71}$$

Therefore, q can be expressed by the sum of them as

$$\begin{aligned} q &= q_0 + q_1 \\ &= q_1 + q_2 + q_3 \end{aligned} \tag{72}$$

Since q_1 follows a χ^2 distribution with a freedom of 1, $q_2 + q_3$ follows a χ^2 distribution with a freedom of $n-1$. We then analyze q_2 and q_3 separately.

Let us consider ς for fixed t of t_0, which is given by

$$\begin{aligned} \varsigma &= \frac{\sum_{i=1}^{n}(t_i - \bar{t})v_i}{T} \\ &= \frac{\sum_{i=1}^{n}(t_i - \bar{t})v_i}{\sqrt{\sum_{i=1}^{n}(t_i - \bar{t})^2}} \\ &= \frac{(t_0 - \bar{t})\sum_{i=1}^{n} v_i}{\sqrt{n(t_0 - \bar{t})^2}} \\ &= \frac{1}{\sqrt{n}}\sum_{i=1}^{n} v_i \end{aligned} \tag{73}$$

This follows a standard normal distribution. Therefore, $q_2(\mathbf{v}, \mathbf{t}_0) = \varsigma^2$ follows a χ^2 distribution with a freedom of 1. Since Eq. (73) does not include t_0, $q_2(\mathbf{v}, \mathbf{t}_0)$ follows a χ^2 distribution with a freedom of 1 for any t. We can conclude that q_2 follows a χ^2 distribution with a freedom of 1. Consequently, q_3 follows a χ^2 distribution with a freedom of $n-2$.

We further consider q_4 as

$$q_4(\mathbf{t}) = T^2 \tag{74}$$

Inspecting Eq. (65), this follows a χ^2 distribution with a freedom of $n-1$.

Finally, we obtain three independent variables q_2, q_3, and q_4 whose corresponding probability distributions we know.

We introduce three variables below instead of using q_2, q_3, and q_4.

We consider a variable λ as

$$\begin{aligned}\lambda &= \varsigma \\ &= \frac{1}{\sqrt{1-\rho^2}}(rU - \rho T)\end{aligned} \tag{75}$$

Since $q_2 = \varsigma^2$ follows a χ^2 distribution with a freedom of 1, λ follows a standard normal distribution, and the corresponding probability distribution is given by

$$f(\lambda) = \frac{1}{\sqrt{2\pi}} \exp\left(-\frac{\lambda^2}{2}\right) \tag{76}$$

Next, we introduce a variable

$$\begin{aligned}\mu &= \frac{1}{2} q_3 \\ &= \frac{1}{2\left(1-\rho^2\right)}\left(1-r^2\right)U^2\end{aligned} \tag{77}$$

Since q_3 follows a χ^2 distribution with a freedom of $n-2$, we obtain

$$f(q_3) = \frac{1}{2^{\frac{n-2}{2}}\Gamma\left(\frac{n-2}{2}\right)} q_3^{\frac{n-4}{2}} \exp\left(-\frac{q_3}{2}\right) \tag{78}$$

Therefore, we can evaluate a probability distribution for μ as follows.

$$d\mu = \frac{1}{2} dq_3 \tag{79}$$

We then have

$$\begin{aligned} f(q_3)dq_3 &= f(\mu)d\mu \\ &= \frac{1}{2^{\frac{n-2}{2}}\Gamma\left(\frac{n-2}{2}\right)}(2\mu)^{\frac{n-4}{2}}\exp(-\mu)\times 2d\mu \\ &= \frac{1}{\Gamma\left(\frac{n-2}{2}\right)}\mu^{\frac{n-4}{2}}\exp(-\mu)d\mu \end{aligned} \tag{80}$$

Finally, we introduce a variable of

$$\begin{aligned} \upsilon &= \frac{1}{2}q_4 \\ &= \frac{1}{2}T^2 \end{aligned} \tag{81}$$

Since q_4 follows a χ^2 distribution with a freedom of $n-1$, we obtain a probability distribution for ν as

$$\begin{aligned} f(\nu)d\nu &= \frac{1}{2^{\frac{n-1}{2}}\Gamma\left(\frac{n-1}{2}\right)}(2\nu)^{\frac{n-3}{2}}\exp(-\nu)\times 2d\mu \\ &= \frac{1}{\Gamma\left(\frac{n-1}{2}\right)}\nu^{\frac{n-3}{2}}\exp(-\nu)d\nu \end{aligned} \tag{82}$$

Therefore, the simultaneous three variable probability distribution is given by

$$\begin{aligned} f(\lambda,\mu,\upsilon) &= \frac{1}{\sqrt{2\pi}}e^{-\frac{\lambda^2}{2}}\frac{\mu^{\frac{n-4}{2}}e^{-\mu}}{\Gamma\left(\frac{n-2}{2}\right)}\frac{\upsilon^{\frac{n-3}{2}}e^{-\upsilon}}{\Gamma\left(\frac{n-1}{2}\right)} \\ &= \frac{1}{\sqrt{2\pi}\Gamma\left(\frac{n-2}{2}\right)\Gamma\left(\frac{n-1}{2}\right)}\mu^{\frac{n-4}{2}}\upsilon^{\frac{n-3}{2}}\exp\left(-\frac{\lambda^2}{2}-\mu-\upsilon\right) \end{aligned} \tag{83}$$

We convert this probability distribution to the one for (r,T,U).
We evaluated the partial differentials below.

$$\begin{aligned}\frac{\partial \lambda}{\partial r} &= \frac{1}{\sqrt{1-\rho^2}}\frac{\partial (rU - \rho T)}{\partial r} \\ &= \frac{1}{\sqrt{1-\rho^2}}U\end{aligned} \tag{84}$$

$$\begin{aligned}\frac{\partial \lambda}{\partial T} &= \frac{1}{\sqrt{1-\rho^2}}\frac{\partial (rU - \rho T)}{\partial T} \\ &= -\frac{1}{\sqrt{1-\rho^2}}\rho\end{aligned} \tag{85}$$

$$\begin{aligned}\frac{\partial \lambda}{\partial U} &= \frac{1}{\sqrt{1-\rho^2}}\frac{\partial (rU - \rho T)}{\partial U} \\ &= \frac{1}{\sqrt{1-\rho^2}}r\end{aligned} \tag{86}$$

$$\begin{aligned}\frac{\partial \mu}{\partial r} &= \frac{1}{2(1-\rho^2)}\frac{\partial}{\partial r}\left[(1-r^2)U^2\right] \\ &= -\frac{1}{(1-\rho^2)}rU^2\end{aligned} \tag{87}$$

$$\begin{aligned}\frac{\partial \mu}{\partial T} &= \frac{1}{2(1-\rho^2)}\frac{\partial}{\partial T}\left[(1-r^2)U^2\right] \\ &= 0\end{aligned} \tag{88}$$

$$\begin{aligned}\frac{\partial \mu}{\partial U} &= \frac{1}{2(1-\rho^2)}\frac{\partial}{\partial U}\left[(1-r^2)U^2\right] \\ &= \frac{1}{1-\rho^2}(1-r^2)U\end{aligned} \tag{89}$$

$$\begin{aligned}\frac{\partial \upsilon}{\partial r} &= \frac{1}{2}\frac{\partial}{\partial r}T^2 \\ &= 0\end{aligned} \tag{90}$$

$$\begin{aligned}\frac{\partial \upsilon}{\partial T} &= \frac{1}{2}\frac{\partial}{\partial T}T^2 \\ &= T\end{aligned} \tag{91}$$

$$\begin{aligned}\frac{\partial \upsilon}{\partial U} &= \frac{1}{2}\frac{\partial}{\partial U}T^2 \\ &= 0\end{aligned} \tag{92}$$

We then obtain corresponding Jacobian as

$$\frac{\partial(\lambda,\mu,\upsilon)}{\partial(r,T,U)}=\begin{vmatrix}\frac{1}{\sqrt{1-\rho^2}}U & -\frac{1}{\sqrt{1-\rho^2}}\rho & \frac{1}{\sqrt{1-\rho^2}}r \\ -\frac{1}{1-\rho^2}rU^2 & 0 & \frac{1}{1-\rho^2}\left(1-r^2\right)U \\ 0 & T & 0\end{vmatrix}$$
$$=-\frac{1}{\left(1-\rho^2\right)^{\frac{3}{2}}}TU^2 \tag{93}$$

Therefore, we obtain a probability distribution with respect to a variable set of (r,T,U) as

$$\begin{aligned}f(r,T,U)&=\frac{1}{\left(1-\rho^2\right)^{\frac{3}{2}}}TU^2\frac{1}{\sqrt{2\pi}\Gamma\left(\frac{n-2}{2}\right)\Gamma\left(\frac{n-1}{2}\right)}\\&\times\left[\frac{1}{2\left(1-\rho^2\right)}\left(1-r^2\right)U^2\right]^{\frac{n-4}{2}}\\&\times\left(\frac{T^2}{2}\right)^{\frac{n-3}{2}}\\&\times\exp\left[-\frac{1}{2\left(1-\rho^2\right)}\left(U^2+T^2-2r\rho TU\right)\right]\\&=\frac{1}{\left(1-\rho^2\right)^{\frac{3}{2}}}TU^2\frac{1}{\sqrt{2\pi}\Gamma\left(\frac{n-2}{2}\right)\Gamma\left(\frac{n-1}{2}\right)}\\&\times\frac{1}{2^{\frac{n-4}{2}}\left(1-\rho^2\right)^{\frac{n-4}{2}}}\left(1-r^2\right)^{\frac{n-4}{2}}U^{n-4}\\&\times\frac{1}{2^{\frac{n-3}{2}}}T^{n-3}\\&\times\exp\left[-\frac{1}{2\left(1-\rho^2\right)}\left(U^2+T^2-2r\rho TU\right)\right]\\&=\frac{1}{\sqrt{2\pi}\Gamma\left(\frac{n-2}{2}\right)\Gamma\left(\frac{n-1}{2}\right)}\frac{1}{2^{n-\frac{7}{2}}}\frac{1}{\left(1-\rho^2\right)^{\frac{n-1}{2}}}\left(1-r^2\right)^{\frac{n-4}{2}}T^{n-2}U^{n-2}\\&\times\exp\left[-\frac{1}{2\left(1-\rho^2\right)}\left(U^2+T^2-2r\rho TU\right)\right]\end{aligned} \tag{94}$$

T and U have value ranges between 0 and ∞. We further convert variables of T and U to α and β given by

$$\begin{cases} T = \sqrt{\alpha} e^{\frac{\beta}{2}} \\ U = \sqrt{\alpha} e^{-\frac{\beta}{2}} \end{cases} \tag{95}$$

Modifying these equations, we obtain

$$\begin{cases} \alpha = TU \\ e^{\beta} = \dfrac{U}{T} \end{cases} \tag{96}$$

The corresponding partial differentials are evaluated as

$$\frac{\partial T}{\partial \alpha} = \frac{1}{2\sqrt{\alpha}} e^{\frac{\beta}{2}} \tag{97}$$

$$\frac{\partial T}{\partial \beta} = \frac{1}{2} \sqrt{\alpha} e^{\frac{\beta}{2}} \tag{98}$$

$$\frac{\partial U}{\partial \alpha} = \frac{1}{2\sqrt{\alpha}} e^{-\frac{\beta}{2}} \tag{99}$$

$$\frac{\partial U}{\partial \beta} = -\frac{1}{2} \sqrt{\alpha} e^{-\frac{\beta}{2}} \tag{100}$$

The corresponding Jacobian is given by

$$\begin{aligned} \frac{\partial(T,U)}{\partial(\alpha,\beta)} &= \begin{vmatrix} \frac{1}{2\sqrt{\alpha}} e^{\frac{\beta}{2}} & \frac{1}{2}\sqrt{\alpha} e^{\frac{\beta}{2}} \\ \frac{1}{2\sqrt{\alpha}} e^{-\frac{\beta}{2}} & -\frac{1}{2}\sqrt{\alpha} e^{-\frac{\beta}{2}} \end{vmatrix} \\ &= -\frac{1}{2\sqrt{\alpha}} e^{\frac{\beta}{2}} \frac{1}{2}\sqrt{\alpha} e^{-\frac{\beta}{2}} - \frac{1}{2\sqrt{\alpha}} e^{-\frac{\beta}{2}} \frac{1}{2}\sqrt{\alpha} e^{\frac{\beta}{2}} \\ &= -\frac{1}{4} - \frac{1}{4} \\ &= -\frac{1}{2} \end{aligned} \tag{101}$$

Therefore, the term $Q(U,T)$ including T,U is given by

$$\begin{aligned} Q(T,U) &= T^{n-2}U^{n-2}\exp\left[-\frac{1}{2\left(1-\rho^2\right)}\left(U^2+T^2-2r\rho TU\right)\right] \\ &= \left(\sqrt{\alpha}e^{\frac{\beta}{2}}\right)^{n-2}\left(\sqrt{\alpha}e^{-\frac{\beta}{2}}\right)^{n-2}\exp\left[-\frac{1}{2\left(1-\rho^2\right)}\left(\alpha e^{\beta}+\alpha e^{-\beta}-2r\rho\alpha\right)\right] \\ &= \alpha^{n-2}\exp\left[-\frac{\alpha}{1-\rho^2}\left(\cosh\beta-r\rho\right)\right] \end{aligned} \tag{102}$$

α changes from 0 to ∞, and β changes from $-\infty$ to ∞. Therefore, the corresponding integration is given by

$$\begin{aligned} \iint Q(T,U)dTdU &= \frac{1}{2}\int_{-\infty}^{\infty}\int_{0}^{\infty}\alpha^{n-2}\exp\left[-\frac{\alpha}{1-\rho^2}\left(\cosh\beta-r\rho\right)\right]d\alpha d\beta \\ &= \Gamma(n-1)\left(1-\rho^2\right)^{n-1}\int_0^{\infty}\frac{1}{\left(\cosh\beta-r\rho\right)^{n-1}}d\beta \end{aligned} \tag{103}$$

We set a variable

$$t=\frac{\alpha}{1-\rho^2}\left(\cosh\beta-r\rho\right) \tag{104}$$

and obtain

$$\begin{aligned} \int_0^{\infty}\alpha^{n-2}\exp\left[-\frac{\alpha}{1-\rho^2}\left(\cosh\beta-r\rho\right)\right]d\alpha &= \left(\frac{1-\rho^2}{\cosh\beta-r\rho}\right)^{n-2}\int_0^{\infty}t^{n-2}\exp(-t)\frac{1-\rho^2}{\cosh\beta-r\rho}dt \\ &= \left(\frac{1-\rho^2}{\cosh\beta-r\rho}\right)^{n-1}\int_0^{\infty}t^{n-2}\exp(-t)dt \\ &= \Gamma(n-1)\left(\frac{1-\rho^2}{\cosh\beta-r\rho}\right)^{n-1} \end{aligned} \tag{105}$$

Since $\cosh\beta$ is an even function, and hence we change the integration region from 0 to ∞, and multiply it by 2 and obtain

$$\begin{aligned} f(r) &= \frac{1}{\sqrt{2\pi}\Gamma\left(\frac{n-2}{2}\right)\Gamma\left(\frac{n-1}{2}\right)} \frac{1}{2^{n-\frac{7}{2}}} \frac{1}{\left(1-\rho^2\right)^{\frac{n-1}{2}}} \left(1-r^2\right)^{\frac{n-4}{2}} \\ &\times\Gamma(n-1)\left(1-\rho^2\right)^{n-1} \int_0^\infty \frac{1}{(\cosh\beta - r\rho)^{n-1}} d\beta \\ &= \frac{\Gamma(n-1)}{\sqrt{\pi}\Gamma\left(\frac{n-2}{2}\right)\Gamma\left(\frac{n-1}{2}\right)} \frac{1}{2^{n-3}} \left(1-\rho^2\right)^{\frac{n-1}{2}} \left(1-r^2\right)^{\frac{n-4}{2}} \int_0^\infty \frac{1}{(\cosh\beta - r\rho)^{n-1}} d\beta \end{aligned} \tag{106}$$

Since the term in Eq. (106) is performed further as

$$\Gamma\left(\frac{n-2}{2}\right)\Gamma\left(\frac{n-1}{2}\right)2^{n-3} = \sqrt{\pi}\Gamma(n-2), \frac{\Gamma(n-1)}{\Gamma(n-2)} = n-2 \tag{107}$$

Eq. (106) is reduced to

$$\begin{aligned} f(r) &= \frac{\Gamma(n-1)}{\sqrt{\pi}\Gamma\left(\frac{n-2}{2}\right)\Gamma\left(\frac{n-1}{2}\right)} \frac{1}{2^{n-3}} \left(1-\rho^2\right)^{\frac{n-1}{2}} \left(1-r^2\right)^{\frac{n-4}{2}} \int_0^\infty \frac{1}{(\cosh\beta - r\rho)^{n-1}} d\beta \\ &= \frac{\Gamma(n-1)}{\sqrt{\pi}\sqrt{\pi}\Gamma(n-2)} \left(1-\rho^2\right)^{\frac{n-1}{2}} \left(1-r^2\right)^{\frac{n-4}{2}} \int_0^\infty \frac{1}{(\cosh\beta - r\rho)^{n-1}} d\beta \\ &= \frac{n-2}{\pi} \left(1-\rho^2\right)^{\frac{n-1}{2}} \left(1-r^2\right)^{\frac{n-4}{2}} \int_0^\infty \frac{1}{(\cosh\beta - r\rho)^{n-1}} d\beta \end{aligned} \tag{108}$$

We further perform integration in Eq. (108) as follows.
We introduce a variable N as

$$n-1=N \tag{109}$$

The term is then expanded as

$$\begin{aligned} \frac{1}{(\cosh\beta - r\rho)^{n-1}} &= \frac{1}{(\cosh\beta - r\rho)^{N}} \\ &= \frac{1}{\cosh^N\beta}\left(1-\frac{r\rho}{\cosh\beta}\right)^{-N} \\ &= \frac{1}{\cosh^N\beta}\sum_{m=0}^{\infty}\frac{N(N+1)\cdots(N+m-1)}{m!}\frac{(r\rho)^m}{\cosh^m\beta} \end{aligned} \tag{110}$$

where we utilize a Taylor series below.

$$\begin{aligned} f(x) &= (1-x)^{-N} \\ &= f(1) + \left.\frac{\partial f}{\partial x}\right|_{x=0} \times x + \frac{1}{2!}\left.\frac{\partial^2 f}{\partial x^2}\right|_{x=0} \times x^2 + \frac{1}{3!}\left.\frac{\partial^3 f}{\partial x^3}\right|_{x=0} \times x^3 + \cdots \\ &= 1 + Nx + \frac{1}{2!}N(N-1)x^2 + \frac{1}{3!}N(N-1)x^3 + \cdots \\ &= \sum_{m=0}^{\infty} \frac{N(N+1)\cdots(N+m-1)}{m!} x^m \end{aligned} \tag{111}$$

where

$$x = \frac{r\rho}{\cosh\beta} \tag{112}$$

in this derivation process.

Therefore, we can focus on the integration of the form given by

$$\int_0^{\infty} \frac{1}{\cosh^k u} du \tag{113}$$

Performing a variable conversion as

$$x = \frac{1}{\cosh u} \tag{114}$$

we then obtain

$$\begin{aligned} dx &= \frac{-\sinh u}{\cosh^2 u} du \\ &= \frac{-\sqrt{\cosh^2 u - 1}}{\cosh^2 u} du \\ &= -x^2\sqrt{\frac{1}{x^2} - 1} du \\ &= -x\sqrt{1-x^2} du \end{aligned} \tag{115}$$

The integration of Eq. (113) is then reduced to

$$\begin{aligned}\int_0^\infty \frac{1}{\cosh^k u} du &= -\int_1^0 x^k \frac{1}{x\sqrt{1-x^2}} dx \\ &= \int_0^1 \frac{x^{k-1}}{\sqrt{1-x^2}} dx\end{aligned} \tag{116}$$

Further, we convert the variable as

$$x = \sqrt{y} \tag{117}$$

and obtain

$$\frac{dx}{dy} = \frac{1}{2\sqrt{y}} \tag{118}$$

Therefore, the integration is reduced to

$$\begin{aligned}\int_0^1 \frac{x^{k-1}}{\sqrt{1-x^2}} dx &= \int_0^1 \frac{y^{\frac{k-1}{2}}}{\sqrt{1-y}} \frac{1}{2\sqrt{y}} dy \\ &= \frac{1}{2}\int_0^1 \frac{y^{\frac{k}{2}-1}}{\sqrt{1-y}} dy \\ &= \frac{1}{2} B\left(\frac{k}{2}, \frac{1}{2}\right) \\ &= \frac{1}{2} \frac{\Gamma\left(\frac{k}{2}\right)\sqrt{\pi}}{\Gamma\left(\frac{k+1}{2}\right)}\end{aligned} \tag{119}$$

where Γ is a Gamma function and B is a Beta function (see Appendix 1-2 of volume 3). The integration is then given by

$$\begin{aligned}
&\int_0^\infty \frac{1}{(\cosh\beta - r\rho)^{n-1}} d\beta \\
&= \int_0^\infty \frac{1}{\cosh^N \beta} \sum_{m=0}^{\infty} \frac{N(N+1)\cdots(N+m-1)}{m!} \frac{(r\rho)^m}{\cosh^m \beta} d\beta \\
&= \sum_{m=0}^{\infty} \frac{N(N+1)\cdots(N+m-1)}{m!} (r\rho)^m \int_0^\infty \frac{1}{\cosh^{N+m} \beta} d\beta \\
&= \sum_{m=0}^{\infty} \frac{N(N+1)\cdots(N+m-1)}{m!} (r\rho)^m \frac{1}{2} \frac{\Gamma\left(\frac{N+m}{2}\right)\sqrt{\pi}}{\Gamma\left(\frac{N+m+1}{2}\right)} \\
&= \frac{1}{\Gamma\left(\frac{N}{2}\right)\Gamma\left(\frac{N-1}{2}\right)} \\
&\times \sum_{m=0}^{\infty} \frac{N(N+1)\cdots(N+m-1)}{m!} (r\rho)^m \frac{1}{2} \frac{\Gamma\left(\frac{N}{2}\right)\Gamma\left(\frac{N-1}{2}\right)}{\Gamma\left(\frac{N+m}{2}\right)\Gamma\left(\frac{N+m+1}{2}\right)} \Gamma^2\left(\frac{N+m}{2}\right)\sqrt{\pi}
\end{aligned} \tag{120}$$

Since we have relationships given by

$$\Gamma\left(\frac{N}{2}\right)\Gamma\left(\frac{N-1}{2}\right) = 2^{2-N}\sqrt{\pi}(N-2)! \tag{121}$$

$$\Gamma\left(\frac{N+m}{2}\right)\Gamma\left(\frac{N+m+1}{2}\right) = 2^{1-N-m}\sqrt{\pi}\Gamma(N+m) \tag{122}$$

the term in Eq. (120) is reduced to

$$\begin{aligned}
&\frac{1}{2} \frac{\Gamma\left(\frac{N}{2}\right)\Gamma\left(\frac{N-1}{2}\right)}{\Gamma\left(\frac{N+m}{2}\right)\Gamma\left(\frac{N+m+1}{2}\right)} \Gamma^2\left(\frac{N+m}{2}\right)\sqrt{\pi} \\
&= \frac{1}{2} \frac{2^{2-N}\sqrt{\pi}(N-2)!}{2^{1-N-m}\sqrt{\pi}\Gamma(N+m)} \Gamma^2\left(\frac{N+m}{2}\right)\sqrt{\pi} \\
&= 2^m \sqrt{\pi}(N-2)! \frac{\Gamma^2\left(\frac{N+m}{2}\right)}{\Gamma(N+m)} \\
&= 2^m \sqrt{\pi}(N-2)! \frac{\Gamma^2\left(\frac{N+m}{2}\right)}{(N+m-1)!}
\end{aligned} \tag{123}$$

Therefore, we substitute this into Eq. (120), we obtain

$$\int_0^{\infty} \frac{1}{\left(\cosh\beta - r\rho\right)^{n-1}} d\beta$$
$$= \frac{1}{\Gamma\left(\frac{N}{2}\right)\Gamma\left(\frac{N-1}{2}\right)}$$
$$\times \sum_{m=0}^{\infty} \frac{N(N+1)\cdots(N+m-1)}{m!} (r\rho)^m 2^m \sqrt{\pi}(N-2)! \frac{\Gamma^2\left(\frac{N+m}{2}\right)}{(N+m-1)!}$$
$$= \frac{1}{\Gamma\left(\frac{N}{2}\right)\Gamma\left(\frac{N-1}{2}\right)}$$
$$\times \sum_{m=0}^{\infty} \frac{1}{m!} (r\rho)^m 2^m \sqrt{\pi}(N-2)! \frac{\Gamma^2\left(\frac{N+m}{2}\right)}{(N-1)!}$$
$$= \frac{1}{\Gamma\left(\frac{N}{2}\right)\Gamma\left(\frac{N-1}{2}\right)} \times \sum_{m=0}^{\infty} \sqrt{\pi} \frac{(r\rho)^m 2^m}{m!} \frac{\Gamma^2\left(\frac{N+m}{2}\right)}{(N-1)} \tag{124}$$
$$= \frac{1}{\Gamma\left(\frac{n-1}{2}\right)\Gamma\left(\frac{n-2}{2}\right)} \times \sum_{m=0}^{\infty} \sqrt{\pi} \frac{(r\rho)^m 2^m}{m!} \frac{\Gamma^2\left(\frac{n-1+m}{2}\right)}{(n-2)}$$

Therefore, the probability distribution for sample correlation factor r is given by

$$\begin{aligned} f(r) &= \frac{n-2}{\pi}\left(1-\rho^2\right)^{\frac{n-1}{2}}\left(1-r^2\right)^{\frac{n-4}{2}} \int_0^{\infty} \frac{1}{\left(\cosh\beta - r\rho\right)^{n-1}} d\beta \\ &= \frac{n-2}{\pi}\left(1-\rho^2\right)^{\frac{n-1}{2}}\left(1-r^2\right)^{\frac{n-4}{2}} \frac{1}{\Gamma\left(\frac{n-1}{2}\right)\Gamma\left(\frac{n-2}{2}\right)} \times \sum_{m=0}^{\infty} \sqrt{\pi} \frac{(r\rho)^m 2^m}{m!} \frac{\Gamma^2\left(\frac{n-1+m}{2}\right)}{(n-2)} \\ &= \frac{\left(1-\rho^2\right)^{\frac{n-1}{2}}\left(1-r^2\right)^{\frac{n-4}{2}}}{\sqrt{\pi}\Gamma\left(\frac{n-1}{2}\right)\Gamma\left(\frac{n-2}{2}\right)} \times \sum_{m=0}^{\infty} \frac{(2r\rho)^m}{m!} \Gamma^2\left(\frac{n-1+m}{2}\right) \end{aligned} \tag{125}$$

This is a rigorous model for a probability distribution for correlation factor. However, it is rather complicated and is difficult to use. Therefore, we consider a variable conversion and derive an approximated normal distribution, and do range analysis with the approximated normal distribution.

5. Approximated Normal Distribution for Converted Variable

We convert the variable r to ξ as

$$\xi = \frac{1}{2}\ln\left(\frac{1+r}{1-r}\right) \tag{126}$$

This can be expanded as

$$\begin{aligned}\xi &= \frac{1}{2}\ln\left(\frac{1+r}{1-r}\right) \\ &= r + \frac{1}{3}r^3 + \frac{1}{5}r^5 + \cdots \\ &= \sum_{k=1}^{\infty}\frac{1}{2k-1}r^{2k-1}\end{aligned} \tag{127}$$

This changes a variable range from $-\infty$ to ∞ when r changes from -1 to +1 as shown in Figure 1. Let us consider the region for $r>0$. When r is less than0.5, ξ is almost equal to r, and then it deviate from r and become much larger than r when r approaches to 1, which is the required feature.

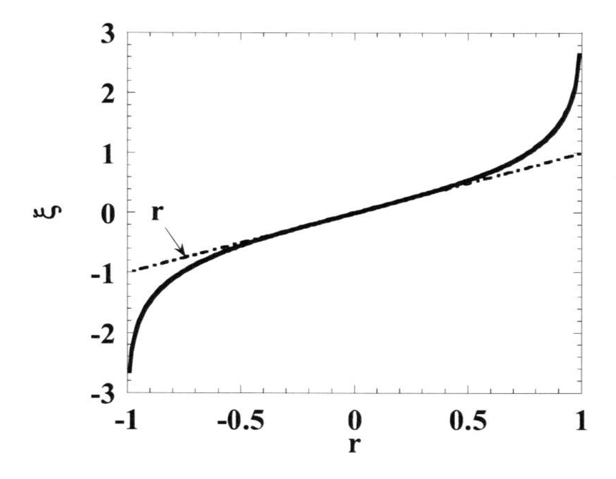

Figure 1. The relationship between ξ and sample correlation factor r.

We then derive a probability distribution for ξ.
We solve Eq. (126) with respect to r and obtain

$$r = \tanh \xi \tag{128}$$

Each increment is related to each other as

$$dr = \frac{1}{\cosh^2 \xi} d\xi \tag{129}$$

Therefore, the corresponding probability distribution is given by

$$f(\xi) = \frac{\left(1-\rho^2\right)^{\frac{n-1}{2}}}{\sqrt{\pi}\Gamma\left(\frac{n-1}{2}\right)\Gamma\left(\frac{n-2}{2}\right)} \times \frac{\left(1-\tanh^2 \xi\right)^{\frac{n-4}{2}}}{\cosh^2 \xi} \sum_{m=0}^{\infty} \frac{\left(2\rho \tanh \xi\right)^m}{m!} \Gamma^2\left(\frac{n-1+m}{2}\right) \tag{130}$$

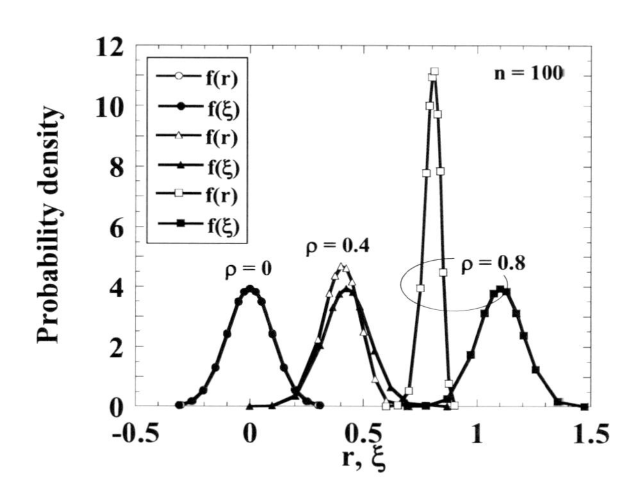

Figure 2. Probability distribution function for r and ξ.

Figure 2 shows a probability distribution for r and ξ with a sample number of 100. $f(r)$ becomes narrow and asymmetrical with increasing ρ, which is due to the fact the sample correlation factor has a upper limit of 1.

On the other hand, $f(\xi)$ has the almost identical shape and move parallel. $f(\xi)$ and $f(r)$ becomes same as the absolute value of ρ decreases.

In the small ρ, we can expect a small r. We can then approximate ξ as

$$\xi = r + \frac{1}{3}r^3 + \frac{1}{5}r^5 + \cdots$$
$$\approx r \tag{131}$$

Therefore, $f(r)$ and $f(\xi)$ are the same.

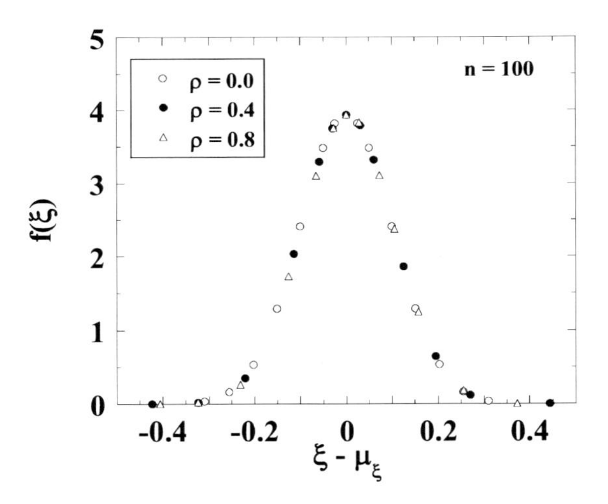

Figure 3. The probability distribution with respect to the deviation of ξ .

We study the average of the ξ .

We guess that the average of ξ can be obtained by substituting ρ into Eq. (126) as

$$\mu_\xi = \frac{1}{2}\ln\left(\frac{1+\rho}{1-\rho}\right) \tag{132}$$

Therefore, we replot by changing the lateral axis in Figure 2 from ξ to $\xi - \mu_\xi$, which is shown in Figure 3. All data for $\rho = 0, 0.4, 0.8$ become almost the same. Therefore, the average is well expressed by Eq. (132), and all the profiles are the same if we plot them with a lateral axis of $\xi - \mu_\xi$.

Next, we consider the variance of the profile.

When we use a lateral axis as $\xi - \mu_\xi$, the profiles do not have ρ dependence. Therefore, we set $\rho = 0$. In this case, $f(\xi) \approx f(r)$, and the probability distribution $f(t)$ converted from $f(r)$ is a t distribution with a freedom of $n-2$. The variance $\sigma_{t(n-2)}^{(2)}$ of the t distribution with a freedom of $n-2$ is given by

$$\sigma_{t(n-2)}^{(2)} = \frac{n-2}{n-4} \tag{133}$$

We re-express the relationship between t and r as

$$t = \sqrt{n-2}\frac{r}{\sqrt{1-r^2}} \tag{134}$$

This is not so simple relationship, but we approximate that r is small, and obtain

$$t \approx \sqrt{n-2}r \tag{135}$$

We then have

$$r \approx \frac{t}{\sqrt{n-2}} \tag{136}$$

In this case, the variance for r expressed by $\sigma_{r(n-2)}^{(2)}$ is given by

$$\begin{aligned} \sigma_r^{(2)} &= \frac{1}{\left(\sqrt{n-2}\right)^2}\sigma_{t(n-2)}^{(2)} \\ &= \frac{1}{n-2}\frac{n-2}{n-4} \\ &= \frac{1}{n-4} \end{aligned} \tag{137}$$

This is the approximated one. When n is large, we can expect that this treatment is accurate. However, we may not expect accurate one for small n. Therefore, we propose that we modify the model of Eq. (137) and change the variable as

$$\sigma_\xi^{(2)} = \frac{1}{n-\alpha} \tag{138}$$

This model includes a parameter α. We decide the value by comparing the calculated results. In the standard text book, α is set at 3.

The model and numerical data agree well with $\alpha = 2.5$ as shown in Figure 4.

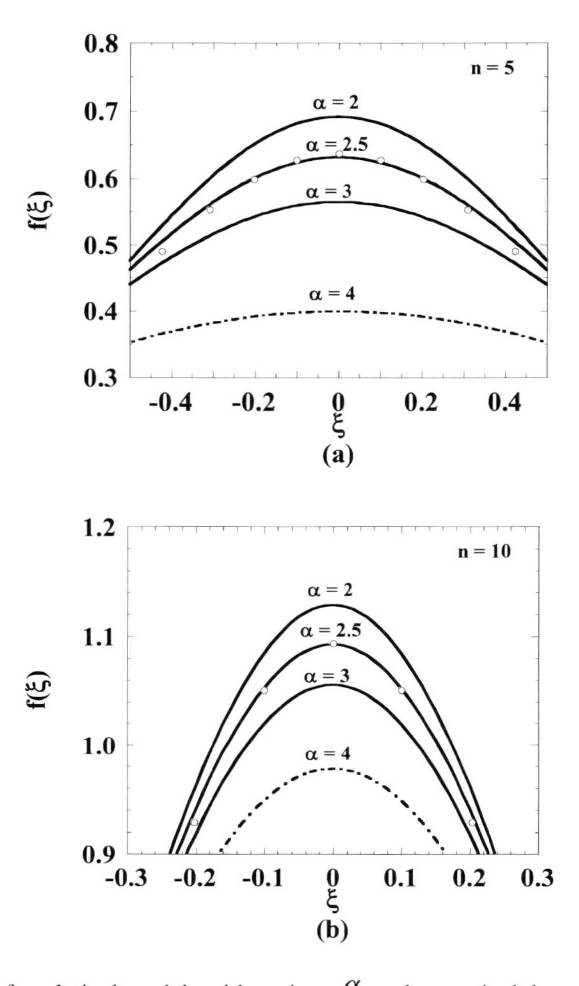

Figure 4. Comparison of analytical models with various α and numerical data.

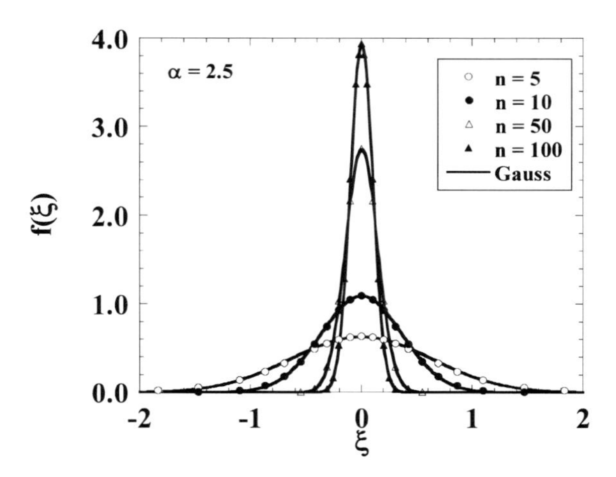

Figure 5. Comparison between analytical model and numerical model for various n.

Figure 5 compares an analytical model with $\alpha = 2.5$ and a numerical model for various n. An analytical model well reproduces the numerical data for any n.

Therefore, the probability distribution for ξ can be approximated with a normal distribution as

$$
\begin{aligned}
f(\xi) &= \frac{\left(1-\rho^2\right)^{\frac{n-1}{2}}}{\sqrt{\pi}\,\Gamma\left(\frac{n-1}{2}\right)\Gamma\left(\frac{n-2}{2}\right)} \times \frac{\left(1-\tanh^2\xi\right)^{\frac{n-4}{2}}}{\cosh^2\xi} \sum_{m=0}^{\infty} \frac{\left(2\rho\tanh\xi\right)^m}{m!}\Gamma^2\left(\frac{n-1+m}{2}\right) \\
&\approx \frac{1}{\sqrt{2\pi}\frac{1}{\sqrt{n-2.5}}} \exp\left[-\frac{\left(\xi - \frac{1}{2}\ln\left(\frac{1+\rho}{1-\rho}\right)\right)^2}{2\frac{1}{n-2.5}}\right]
\end{aligned}
\tag{139}
$$

When we perform the conversion given by

$$
z = \frac{\xi - \frac{1}{2}\ln\left(\frac{1+\rho}{1-\rho}\right)}{\sqrt{\frac{1}{n-2.5}}}
\tag{140}
$$

z follows a standard normal distribution as

$$
f(z) = \frac{1}{\sqrt{2\pi}}\exp\left(-\frac{z^2}{2}\right)
\tag{141}
$$

We finally evaluate the normal distribution approximation for $\rho \neq 0$.

We use a sample number $n = 5$ where the accuracy is sensitive. The analytical model reproduces the numerical data with $\alpha = 2.5$ as shown in Figure 6.

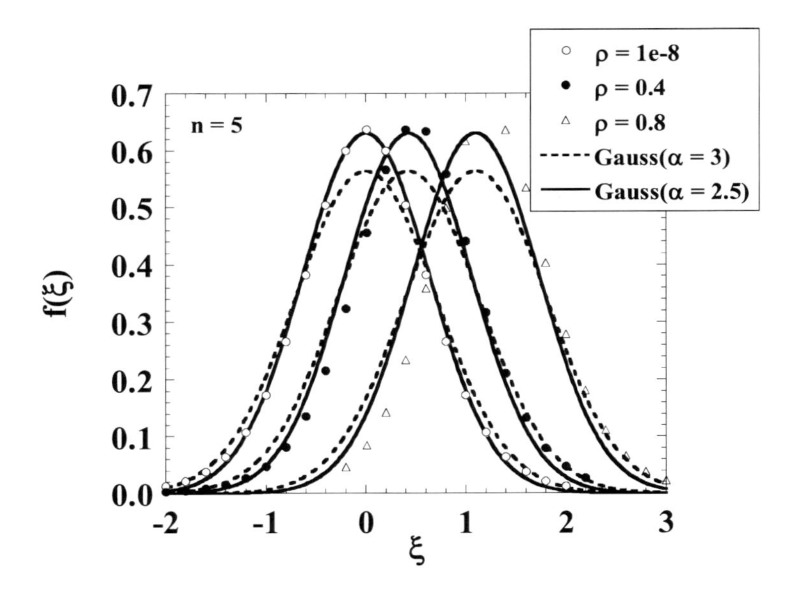

Figure 6. Comparison of numerical and analytical model for various ρ at the sample number of 5. The parameters for the analytical model are set at 3 and 2.5.

6. Prediction of Correlation Factor

We showed that the converted variable follows a normal distribution and we further obtained approximated averages and variances. Therefore, related prediction range is given by

$$\frac{1}{2}\ln\left(\frac{1+r_{xy}}{1-r_{xy}}\right)-z_P\frac{1}{\sqrt{n-2.5}}\leq\frac{1}{2}\ln\left(\frac{1+\rho}{1-\rho}\right)\leq\frac{1}{2}\ln\left(\frac{1+r_{xy}}{1-r_{xy}}\right)+z_P\frac{1}{\sqrt{n-2.5}} \tag{142}$$

We then obtain the following form

$$\rho_{\min}\leq\rho\leq\rho_{\max} \tag{143}$$

where $\rho_{\max}$ is evaluated from the equation of

$$\frac{1}{2}\ln\left(\frac{1+\rho_{\max}}{1-\rho_{\max}}\right)=\frac{1}{2}\ln\left(\frac{1+r_{xy}}{1-r_{xy}}\right)+z_P\frac{1}{\sqrt{n-2.5}} \tag{144}$$

We set G_H as

$$G_H \equiv \frac{1}{2}\ln\left(\frac{1+r_{xy}}{1-r_{xy}}\right)+z_P\frac{1}{\sqrt{n-2.5}} \tag{145}$$

and obtain $\rho_{\max}$ as a function of G_H as below.

$$\begin{aligned}
&\ln\left(\frac{1+\rho_{\max}}{1-\rho_{\max}}\right)=2G_H\\
&\frac{1+\rho_{\max}}{1-\rho_{\max}}=e^{2G_H}\\
&1+\rho_{\max}=\left(1-\rho_{\max}\right)e^{2G_H}\\
&\left(1+e^{2G_H}\right)\rho_{\max}=e^{2G_H}-1\\
&\rho_{\max}=\frac{e^{2G_H}-1}{e^{2G_H}+1}
\end{aligned} \tag{146}$$

Similarly, we obtain $\rho_{\min}$ as

$$\rho_{\min}=\frac{e^{2G_L}-1}{e^{2G_L}+1} \tag{147}$$

where

$$G_L \equiv \frac{1}{2}\ln\left(\frac{1+r_{xy}}{1-r_{xy}}\right)-z_P\frac{1}{\sqrt{n-2.5}} \tag{148}$$

Figure 7 shows the prediction range for a sample correlation factor. The range is very wide, and hence we need a sample number at least 50 to ensure the range is less than 0.5.

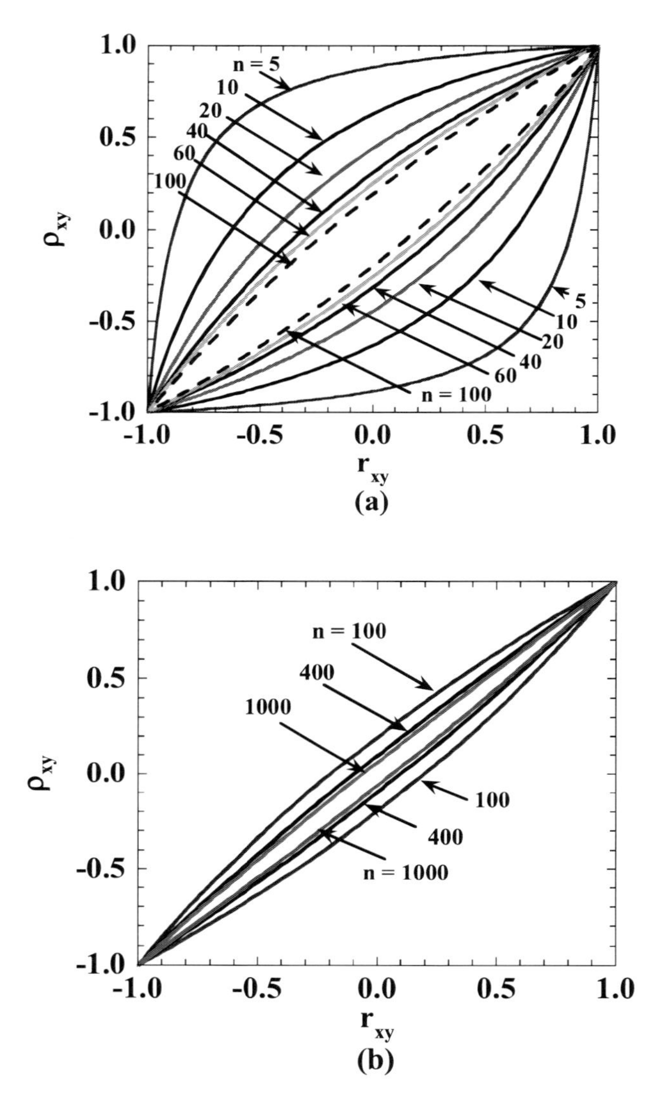

Figure 7. Prediction range for sample correlation factor. (a) $n \leq 100$, (b) $n \geq 100$.

Let us study Eq. (143) in more detail.

When n is sufficiently large, we can neglect the second term in G_H , and it is reduced to

$$G_H \approx \frac{1}{2}\ln\left(\frac{1+r_{xy}}{1-r_{xy}}\right) \tag{149}$$

Then the ρ_{max} is also reduced to

$$\rho_{\max} = \frac{e^{\ln\left(\frac{1+r_{xy}}{1-r_{xy}}\right)} - 1}{e^{\ln\left(\frac{1+r_{xy}}{1-r_{xy}}\right)} + 1} = \frac{\frac{1+r_{xy}}{1-r_{xy}} - 1}{\frac{1+r_{xy}}{1-r_{xy}} + 1} = \frac{2r_{xy}}{2} = r_{xy} \tag{150}$$

Then, the sample correlation factor becomes that of population.
When n is not so sufficiently large, we can approximate as follows.

$$\begin{aligned} \exp(2G_H) &= \exp\left(\ln\left(\frac{1+r_{xy}}{1-r_{xy}}\right) + z_P \frac{2}{\sqrt{n-2.5}}\right) \\ &= \frac{1+r_{xy}}{1-r_{xy}} \exp\left(z_P \frac{2}{\sqrt{n-2.5}}\right) \\ &\approx \frac{1+r_{xy}}{1-r_{xy}}\left(1 + z_P \frac{2}{\sqrt{n-2.5}}\right) \end{aligned} \tag{151}$$

Therefore, we obtain

$$\rho_{\max} = \frac{\frac{1+r_{xy}}{1-r_{xy}}\left(1 + z_P \frac{2}{\sqrt{n-2.5}}\right) - 1}{\frac{1+r_{xy}}{1-r_{xy}}\left(1 + z_P \frac{2}{\sqrt{n-2.5}}\right) + 1} = \frac{r_{xy} + z_P \frac{1+r_{xy}}{\sqrt{n-2.5}}}{1 + z_P \frac{1+r_{xy}}{\sqrt{n-2.5}}} \tag{152}$$

Similarly, we obtain an approximated $\rho_{\min}$ as

$$\rho_{\min} = \frac{r_{xy} - z_P \frac{1 + r_{xy}}{\sqrt{n - 2.5}}}{1 - z_P \frac{1 + r_{xy}}{\sqrt{n - 2.5}}} \tag{153}$$

The approximated model is compared with the rigorous one as shown in Figure 8. The agreement is not good for n less than 50, but the simple model does work for n larger than 100. The accuracy of the model is related to the Taylor expansion, and it is expressed by

$$n \gg 3 + 4z_p^2 \tag{154}$$

We decide the prediction probability as 0.95, and corresponding z_p is 1.96. Therefore, Eq. (154) is reduced to

$$n \gg 3 + 4 \times 1.96^2 = 18.4 \tag{155}$$

This well explains the results in Figure 8.

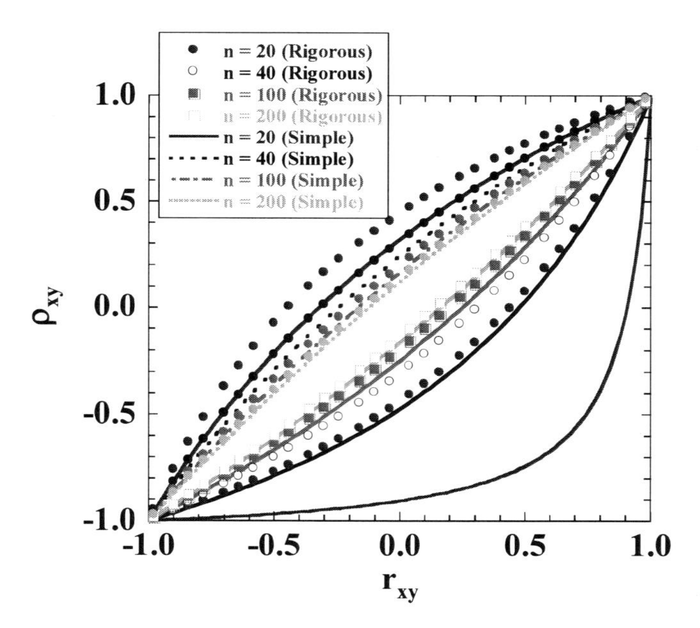

Figure 8. Comparison between simple and rigorous correlation factor models.

7. PREDICTION WITH MORE RIGOROUS AVERAGE MODEL

Let us inspect the Figure 6 in more detail. The analytical normal distribution deviates from the numerical data for large ρ. The analytical models shift left slightly with increasing ρ. This is due to the inaccurate approximation of the model for an average.

This is can be improved by using a model given by

$$\mu_\xi = \frac{1}{2}\ln\left(\frac{1+\rho}{1-\rho}\right) + \frac{\rho}{2(n-1)} \tag{156}$$

Figure 9 compares the modified analytical model and numerical data, where both agree well for various ρ.

Finally, we express a probability distribution function as

$$\begin{aligned} f(\xi) &= \frac{\left(1-\rho^2\right)^{\frac{n-1}{2}}}{\sqrt{\pi}\Gamma\left(\frac{n-1}{2}\right)\Gamma\left(\frac{n-2}{2}\right)} \times \frac{\left(1-\tanh^2\xi\right)^{\frac{n-4}{2}}}{\cosh^2\xi} \sum_{m=0}^{\infty} \frac{(2\rho\tanh\xi)^m}{m!}\Gamma^2\left(\frac{n-1+m}{2}\right) \\ &\approx \frac{1}{\sqrt{2\pi}\frac{1}{\sqrt{n-2.5}}} \exp\left[-\frac{\left(\xi-\left[\frac{1}{2}\ln\left(\frac{1+\rho}{1-\rho}\right)+\frac{\rho}{2(n-1)}\right]\right)^2}{\frac{2}{n-2.5}}\right] \end{aligned} \tag{157}$$

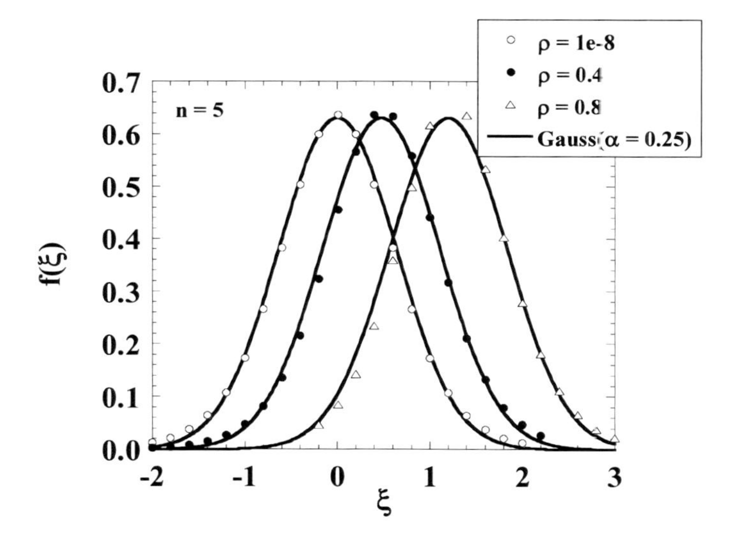

Figure 9. Comparison with analytical data with modified average model with numerical data.

Therefore, we use a normal distribution with an average of $\frac{1}{2}\ln\left(\frac{1+\rho}{1-\rho}\right)+\frac{\rho}{2(n-1)}$, and a standard deviation of $1/\sqrt{n-2.5}$.

Consequently, we set the normalized variable as

$$z=\frac{\frac{1}{2}\ln\left(\frac{1+r_{xy}}{1-r_{xy}}\right)-\left[\frac{1}{2}\ln\left(\frac{1+\rho}{1-\rho}\right)+\frac{1}{2(n-1)}\right]}{\frac{1}{\sqrt{n-2.5}}} \tag{158}$$

Then, it follows a standard normal distribution given by

$$f(z)=\frac{1}{\sqrt{2\pi}}\exp\left(-\frac{z^2}{2}\right) \tag{159}$$

Based on the above analysis, we predict the range of correlation factor.

First of all, we want to obtain the central value of ρ denoted as ρ_{cent}, which is commonly regarded as r_{xy}. We can improve the accuracy as follows.

ρ_{cent} should satisfy the relationship given by

$$\frac{1}{2}\ln\left(\frac{1+\rho_{cent}}{1-\rho_{cent}}\right)+\frac{\rho_{cent}}{2(n-1)}=\frac{1}{2}\ln\left(\frac{1+r_{xy}}{1-r_{xy}}\right) \tag{160}$$

If there is no the second term in the left side of Eq. (160), we obtain

$$\rho_{cent}=r_{xy} \tag{161}$$

We construct a function given by

$$f_{cent}(\rho)=\frac{1}{2}\ln\left(\frac{1+\rho}{1-\rho}\right)+\frac{\rho}{2(n-1)}-\frac{1}{2}\ln\left(\frac{1+r_{xy}}{1-r_{xy}}\right) \tag{162}$$

ρ_{cent} is a root of the equation given by

$$f_{cent}(\rho_{cent}) = 0 \tag{163}$$

We can expect this ρ_{cent} is not far from the r_{xy} from Figure 6 and Figure 9. Therefore, we can expect

$$f_{cent}(\rho) = f(r_{xy}) + f_{cent}'(r_{xy})(\rho - r_{xy}) \tag{164}$$

Therefore, we obtain

$$0 = f(r_{xy}) + f'(r_{xy})(\rho_{cent} - r_{xy}) \tag{165}$$

Therefore, we obtain ρ_{cent} as

$$\rho_{cent} = r_{xy} - \frac{f_{cent}(r_{xy})}{f_{cent}'(r_{xy})} \tag{166}$$

where

$$\begin{aligned} f_{cent}(r_{xy}) &= \frac{1}{2}\ln\ln\left(\frac{1+r_{xy}}{1-r_{xy}}\right) + \frac{r_{xy}}{2(n-1)} - \frac{1}{2}\ln\left(\frac{1+r_{xy}}{1-r_{xy}}\right) \\ &= \frac{r_{xy}}{2(n-1)} \end{aligned} \tag{167}$$

$$\begin{aligned} f_{cent}'(r_{xy}) &= \frac{1}{2}\left[\frac{1}{1+r_{xy}} + \frac{1}{1-r_{xy}}\right] + \frac{1}{2(n-1)} \\ &= \frac{1}{1-r_{xy}^2} + \frac{1}{2(n-1)} \end{aligned} \tag{168}$$

Therefore, we obtain

$$\rho_{cent} = r_{xy} - \frac{\frac{1}{2(n-1)}}{\frac{1}{1-r_{xy}^2} + \frac{1}{2(n-1)}} r_{xy}$$
$$= \frac{\frac{1}{1-r_{xy}^2}}{\frac{1}{1-r_{xy}^2} + \frac{1}{2(n-1)}} r_{xy} \tag{169}$$

Next, we evaluate the range of the correlation factor range.

The related estimated range is expressed by

$$\frac{1}{2}\ln\left(\frac{1+r_{xy}}{1-r_{xy}}\right) - z_P \frac{1}{\sqrt{n-2.5}} \le \frac{1}{2}\ln\left(\frac{1+\rho}{1-\rho}\right) + \frac{\rho}{2(n-1)} \le \frac{1}{2}\ln\left(\frac{1+r_{xy}}{1-r_{xy}}\right) + z_P \frac{1}{\sqrt{n-2.5}} \tag{170}$$

Form this equation, we want to obtain the final form below.

$$\rho_{\min} \le \rho \le \rho_{\max} \tag{171}$$

We cannot obtain analytical forms for $\rho_{\min}$ and $\rho_{\max}$ from the Eq. (170).

Therefore, we show approximated analytical forms and a numerical procedure to obtain rigorous one.

We first neglect a correlation term of $\rho/[2(n-1)]$, and set $\rho_{\max}, \rho_{\min}$ under this treatment as $\rho_{\max 0}, \rho_{\min 0}$.

$\rho_{\max 0}$ can be evaluated from the equation given by

$$\frac{1}{2}\ln\left(\frac{1+\rho_{\max 0}}{1-\rho_{\max 0}}\right) = \frac{1}{2}\ln\left(\frac{1+r_{xy}}{1-r_{xy}}\right) + z_P \frac{1}{\sqrt{n-2.5}} \tag{172}$$

We introduce a variable as

$$G_H \equiv \frac{1}{2}\ln\left(\frac{1+r_{xy}}{1-r_{xy}}\right) + z_P \frac{1}{\sqrt{n-2.5}} \tag{173}$$

We then have an analytical form for $\rho_{\max 0}$ as below.

$$\begin{aligned}
&\ln\left(\frac{1+\rho_{\max 0}}{1-\rho_{\max 0}}\right) = 2G_H \\
&\frac{1+\rho_{\max 0}}{1-\rho_{\max 0}} = e^{2G_H} \\
&1+\rho_{\max 0} = \left(1-\rho_{\max 0}\right)e^{2G_H} \\
&\left(1+e^{2G_H}\right)\rho_{\max 0} = e^{2G_H} - 1 \\
&\rho_{\max 0} = \frac{e^{2G_H}-1}{e^{2G_H}+1}
\end{aligned} \tag{174}$$

Similarly, we obtain

$$\rho_{\min 0} = \frac{e^{2G_L}-1}{e^{2G_L}+1} \tag{175}$$

where

$$G_L \equiv \frac{1}{2}\ln\left(\frac{1+r_{xy}}{1-r_{xy}}\right) - z_P\frac{1}{\sqrt{n-2.5}} \tag{176}$$

Next, we include the correlation factor.

$\rho_{\max}$ can be evaluated from the equation given by

$$\frac{1}{2}\ln\left(\frac{1+\rho_{\max}}{1-\rho_{\max}}\right) + \frac{\rho_{\max}}{2(n-1)} = \frac{1}{2}\ln\left(\frac{1+r_{xy}}{1-r_{xy}}\right) + z_P\frac{1}{\sqrt{n-2.5}} \tag{177}$$

We construct a function given by

$$f(\rho) = \frac{1}{2}\ln\left(\frac{1+\rho}{1-\rho}\right) + \frac{\rho}{2(n-1)} - \frac{1}{2}\ln\left(\frac{1+r_{xy}}{1-r_{xy}}\right) - z_P\frac{1}{\sqrt{n-2.5}} \tag{178}$$

ρ_{max} is a root of the equation given by

$$f\left(\rho_{max}\right)=0 \tag{179}$$

We can expect that this ρ_{max} is not far from the $\rho_{max\,0}$ from Figure 6 and Figure 9. Therefore, we can expect

$$f\left(\rho\right)=f\left(\rho_{max\,0}\right)+f'\left(\rho_{max\,0}\right)\left(\rho-\rho_{max\,0}\right) \tag{180}$$

We then obtain

$$0=f\left(\rho_{max\,0}\right)+f'\left(\rho_{max\,0}\right)\left(\rho_{max}-\rho_{max\,0}\right) \tag{181}$$

Therefore, we obtain ρ_{max} as

$$\rho_{max}=\rho_{max\,0}-\frac{f\left(\rho_{max\,0}\right)}{f'\left(\rho_{max\,0}\right)} \tag{182}$$

where

$$\begin{aligned} f\left(\rho_{max\,0}\right) &= \frac{1}{2}\ln\left(\frac{1+\rho_{max\,0}}{1-\rho_{max\,0}}\right)+\frac{\rho_{max\,0}}{2\left(n-1\right)}-\frac{1}{2}\ln\left(\frac{1+r_{xy}}{1-r_{xy}}\right)-z_P\frac{1}{\sqrt{n-2.5}} \\ &= \frac{\rho_{max\,0}}{2\left(n-1\right)} \end{aligned} \tag{183}$$

$$\begin{aligned} f'\left(\rho_{max\,0}\right) &= \frac{1}{2}\left[\frac{1}{1+\rho_{max\,0}}+\frac{1}{1-\rho_{max\,0}}\right]+\frac{1}{2\left(n-1\right)} \\ &= \frac{1}{1-\rho^2_{max\,0}}+\frac{1}{2\left(n-1\right)} \end{aligned} \tag{184}$$

Therefore, we obtain

$$\rho_{\max} = \rho_{\max 0} - \frac{\frac{1}{2(n-1)}}{\frac{1}{1-\rho^2_{\max 0}} + \frac{1}{2(n-1)}} \rho_{\max 0}$$

$$= \frac{\frac{1}{1-\rho^2_{\max 0}}}{\frac{1}{1-\rho^2_{\max 0}} + \frac{1}{2(n-1)}} \rho_{\max 0} \tag{185}$$

Similarly, we can obtain $\rho_{\min}$. $\rho_{\min}$ can be evaluated from the equation given by

$$\frac{1}{2}\ln\left(\frac{1+\rho_{\min}}{1-\rho_{\min}}\right) + \frac{\rho_{\min}}{2(n-1)} = \frac{1}{2}\ln\left(\frac{1+r_{xy}}{1-r_{xy}}\right) - z_P \frac{1}{\sqrt{n-2.5}} \tag{186}$$

We introduce a function for $\rho_{\min}$ given by

$$g(\rho) = \frac{1}{2}\ln\left(\frac{1+\rho}{1-\rho}\right) + \frac{\rho}{2(n-1)} - \frac{1}{2}\ln\left(\frac{1+r_{xy}}{1-r_{xy}}\right) + z_P \frac{1}{\sqrt{n-2.5}} \tag{187}$$

$\rho_{\min}$ is given by

$$\rho_{\min} = \rho_{\min 0} - \frac{g(\rho_{\min 0})}{g'(\rho_{\min 0})} \tag{188}$$

where

$$g'(\rho_{\min 0}) = \frac{1}{1-\rho^2_{\min 0}} + \frac{1}{2(n-1)} \tag{189}$$

We then obtain

$$\rho_{\min} = \frac{\frac{1}{1-\rho^2_{\min 0}}}{\frac{1}{1-\rho^2_{\min 0}} + \frac{1}{2(n-1)}} \rho_{\min 0} \tag{190}$$

8. Testing of Correlation Factor

The sample correlation factor is not 0 even when correlation factor of the population is 0. We can test it setting $\rho = 0$ in Eq. (125), and obtain

$$f(r) = \frac{\Gamma\left(\frac{n-1}{2}\right)}{\sqrt{\pi}\Gamma\left(\frac{n-2}{2}\right)}\left(1-r^2\right)^{\frac{n-4}{2}} \tag{191}$$

We introduce a variable

$$t = \sqrt{n-2}\frac{r}{\sqrt{1-r^2}} \tag{192}$$

The corresponding increment is

$$\begin{aligned} dt &= \sqrt{n-2}\frac{\sqrt{1-r^2} - r\frac{-2r}{2\sqrt{1-r^2}}}{1-r^2}dr \\ &= \sqrt{n-2}\frac{\frac{1-r^2+r^2}{\sqrt{1-r^2}}}{1-r^2}dr \\ &= \sqrt{n-2}\frac{1}{\left(1-r^2\right)^{\frac{3}{2}}}dr \end{aligned} \tag{193}$$

The relationship between t and r is modified from Eq. (192) as

$$t^2\left(1-r^2\right) = (n-2)r^2 \tag{194}$$

We solve Eq. (194) with respect to r^2, and obtain

$$r^2 = \frac{t^2}{(n-2)+t^2} \tag{195}$$

Therefore, the probability distribution for r is given by

$$
\begin{aligned}
f(r)dr &= \frac{\Gamma\left(\frac{n-1}{2}\right)}{\sqrt{\pi}\Gamma\left(\frac{n-2}{2}\right)}\left(1-r^2\right)^{\frac{n-4}{2}} dr \\
&= \frac{\Gamma\left(\frac{n-1}{2}\right)}{\sqrt{\pi}\Gamma\left(\frac{n-2}{2}\right)}\left(1-r^2\right)^{\frac{n-4}{2}} \frac{1}{\frac{\sqrt{n-2}}{\left(1-r^2\right)^{\frac{3}{2}}}} dt \\
&= \frac{1}{\sqrt{n-2}} \frac{\Gamma\left(\frac{n-1}{2}\right)}{\sqrt{\pi}\Gamma\left(\frac{n-2}{2}\right)}\left(1-r^2\right)^{\frac{n-1}{2}} dt \\
&= \frac{1}{\sqrt{n-2}} \frac{\Gamma\left(\frac{n-1}{2}\right)}{\sqrt{\pi}\Gamma\left(\frac{n-2}{2}\right)}\left(1-\frac{t^2}{(n-2)+t^2}\right)^{\frac{n-1}{2}} dt \\
&= \frac{1}{\sqrt{n-2}} \frac{\Gamma\left(\frac{n-1}{2}\right)}{\sqrt{\pi}\Gamma\left(\frac{n-2}{2}\right)}\left(\frac{n-2}{(n-2)+t^2}\right)^{\frac{n-1}{2}} dt \\
&= \frac{1}{\sqrt{n-2}} \frac{\Gamma\left(\frac{n-1}{2}\right)}{\sqrt{\pi}\Gamma\left(\frac{n-2}{2}\right)}\left(\frac{1}{1+\frac{t^2}{n-2}}\right)^{\frac{n-1}{2}} dt \\
&= \frac{1}{\sqrt{n-2}} \frac{\Gamma\left(\frac{n-1}{2}\right)}{\sqrt{\pi}\Gamma\left(\frac{n-2}{2}\right)}\left(1+\frac{t^2}{n-2}\right)^{-\frac{n-1}{2}} dt \\
&= \frac{1}{\sqrt{n-2}} \frac{\Gamma\left(\frac{(n-2)+1}{2}\right)}{\sqrt{\pi}\Gamma\left(\frac{n-2}{2}\right)}\left(1+\frac{t^2}{n-2}\right)^{-\frac{(n-2)+1}{2}} dt
\end{aligned}
\tag{196}
$$

Therefore, the probability distribution function $f(r)$ is converted to

$$
f(t) = \frac{1}{\sqrt{n-2}} \frac{\Gamma\left(\frac{(n-2)+1}{2}\right)}{\sqrt{\pi}\Gamma\left(\frac{n-2}{2}\right)}\left(1+\frac{t^2}{n-2}\right)^{-\frac{(n-2)+1}{2}} \tag{197}
$$

This is a t distribution with a freedom of $n-2$.
When we obtain sample correlation factor r from n samples, we evaluate t given by

$$
t = \sqrt{n-2}\frac{r}{\sqrt{1-r^2}} \tag{198}
$$

We decide a prediction probability P and obtain the corresponding P point of t_P with a t distribution with a freedom of $n-2$, and compare this with Eq.(198), and perform a test as

$$\begin{cases} t \le t_p & \text{Two variables do not dependent each other} \\ t > t_p & \text{Two variables depend on each other} \end{cases} \tag{199}$$

9. Testing of Two Correlation Factor Difference

We want to test that the two groups correlation factors are the same or not.

We treat two groups A and B and corresponding correlation factors and sample numbers are r_A and r_B, and n_A and n_B. We introduce variables given by

$$z_A = \frac{1}{2}\ln\left(\frac{1+r_A}{1-r_A}\right) \tag{200}$$

$$z_B = \frac{1}{2}\ln\left(\frac{1+r_B}{1-r_B}\right) \tag{201}$$

The average of z_A and z_B are 0. The corresponding variances are given by

$$\sigma_A^{(2)} = \frac{1}{n_A - 2.5} \tag{202}$$

$$\sigma_B^{(2)} = \frac{1}{n_B - 2.5} \tag{203}$$

We introduce a variable as

$$\begin{aligned} Z &= z_A - z_B \\ &= \frac{1}{2}\ln\left(\frac{1+r_A}{1-r_A}\right) - \frac{1}{2}\ln\left(\frac{1+r_B}{1-r_B}\right) \\ &= \frac{1}{2}\ln\left(\frac{1+r_A}{1+r_B}\frac{1-r_B}{1-r_A}\right) \end{aligned} \tag{204}$$

The $r_A = r_B$ is identical to $Z = 0$.

The average of Z is zero, and variance $\sigma^{(2)}$ is

$$\sigma^{(2)} = \sigma_A^{(2)} + \sigma_B^{(2)} \tag{205}$$

Therefore, we can evaluate corresponding z_P for a normal distribution and we can judge whether the correlation factor is the same if the following is valid.

$$|Z| < z_p \tag{206}$$

SUMMARY

The probability distribution function of two variables is given by

$$f(x_1, x_2) = \frac{1}{2\pi\sqrt{(1-\rho^2)\sigma_1^{(2)}\sigma_2^{(2)}}} \exp\left[-\frac{1}{2(1-\rho^2)}\left(\frac{x_1^{\ 2}}{\sigma_1^{(2)}} - 2\rho\frac{x_1 x_2}{\sqrt{\sigma_1^{(2)}\sigma_2^{(2)}}} + \frac{x_2^{\ 2}}{\sigma_2^{(2)}}\right)\right]$$

This can be extended to multi variables given by

$$f(x_1, x_2, \cdots, x_m) = \frac{1}{(2\pi)^{\frac{m}{2}}\sqrt{|\Sigma|}} \exp\left(-\frac{D^2}{2}\right)$$

where

$$\Sigma = \begin{pmatrix} \sigma_{11}^{(2)} & \sigma_{12}^{(2)} & \cdots & \sigma_{1m}^{(2)} \\ \sigma_{21}^{(2)} & \sigma_{22}^{(2)} & \cdots & \sigma_{2m}^{(2)} \\ \vdots & \vdots & \ddots & \vdots \\ \sigma_{m1}^{(2)} & \sigma_{m2}^{(2)} & \cdots & \sigma_{mm}^{(2)} \end{pmatrix}$$

and D is given by

$$D^2 = \left(x_1 - \mu_1, x_2 - \mu_2, \cdots, x_m - \mu_m\right) \begin{pmatrix} \sigma^{11(2)} & \sigma^{12(2)} & \cdots & \sigma^{1m(2)} \\ \sigma^{21(2)} & \sigma^{22(2)} & \cdots & \sigma^{2m(2)} \\ \vdots & \vdots & \ddots & \vdots \\ \sigma^{m1(2)} & \sigma^{m2(2)} & \cdots & \sigma^{mm(2)} \end{pmatrix} \begin{pmatrix} x_1 - \mu_1 \\ x_2 - \mu_2 \\ \vdots \\ x_m - \mu_m \end{pmatrix}$$

The probability distribution for a sample correlation factor is given by

$$f(r) = \frac{\left(1-\rho^2\right)^{\frac{n-1}{2}} \left(1-r^2\right)^{\frac{n-4}{2}}}{\sqrt{\pi}\Gamma\left(\frac{n-1}{2}\right)\Gamma\left(\frac{n-2}{2}\right)} \times \sum_{m=0}^{\infty} \frac{(2r\rho)^m}{m!} \Gamma^2\left(\frac{n-1+m}{2}\right)$$

Introducing a variable ξ, the distribution function is converted to

$$\xi = \frac{1}{2}\ln\left(\frac{1+r}{1-r}\right)$$

Performing numerical study, we show that the average and variance of this function are given by

$$\mu_\xi = \frac{1}{2}\ln\left(\frac{1+\rho}{1-\rho}\right) + \frac{\rho}{2(n-1)}$$

$$\sigma_\xi^{(2)} = \frac{1}{n-2.5}$$

Therefore, the distribution is approximately expressed by

$$f(\xi) = \frac{1}{\sqrt{2\pi\sigma_\xi^{(2)}}} \exp\left[-\frac{\left(\xi - \mu_\xi\right)^2}{2\sigma_\xi^{(2)}}\right]$$

The central correlation factor in the population ρ_{cent} is given by

$$\rho_{cent} = \frac{\frac{1}{1-r_{xy}^2}}{\frac{1}{1-r_{xy}^2} + \frac{1}{2(n-1)}} r_{xy}$$

We then obtain an error range as

$$\rho_{\min} \le \rho \le \rho_{\max}$$

where

$$\rho_{\max} = \frac{\dfrac{1}{1-\rho_{\max 0}^2}}{\dfrac{1}{1-\rho_{\max 0}^2} + \dfrac{1}{2(n-1)}} \rho_{\max 0}$$

$$\rho_{\min} = \frac{\dfrac{1}{1-\rho_{\min 0}^2}}{\dfrac{1}{1-\rho_{\min 0}^2} + \dfrac{1}{2(n-1)}} \rho_{\min 0}$$

$$\rho_{\max 0} = \frac{e^{2G_H} - 1}{e^{2G_H} + 1}$$

$$\rho_{\min 0} = \frac{e^{2G_L} - 1}{e^{2G_L} + 1}$$

$$G_H \equiv \frac{1}{2}\ln\left(\frac{1+r_{xy}}{1-r_{xy}}\right) + z_P \frac{1}{\sqrt{n-2.5}}$$

$$G_L \equiv \frac{1}{2}\ln\left(\frac{1+r_{xy}}{1-r_{xy}}\right) - z_P \frac{1}{\sqrt{n-2.5}}$$

The correlation factor distribution for $\rho = 0$ is given by

$$f(r) = \frac{\Gamma\left(\dfrac{n-1}{2}\right)}{\sqrt{\pi}\,\Gamma\left(\dfrac{n-2}{2}\right)} \left(1-r^2\right)^{\frac{n-4}{2}}$$

Introducing a variable

$$t = \sqrt{n-2}\frac{r}{\sqrt{1-r^2}}$$

We modify the distribution as

$$f(t) = \frac{1}{\sqrt{n-2}} \frac{\Gamma\left(\frac{(n-2)+1}{2}\right)}{\sqrt{\pi}\Gamma\left(\frac{n-2}{2}\right)} \left(1 + \frac{t^2}{n-2}\right)^{-\frac{(n-2)+1}{2}}$$

This is a t distribution with a freedom of $n-2$. We can hence compare this with the P point t_P with a freedom of $n-2$.

Chapter 3

REGRESSION ANALYSIS

ABSTRACT

We studied the significance of relationship between two variables by evaluating a correlation factor in the previous two chapters. We then predict the objective variable value for given explanatory variable values, which is called as the regression analysis. We also evaluate the accuracy of the regression line, and predictive a range of regression and objective value. We further evaluate the predictive range of factors of regression line, the predictive range of regression, and the range of object values. We first focus on the linear regression to express the relationship, and then extend it to various functions.

Keywords: regression line, linear function, logarithmic regression, power law function, multi-power law function, leverage ratio, logistic curve

1. INTRODUCTION

We can evaluate two variables such as smoking and life expectancy. We can obtain a corresponding correlation factor. The correlation factor value expresses the strength of the relationship between the two variables. We further want to predict the life expectancy of a person who takes one box tobacco per a day. We should express the relationship between two variables quantitatively, which we treat in this chapter. The relationship is expressed by a line in general, and it is called as regression line. Further, the relationship is not expressed with only a line, but may be with an arbitrary function.

We perform matrix operation for the analysis in this chapter. The matrix operation is summarized in Chapter 15 of volume 3.

2. PATH DIAGRAM

It is convenient to express the relationship between variables qualitatively. A path diagram is sometimes used to do it.

The evaluated variables are expressed by a square, and the correlation is expressed by an arrowed arch line, and the objective and explanatory variables are connected by an arrowed line where the arrow is terminated to the objected variable. The error is expressed by a circle.

Therefore, the objective variable y and the explanatory variable x are expressed as shown in Figure 1.

Figure 1. Path diagram for regression.

3. REGRESSION EQUATION FOR A POPULATION

First, we consider a regression for a population, where we obtain N sets of data $(x_1, y_1), (x_2, y_2), \cdots, (x_N, y_N)$. We obtain a scattered plot as shown in the previous chapter. We want to obtain a corresponding theory that expresses relationship between the two variables.

Table 1. Relationship between secondary industry number and detached house number

Prefecture ID	Scondary industry number	Detached house number
1	630	680
2	1050	1300
3	350	400
4	820	1240
5	1400	1300
6	1300	1500
7	145	195
8	740	1050
9	1200	1250
10	200	320
11	450	630
12	1500	1750

We obtain the relationship between a secondary industry number and a detached house number as shown in Table 1. Figure 2 shows the scattered plot of the Table 1. We can estimate the correlation factor as shown in the previous chapter, and it is given by

$$\rho_{xy} = 0.959 \tag{1}$$

That is, we have a strong relationship between the secondary industry number and the detached house number.

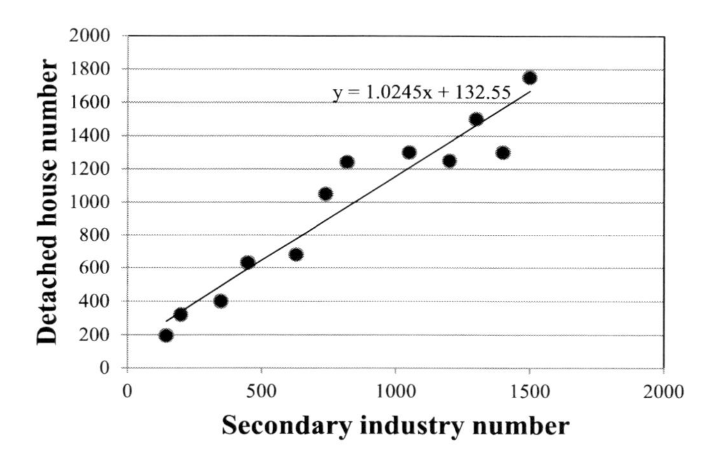

Figure 2. Dependence of detached house number on secondary industry number.

We assume that the detached house number data y_i is related to the secondary industry data x_i as

$$\begin{aligned} y_i &= a_0 + a_1 x_i + e_i \\ &= Y_i + e_i \end{aligned} \tag{2}$$

where a_0 and a_1 are a constant number independent on data i. e_i is the error associated with the data number i. Y_i is the regression value and is given by

$$Y_i = a_0 + a_1 x_i \tag{3}$$

We assume that the e_i follows the normal distribution with $N\left(0, \sigma_e^{(2)}\right)$, where

$$\sigma_e^{(2)} = \frac{1}{N}\sum_{i=1}^{N}(y_i - Y_i)^2 \tag{4}$$

and it is independent on variable x_i , that is

$$Cov[x_i, e_i] = 0 \tag{5}$$

We show how we can derive the regression line.

The deviation from the regression line and data e_i is expressed by

$$\begin{aligned} e_i &= y_i - Y_i \\ &= y_i - (a_0 + a_1 x_i) \end{aligned} \tag{6}$$

A variance of e_i is denoted as $\sigma_e^{(2)}$ and is given by

$$\begin{aligned} \sigma_e^{(2)} &= \frac{1}{N}\left(e_1^2 + e_2^2 + \cdots + e_N^2\right) \\ &= \frac{1}{N}\left\{\left[y_1 - (a_0 + a_1 x_1)\right]^2 + \left[y_2 - (a_0 + a_1 x_2)\right]^2 + \cdots + \left[y_N - (a_0 + a_1 x_N)\right]^2\right\} \end{aligned} \tag{7}$$

We decide variables a_0 and a_1 so that $\sigma_e^{(2)}$ has the minimum value.

We first differentiate $\sigma_e^{(2)}$ with respect to a_0 and set it to be zero as follows

$$\begin{aligned} \frac{\partial \sigma_e^{(2)}}{\partial a_0} &= \frac{\partial}{\partial a_0}\frac{1}{N}\sum_{i=1}^{N}\left[y_i - (a_0 + a_1 x_i)\right]^2 \\ &= -\frac{1}{N}\sum_{i=1}^{N} 2\left[y_i - (a_0 + a_1 x_i)\right] \\ &= -2\left\{\frac{1}{N}\sum_{i=1}^{N} y_i - a_1 \frac{1}{N}\sum_{i=1}^{N} x_i - a_0\right\} \\ &= -2\left\{\mu_y - a_1 \mu_x - a_0\right\} \\ &= 0 \end{aligned} \tag{8}$$

where

$$\mu_y = \frac{1}{N}\sum_{i=1}^{N} y_i \tag{9}$$

$$\mu_x = \frac{1}{N}\sum_{i=1}^{N} x_i \tag{10}$$

Therefore, we obtain

$$\mu_y = a_0 + a_1\mu_x \tag{11}$$

Differentiating $\sigma_e^{(2)}$ with respect to a_1, we obtain

$$\begin{aligned}\frac{\partial \sigma_e^{(2)}}{\partial a_1} &= \frac{\partial}{\partial a_1}\frac{1}{N}\sum_{i=1}^{N}\left[y_i - \left(a_0 + a_1 x_i\right)\right]^2 \\ &= -\frac{1}{N}\sum_{i=1}^{N} 2\left[y_i - \left(a_0 + a_1 x_i\right)\right]x_i \\ &= 0\end{aligned} \tag{12}$$

Substituting Eq. (11) of a_0 into Eq. (12), we obtain

$$\begin{aligned}&\frac{1}{N}\sum_{i=1}^{N} x_i\left[y_i - \left(a_0 + a_1 x_i\right)\right] \\ &= \frac{1}{N}\sum_{i=1}^{N} x_i\left[y_i - \left(a_1 x_i + \mu_y - a_1\mu_x\right)\right] \\ &= \frac{1}{N}\sum_{i=1}^{N} x_i\left[\left(y_i - \mu_y\right) - a_1\left(x_i - \mu_x\right)\right] = 0\end{aligned} \tag{13}$$

It should also be noted that

$$\frac{1}{N}\sum_{i=1}^{N} \mu_x\left[\left(y_i - \mu_y\right) - a_1\left(x_i - \mu_x\right)\right] = 0 \tag{14}$$

Subtracting Eq. (14) from Eq. (13), we obtain

$$\frac{1}{N}\sum_{i=1}^{N}\left(x_i - \mu_x\right)\left[\left(y_i - \mu_y\right) - a_1\left(x_i - \mu_x\right)\right] = 0 \tag{15}$$

We then obtain

$$a_1 = \frac{\sigma_{xy}^{(2)}}{\sigma_{xx}^{(2)}} \tag{16}$$

where

$$\sigma_{xx}^{(2)} = \frac{1}{N}\sum_{i=1}^{N}\left(x_i - \mu_x\right)^2 \tag{17}$$

$$\sigma_{xy}^{(2)} \frac{1}{N}\sum_{i=1}^{N}\left(y_i - \mu_y\right)\left(x_i - \mu_x\right) \tag{18}$$

Eq. (16) can be expressed as a function of a correlation factor ρ_{xy} as

$$\begin{aligned} a_1 &= \frac{\sigma_{xy}^{(2)}}{\sigma_{xx}^{(2)}} \\ &= \rho_{xy}\frac{\sqrt{\sigma_{xx}^{(2)}\sigma_{yy}^{(2)}}}{\sigma_{xy}^{(2)}}\frac{\sigma_{xy}^{(2)}}{\sigma_{xx}^{(2)}} \\ &= \rho_{xy}\sqrt{\frac{\sigma_{yy}^{(2)}}{\sigma_{xx}^{(2)}}} \end{aligned} \tag{19}$$

where

$$\rho_{xy} = \frac{\sigma_{xy}^{(2)}}{\sqrt{\sigma_{xx}^{(2)}\sigma_{yy}^{(2)}}} \tag{20}$$

a_0 can be determined from Eq. (11) as

$$\begin{aligned} a_0 &= \mu_y - a_1\mu_x \\ &= \mu_y - \rho_{xy}\sqrt{\frac{\sigma_{yy}^{(2)}}{\sigma_{xx}^{(2)}}}\mu_x \end{aligned} \tag{21}$$

Finally, we obtain a regression line as

$$
\begin{aligned}
Y &= \mu_y - \rho_{xy}\sqrt{\frac{\sigma_{yy}^{(2)}}{\sigma_{xx}^{(2)}}}\mu_x + \rho_{xy}\sqrt{\frac{\sigma_{yy}^{(2)}}{\sigma_{xx}^{(2)}}}x \\
&= \rho_{xy}\sqrt{\frac{\sigma_{yy}^{(2)}}{\sigma_{xx}^{(2)}}}\left(x-\mu_x\right)+\mu_y
\end{aligned}
\tag{22}
$$

This can be modified as a normalized form as

$$
\frac{Y-\mu_y}{\sqrt{\sigma_{yy}^{(2)}}} = \rho_{xy}\left(\frac{x-\mu_x}{\sqrt{\sigma_{xx}^{(2)}}}\right) \tag{23}
$$

Eq. (22) can then be expressed with a normalized variable as

$$
z_Y = \rho_{xy} z_x \tag{24}
$$

where

$$
z_Y = \frac{Y-\mu_y}{\sqrt{\sigma_{yy}^{(2)}}} \tag{25}
$$

$$
z_x = \frac{x-\mu_x}{\sqrt{\sigma_{xx}^{(2)}}} \tag{26}
$$

Now, we can evaluate the assumption for e.

The average of e is denoted as μ_e and can be evaluated as

$$
\begin{aligned}
\mu_e &= \frac{1}{N}\sum_{i=1}^{N}\left(y_i - Y_i\right) \\
&= \frac{1}{N}\sum_{i=1}^{N}\left[y_i - \left(a_0 + a_1 x_i\right)\right] \\
&= \frac{1}{N}\sum_{i=1}^{N}\left[\left(y_i - \mu_y\right) - a_1\left(x_i - \mu_x\right)\right] \\
&= 0
\end{aligned}
\tag{27}
$$

This is zero as was assumed.

The covariance between x and e is given by

$$
\begin{aligned}
Cov[x,e] &= \frac{1}{N}\sum_{i=1}^{N}(x_i-\mu_x)e_i \\
&= \frac{1}{N}\sum_{i=1}^{N}(x_i-\mu_x)\left[y_i-(a_0+a_1x_i)\right] \\
&= \frac{1}{N}\sum_{i=1}^{N}(x_i-\mu_x)\left[(y_i-\mu_y)-a_1(x_i-\mu_x)\right] \\
&= \sigma_{xy}^{(2)}-a_1\sigma_{xx}^{(2)} \\
&= \sigma_{xy}^{(2)}-\frac{\sigma_{xy}^{(2)}}{\sigma_{xx}^{(2)}}\sigma_{xx}^{(2)} \\
&= 0
\end{aligned}
\tag{28}
$$

Therefore, both variables are independent as is assumed.

The variance of e can be also evaluated as

$$
\begin{aligned}
&\sigma_e^{(2)} \\
&= \frac{e_1^2+e_2^2+\cdots+e_N^2}{N} \\
&= \frac{1}{N}\left\{(y_1-a_0-a_1x_1)^2+(y_2-a_0-a_1x_2)^2+\cdots+(y_n-a_0-a_1x_N)^2\right\} \\
&= \frac{1}{N}\left\{\left[(y_1-\mu_y)-a_1(x_1-\mu_x)\right]^2+\left[(y_2-\mu_y)-a_1(x_2-\mu_x)\right]^2+\cdots+\left[(y_n-\mu_y)-a_1(x_n-\mu_x)\right]^2\right\} \\
&= \frac{1}{N}\left[(y_1-\mu_y)^2+(y_2-\mu_y)^2+\cdots+(y_N-\mu_y)^2\right] \\
&\quad -\frac{2a_1}{N}\left[(y_1-\mu_y)(x_1-\mu_x)+(y_2-\mu_y)(x_2-\mu_x)+\cdots+(y_N-\mu_y)(x_N-\mu_x)\right] \\
&\quad +\frac{a_1^2}{N}\left[(x_1-\mu_x)^2+(x_2-\mu_x)^2+\cdots+(x_N-\mu_x)^2\right] \\
&= \sigma_{yy}^{(2)}-2a_1\sigma_{xy}^{(2)}+a_1^2\sigma_{xx}^{(2)} \\
&= \sigma_{yy}^{(2)}-2\frac{\sqrt{\sigma_{yy}^{(2)}}}{\sqrt{\sigma_{xx}^{(2)}}}\rho_{xy}\sigma_{xy}^{(2)}+\left[\frac{\sqrt{\sigma_{yy}^{(2)}}}{\sqrt{\sigma_{xx}^{(2)}}}\rho_{xy}\right]^2\sigma_{xx}^{(2)} \\
&= \sigma_{yy}^{(2)}-2\sigma_{yy}^{(2)}\rho_{xy}\frac{\sigma_{xy}^{(2)}}{\sqrt{\sigma_{xx}^{(2)}}\sqrt{\sigma_{yy}^{(2)}}}+\sigma_{yy}^{(2)}\rho_{xy}^2 \\
&= \sigma_{yy}^{(2)}\left(1-\rho_{xy}^2\right)
\end{aligned}
\tag{29}
$$

4. Different Expression for Correlation Factor

Let us consider the correlation between y_i and the regression line data denoted by Y_i, where

$$Y_i = \frac{\sqrt{\sigma_{yy}{}^{(2)}}}{\sqrt{\sigma_{xx}{}^{(2)}}} \rho_{xy} \left(x_i - \mu_x \right) + \mu_y \tag{30}$$

The average and variance of Y are given by

$$\mu_Y = \mu_y \tag{31}$$

$$\sigma_{YY}^{(2)} = \frac{\sigma_{yy}{}^{(2)}}{\sigma_{xx}{}^{(2)}} \rho_{xy}^2 \sigma_{xx}^{(2)} \tag{32}$$

The correlation between y_i and Y_i is then given by

$$\begin{aligned}
\frac{\frac{1}{N}\sum_{i=1}^{N}\left(y_i - \mu_y\right)\left(Y_i - \mu_y\right)}{\sqrt{\sigma_{yy}^{(2)}\sigma_{YY}^{(2)}}} &= \frac{\frac{1}{N}\sum_{i=1}^{N}\left(y_i - \mu_y\right)\left(\frac{\sqrt{\sigma_{yy}{}^{(2)}}}{\sqrt{\sigma_{xx}{}^{(2)}}} \rho_{xy}\left(x_i - \mu_x\right) + \mu_y - \mu_y\right)}{\sqrt{\sigma_{yy}^{(2)} \frac{\sigma_{yy}{}^{(2)}}{\sigma_{xx}{}^{(2)}} \rho_{xy}^2 \sigma_{xx}^{(2)}}} \\
&= \frac{\frac{\sqrt{\sigma_{yy}{}^{(2)}}}{\sqrt{\sigma_{xx}{}^{(2)}}} \rho_{xy} \frac{1}{N}\sum_{i=1}^{N}\left(y_i - \mu_y\right)\left(x_i - \mu_x\right)}{\frac{\sqrt{\sigma_{yy}{}^{(2)}}}{\sqrt{\sigma_{xx}{}^{(2)}}} \rho_{xy} \sqrt{\sigma_{yy}{}^{(2)}\sigma_{xx}^{(2)}}} \\
&= \frac{\frac{1}{N}\sum_{i=1}^{N}\left(y_i - \mu_y\right)\left(x_i - \mu_x\right)}{\sqrt{\sigma_{yy}{}^{(2)}\sigma_{xx}^{(2)}}} \\
&= \rho_{xy}
\end{aligned} \tag{33}$$

It should be noted that the correlation between y and regression line data Y is equal to the correlation factor between y and x.

5. Forced Regression

In some cases, we want to set some restrictions, that is, we sometimes want to set a constant boundary value at 0 for $x = 0$. In this special case, we can obtain a_1 by setting $a_0 = 0$ in Eq. (12) and obtain

$$\begin{aligned}\frac{\partial \sigma_e^{(2)}}{\partial a_1} &= \frac{\partial}{\partial a_1}\frac{1}{N}\sum_{i=1}^{N}\left[y_i - a_1 x_i\right]^2 \\ &= -\frac{1}{N}\sum_{i=1}^{N}2\left[y_i - a_1 x_i\right]x_i \\ &= 0\end{aligned} \tag{34}$$

This can be modified as

$$\begin{aligned}&\frac{1}{N}\sum_{i=1}^{N}\left[y_i - a_1 x_i\right]x_i \\ &= \frac{1}{N}\sum_{i=1}^{N}\left[y_i - \mu_y + \mu_y - a_1\left(x_i - \mu_x + \mu_x\right)\right]\left(x_i - \mu_x + \mu_x\right) \\ &= \frac{1}{N}\sum_{i=1}^{N}\left[\left(y_i - \mu_y\right) + \mu_y - a_1\left(x_i - \mu_x\right) - a_1\mu_x\right]\left(x_i - \mu_x\right) \\ &+ \frac{1}{N}\sum_{i=1}^{N}\left[\left(y_i - \mu_y\right) + \mu_y - a_1\left(x_i - \mu_x\right) - a_1\mu_x\right]\mu_x \\ &= \sigma_{xy}^{(2)} - a_1\sigma_{xx}^{(2)} + \mu_y\mu_x - a_1{\mu_x}^2 \\ &= 0\end{aligned} \tag{35}$$

Therefore, we obtain

$$\begin{aligned}a_1 &= \frac{\sigma_{xy}^{(2)} + \mu_y\mu_x}{\sigma_{xx}^{(2)} + {\mu_x}^2} \\ &= \frac{\rho_{xy}\dfrac{\sqrt{\sigma_{xx}^{(2)}\sigma_{yy}^{(2)}}}{\sigma_{xy}^{(2)}}\sigma_{xy}^{(2)} + \mu_y\mu_x}{\sigma_{xx}^{(2)} + {\mu_x}^2} \\ &= \frac{\rho_{xy}\sqrt{\sigma_{xx}^{(2)}\sigma_{yy}^{(2)}} + \mu_y\mu_x}{\sigma_{xx}^{(2)} + {\mu_x}^2}\end{aligned} \tag{36}$$

We then obtain a corresponding regression equation as

$$Y = \frac{\rho_{xy}\sqrt{{\sigma_{xx}}^{(2)}}\sqrt{{\sigma_{yy}}^{(2)}} + \mu_x\mu_y}{{\sigma_{xx}}^{(2)} + {\mu_x}^2}x \tag{37}$$

In general, we need not to set a_0 as 0, but set is as a given value, and the corresponding a_1 is then given by

$$a_1 = \frac{\rho_{xy}\sqrt{\sigma_{xx}{}^{(2)}}\sqrt{\sigma_{yy}{}^{(2)}} + \mu_x\mu_y - a_0\mu_x}{\left(\sigma_{xx}{}^{(2)} + \mu_x{}^2\right)} \tag{38}$$

and the regression equation is given by

$$Y = \frac{\rho_{xy}\sqrt{\sigma_{xx}{}^{(2)}}\sqrt{\sigma_{yy}{}^{(2)}} + \mu_x\mu_y - a_0\mu_x}{\left(\sigma_{xx}{}^{(2)} + \mu_x{}^2\right)} x + a_0 \tag{39}$$

6. Accuracy of the Regression

The determination coefficient below is defined and frequently used to evaluate the accuracy of the regression, which is given by

$$R^2 = \frac{\sigma_r^{(2)}}{\sigma_{yy}^{(2)}} \tag{40}$$

where $\sigma_r^{(2)}$ is the variance of Y given by

$$\sigma_r^{(2)} = \frac{1}{N}\sum_{i=1}^{N}\left(Y_i - \mu_y\right)^2 \tag{41}$$

This is the ratio of the variance of the regression line to the variance of y. If whole data are on the regression line, that is, the regression is perfect, $\sigma_{yy}^{(2)}$ is equal to $\sigma_r^{(2)}$, and $R^2 = 1$.

On the other hand, when the deviation of data from the regression line is significant, the determinant ratio is expected to become small.

We further modify the variance of y related to the regression value as

$$\begin{aligned}
\sigma_{yy}^{(2)} &= \frac{1}{N}\sum_{i=1}^{N}\left(y_i - \mu_y\right)^2 \\
&= \frac{1}{N}\sum_{i=1}^{N}\left(y_i - Y_i + Y_i - \mu_y\right)^2 \\
&= \frac{1}{N}\sum_{i=1}^{N}\left(y_i - Y_i\right)^2 + \frac{1}{N}\sum_{i=1}^{N}\left(Y_i - \mu_y\right)^2 + 2\frac{1}{N}\sum_{i=1}^{N}\left(y_i - Y_i\right)\left(Y_i - \mu_y\right) \\
&= \frac{1}{N}\sum_{i=1}^{N}e_i^2 + \frac{1}{N}\sum_{i=1}^{N}\left(Y_i - \mu_y\right)^2 + 2\frac{1}{N}\sum_{i=1}^{N}e_i\left(a_0 + a_1x_i\right) \\
&= \sigma_e^{(2)} + \sigma_r^{(2)}
\end{aligned} \tag{42}$$

Therefore, the determination coefficient is reduced to

$$R^2 = \frac{\sigma_r^{(2)}}{\sigma_e^{(2)} + \sigma_r^{(2)}} = 1 - \frac{\sigma_e^{(2)}}{\sigma_{yy}^{(2)}} \tag{43}$$

Substituting Eq. (29)into Eq. (43), we obtain

$$R^2 = 1 - \frac{\sigma_{yy}^{(2)}}{\sigma_{yy}^{(2)}}\left(1 - \rho_{xy}^{\ 2}\right) = \rho_{xy}^{\ 2} \tag{44}$$

Consequently, the determination coefficient is the square of the correlation factor, and has a value range between 0 and 1.

From Eqs. (43) and (44), we can easily evaluate $\sigma_e^{(2)}$ as

$$\sigma_e^{(2)} = \left(1 - \rho_{xy}^2\right)\sigma_{yy}^{(2)} \tag{45}$$

7. Regression for Sample

We can perform similar analysis for the sample data. We assume n samples.
The regression line can be expressed by

$$\hat{Y} = \hat{a}_0 + \hat{a}_1 x \tag{46}$$

The form of the regression line is the same as the one for population. However, the coefficients $\hat{a}_0$ and $\hat{a}_1$ depend on extracted samples, and are given by

$$\hat{a}_1 = \frac{\sqrt{S_{yy}^{(2)}}}{\sqrt{S_{xx}^{(2)}}} r_{xy} \tag{47}$$

$$a_0 = \mu_y - a_1 \mu_x$$
$$= \bar{y} - \frac{\sqrt{S_{yy}{}^{(2)}}}{\sqrt{S_{xx}{}^{(2)}}} r_{xy} \bar{x} \tag{48}$$

Finally, we obtain a regression line for sample data as

$$Y - \bar{y} = \frac{\sqrt{S_{yy}{}^{(2)}}}{\sqrt{S_{xx}{}^{(2)}}} r_{xy} \left(x - \bar{x} \right) \tag{49}$$

where sample averages and sample variances are given by

$$\bar{x} = \frac{1}{n} \sum x_i \tag{50}$$

$$\bar{y} = \frac{1}{n} \sum y_i \tag{51}$$

$$S_{xx}^{(2)} = \frac{\sum \left(x_i - \bar{x} \right)^2}{n} \tag{52}$$

$$S_{yy}^{(2)} = \frac{\sum \left(y_i - \bar{y} \right)^2}{n} \tag{53}$$

$$S_{xy}^{(2)} = \frac{\sum \left(x_i - \bar{x} \right)\left(y_i - \bar{y} \right)}{n} \tag{54}$$

$$r_{xy} = \frac{S_{xy}^{(2)}}{\sqrt{S_{xx}^{(2)} S_{yy}^{(2)}}} \tag{55}$$

As in the population, the sample correlation factor is expressed by a different way as

$$r_{xy} = \frac{\frac{1}{n} \sum \left(y_i - \mu_y \right)\left(Y_i - \mu_y \right)}{\sqrt{S_{yy}^{(2)} S_{YY}^{(2)}}} \tag{56}$$

The determinant factor for the sample data is given by

$$R^2 = \frac{S_r^{(2)}}{S_{yy}^{(2)}} = 1 - \frac{S_e^{(2)}}{S_{yy}^{(2)}} \tag{57}$$

where

$$S_r^{(2)} = \frac{\sum (y_i - \bar{y})(Y_i - \bar{y})}{n} \tag{58}$$

and it can be proved as the similar way as population as

$$S_{yy}^{(2)} = S_r^{(2)} + S_e^{(2)} \tag{59}$$

The variances are ones for sample data, and hence we replace them by unbiased ones as

$$R^{*2} = 1 - \frac{\frac{n}{\phi_e} S_e^{(2)}}{\frac{n}{\phi_T} S_{yy}^{(2)}} \tag{60}$$

a_0 and a_1 are used for $S_e^{(2)}$ and hence freedom is decreased by two, and the corresponding freedom ϕ_e is given

$$\phi_e = n - 2 \tag{61}$$

$\bar{y}$ is used in $S_{yy}^{(2)}$, and hence the freedom is decreased by one, and the corresponding freedom ϕ_T is given

$$\phi_T = n - 1 \tag{62}$$

R^{*2} is called as a degree of freedom adjusted coefficient of determination.

8. RESIDUAL ERROR AND LEVERAGE RATIO

We study a residual error and a leverage ratio to evaluate the accuracy of the regression. The residual error is given by

$$e_k = y_k - \hat{y}_k \tag{63}$$

The normalized one is then given by

$$z_{ek} = \frac{e_k}{\sqrt{s_e^{(2)}}} \tag{64}$$

This follows a standard normal distribution with $N\left(0,1^2\right)$. We can set a certain critical value for e_k' , and evaluate whether it is outlier or not.

We can evaluate the error of the regression on the different stand point of view, called it as a leverage ratio, and denote it as h_{kk} . This leverage ratio expresses how each data contribute to the regression line, and is evaluated as follows.

The predictive value for k -th sample is given by

$$\begin{aligned} Y_k &= \hat{a}_0 + \hat{a}_1 x_k \\ &= \bar{y} + \hat{a}_1\left(x_k - \bar{x}\right) \end{aligned} \tag{65}$$

We substitute the expression of $\hat{a}_1$ given by,

$$\hat{a}_1 = \frac{S_{xy}^{(2)}}{S_{xx}^{(2)}} \tag{66}$$

into Eq. (66), and obtain

$$
\begin{aligned}
Y_k &= \bar{y} + \frac{S_{xy}^{(2)}}{S_{xx}^{(2)}}\left(x_k - \bar{x}\right) \\
&= \frac{1}{n}\sum_{i=1}^{n} y_i + \frac{\frac{1}{n}\sum_{i=1}^{n}\left(x_i - \bar{x}\right)\left(y_i - \bar{y}\right)}{S_{xx}^{(2)}}\left(x_k - \bar{x}\right) \\
&= \frac{1}{n}\sum_{i=1}^{n} y_i + \frac{\sum_{i=1}^{n}\left(x_i - \bar{x}\right) y_i}{n S_{xx}^{(2)}}\left(x_k - \bar{x}\right)
\end{aligned}
\tag{67}
$$

This can be expressed with a general form as

$$
Y_k = h_{k1} y_1 + h_{k2} y_2 + \cdots + h_{kk} y_k + \cdots + h_{kn} y_n \tag{68}
$$

Comparing Eq. (67) with (68), we obtain h_{kk} as

$$
h_{kk} = \frac{1}{n} + \frac{\left(x_k - \bar{x}\right)^2}{n S_{xx}^{(2)}} \tag{69}
$$

h_{kk} is the variation of the predicted value when y_k changed by 1. The predicted value should be determined by the whole data, and the variation should be small to obtain an accurate regression.

The leverage ratio increases with increasing the deviation of x from the sample average of $\bar{x}$, and it is determined only with x_i independent on y_i. Therefore, the data far from the average of $\bar{x}$ influences much to the regression. This means that we should not use the data much far from the $\bar{x}$

The leverage ratio range can be evaluated below.

It is obvious that the minimum value is given by

$$
h_{kk} \geq \frac{1}{n} \tag{70}
$$

We then consider the maximum value. We set $x_i - \bar{x} = X_i$. h_{kk} must have the maximum value for h_{11} of h_{nn}. We assume h_{nn} as the maximum value, which does not eliminate generality.

We have the relationship given by

$$\sum_{i=1}^{n} X_i = 0 \tag{71}$$

Extracting the term X_n form Eq.(71), we obtain

$$-X_n = \sum_{i=1}^{n-1} X_i \tag{72}$$

We then have

$$\begin{aligned}
1-h_{nn} &= 1-\frac{1}{n}-\frac{(x_n-\bar{x})^2}{nS_{xx}^{(2)}} \\
&= \frac{1}{nS_{xx}^{(2)}}\left[(n-1)S_{xx}^{(2)}-(x_n-\bar{x})^2\right] \\
&= \frac{1}{nS_{xx}^{(2)}}\left[(n-1)S_{xx}^{(2)}-X_n{}^2\right] \\
&= \frac{1}{nS_{xx}^{(2)}}\left[(n-1)\frac{\sum_{i=1}^{n} X_i^2}{n}-X_n{}^2\right] \\
&= \frac{1}{nS_{xx}^{(2)}}\left[(n-1)\frac{\sum_{i=1}^{n-1} X_i^2+X_n{}^2}{n}-X_n{}^2\right] \\
&= \frac{1}{nS_{xx}^{(2)}}\left[\frac{n-1}{n}\sum_{i=1}^{n-1} X_i^2-\frac{X_n{}^2}{n}\right] \\
&= \frac{1}{nS_{xx}^{(2)}}\frac{n-1}{n}\left[\sum_{i=1}^{n-1} X_i^2-\frac{\left(\sum_{i=1}^{n-1} X_i\right)^2}{n-1}\right] \geq 0
\end{aligned} \tag{73}$$

Therefore, we have

$$h_{nn} \leq 1 \tag{74}$$

Consequently, we obtain

$$\frac{1}{n} \leq h_{kk} \leq 1 \tag{75}$$

Sum of the leverage ratio is given by

$$\begin{aligned}\sum_{k=1}^{n} h_{kk} &= \sum_{k=1}^{n} \frac{1}{n} + \sum_{k=1}^{n} \frac{\left(x_k - \bar{x}\right)^2}{n S_{xx}^{(2)}} \\ &= 1 + \frac{S_{xx}^{(2)}}{S_{xx}^{(2)}} \\ &= 2\end{aligned} \tag{76}$$

Therefore, the average value for h_{kk} is denoted as μ_{hkk}

$$\mu_{hkk} = \frac{2}{n} \tag{77}$$

Therefore, we use a critical value for h_{kk} as

$$h_{kk} \leq \kappa \times \mu_{hkk} \tag{78}$$

$\kappa = 2$ is frequently used, and evaluate whether it is outlier or not.

From the above analysis, we can evaluate the error of the regression by z_{ek} or h_{kk}. Merging these parameters, we also evaluate t value of the residual error given by

$$t = \frac{z_{ek}}{\sqrt{1 - h_{kk}}} \tag{79}$$

This includes the both criticisms of parameters of z_{ek} or h_{kk}, and we can set a certain value for t, and evaluate whether it is outlier or not. When h_{kk} approaches to its maximum value of 1, the denominator approaches to 0, and t becomes large. Therefore, the corresponding data is apt to be regarded as an outlier.

9. Prediction of Coefficients of a Regression Line

When we extract samples, the corresponding regression line varies, that is, the factors $\hat{a}_0$ and $\hat{a}_1$ vary depending on the extracted samples. We study averages and variances of $\hat{a}_0$ and $\hat{a}_1$ in this section.

We assume that the population set has a regression line given by

$$y_i = a_0 + a_1 x_i + e_i \quad for\ i = 1, 2, \cdots, n \tag{80}$$

where a_0 and a_1 are established values. e_i is independent on each other and follow a normal distribution with $\left(0, \sigma^{(2)}\right)$. x_i is a given variable. The probability variable in Eq. (80) is then only e_i.

We consider the case where we obtain n set of data $\left(x_i, y_i\right)$. The corresponding $\hat{a}_0$ and $\hat{a}_1$ are given by

$$\begin{aligned}
\hat{a}_1 &= \frac{S_{xy}^{(2)}}{S_{xx}^{(2)}} \\
&= \frac{\sum_{i=1}^{n}\left(x_i - \bar{x}\right)\left(y_i - \bar{y}\right)}{nS_{xx}^{(2)}} \\
&= \sum_{i=1}^{n} \frac{x_i - \bar{x}}{nS_{xx}^{(2)}} y_i \\
&= \sum_{i=1}^{n} \frac{x_i - \bar{x}}{nS_{xx}^{(2)}} \left(a_0 + a_1 x_i + e_i\right)
\end{aligned} \tag{81}$$

$$\begin{aligned}
\hat{a}_0 &= \bar{y} - \hat{a}_1 \bar{x} \\
&= \frac{1}{n}\sum_{i=1}^{n} y_i - \left(\sum_{i=1}^{n} \frac{x_i - \bar{x}}{nS_{xx}^{(2)}} y_i\right)\bar{x} \\
&= \sum_{i=1}^{n}\left(\frac{1}{n} - \frac{x_i - \bar{x}}{nS_{xx}^{(2)}}\bar{x}\right) y_i \\
&= \sum_{i=1}^{n}\left(\frac{1}{n} - \frac{x_i - \bar{x}}{nS_{xx}^{(2)}}\bar{x}\right)\left(a_0 + a_1 x_i + e_i\right)
\end{aligned} \tag{82}$$

We notice that only e_i is a probability variable, and evaluate the expected value as

$$\begin{aligned} E\left[\hat{a}_1\right] &= \sum_{i=1}^{n} \frac{x_i - \overline{x}}{nS_{xx}^{(2)}} \left(a_0 + a_1 x_i\right) + \sum_{i=1}^{n} \frac{x_i - \overline{x}}{nS_{xx}^{(2)}} E\left[e_i\right] \\ &= a_1 \sum_{i=1}^{n} \frac{x_i - \overline{x}}{nS_{xx}^{(2)}} x_i \\ &= a_1 \sum_{i=1}^{n} \frac{x_i - \overline{x}}{nS_{xx}^{(2)}} \left(x_i - \overline{x}\right) \quad for \sum_{i=1}^{n} \frac{x_i - \overline{x}}{nS_{xx}^{(2)}} \overline{x} = 0 \\ &= a_1 \end{aligned} \tag{83}$$

$$\begin{aligned} E\left[\hat{a}_0\right] &= E\left[\overline{y} - \hat{a}_1 \overline{x}\right] \\ &= \overline{y} - E\left[\hat{a}_1\right]\overline{x} \\ &= \overline{y} - a_1 \overline{x} \\ &= a_0 \end{aligned} \tag{84}$$

$$\begin{aligned} V\left[\hat{a}_1\right] &= V\left[\sum_{i=1}^{n} \frac{x_i - \overline{x}}{nS_{xx}^{(2)}} y_i\right] \\ &= \sum_{i=1}^{n} \left(\frac{x_i - \overline{x}}{nS_{xx}^{(2)}}\right)^2 V\left[y_i\right] \\ &= \sum_{i=1}^{n} \left(\frac{x_i - \overline{x}}{nS_{xx}^{(2)}}\right)^2 V\left[e_i\right] \\ &= \sum_{i=1}^{n} \left(\frac{x_i - \overline{x}}{nS_{xx}^{(2)}}\right)^2 \sigma_e^{(2)} \\ &= \frac{\sigma_e^{(2)}}{nS_{xx}^{(2)}} \end{aligned} \tag{85}$$

$$\begin{aligned} V\left[\hat{a}_0\right] &= V\left[\sum_{i=1}^{n} \left(\frac{1}{n} - \frac{x_i - \overline{x}}{nS_{xx}^{(2)}} \overline{x}\right) y_i\right] \\ &= \sum_{i=1}^{n} \left(\frac{1}{n} - \frac{x_i - \overline{x}}{nS_{xx}^{(2)}} \overline{x}\right)^2 V\left[y_i\right] \\ &= \sum_{i=1}^{n} \left(\frac{1}{n} - \frac{x_i - \overline{x}}{nS_{xx}^{(2)}} \overline{x}\right)^2 V\left[e_i\right] \\ &= \sum_{i=1}^{n} \left[\frac{1}{n^2} - 2\frac{x_i - \overline{x}}{n^2 S_{xx}^{(2)}} \overline{x} + \frac{\left(x_i - \overline{x}\right)^2}{n^2 \left(S_{xx}^{(2)}\right)^2} \overline{x}^2\right] \sigma_e^{(2)} \\ &= \left(\frac{1}{n} + \frac{\overline{x}^2}{nS_{xx}^{(2)}}\right) \sigma_e^{(2)} \end{aligned} \tag{86}$$

$$
\begin{aligned}
Cov\left[\hat{a}_1,\hat{a}_0\right] &= Cov\left[\sum_{i=1}^{n}\frac{x_i-\overline{x}}{nS_{xx}^{(2)}}y_i,\sum_{i=1}^{n}\left(\frac{1}{n}-\frac{x_i-\overline{x}}{nS_{xx}^{(2)}}\overline{x}\right)y_i\right] \\
&= E\left[\sum_{i=1}^{n}\frac{x_i-\overline{x}}{nS_{xx}^{(2)}}\left(\frac{1}{n}-\frac{x_i-\overline{x}}{nS_{xx}^{(2)}}\overline{x}\right)y_i^2\right] \\
&= \sum_{i=1}^{n}\frac{x_i-\overline{x}}{nS_{xx}^{(2)}}\left(\frac{1}{n}-\frac{x_i-\overline{x}}{nS_{xx}^{(2)}}\overline{x}\right)E\left[y_i^2\right] \\
&= \sum_{i=1}^{n}\frac{x_i-\overline{x}}{nS_{xx}^{(2)}}\left(\frac{1}{n}-\frac{x_i-\overline{x}}{nS_{xx}^{(2)}}\overline{x}\right)E\left[e_i^2\right] \\
&= \sum_{i=1}^{n}\frac{x_i-\overline{x}}{nS_{xx}^{(2)}}\left(\frac{1}{n}-\frac{x_i-\overline{x}}{nS_{xx}^{(2)}}\overline{x}\right)\sigma_e^{(2)} \\
&= -\sum_{i=1}^{n}\frac{x_i-\overline{x}}{nS_{xx}^{(2)}}\frac{x_i-\overline{x}}{nS_{xx}^{(2)}}\overline{x}\sigma_e^{(2)} \\
&= -\frac{\overline{x}}{nS_{xx}^{(2)}}\sigma_e^{(2)}
\end{aligned}
\tag{87}
$$

We assume that $\hat{a}_1$ and $\hat{a}_0$ follow normal distributions, and hence assume the normalized variables given by

$$
z=\frac{\hat{a}_i-a_i}{\sqrt{V\left[a_i\right]}} \quad for\ i=0,1
\tag{88}
$$

This follows a standard normal distribution. Therefore, the predictive range is given by

$$
\hat{a}_1 - z_p\frac{\sqrt{\sigma_e^{(2)}}}{\sqrt{nS_{xx}^{(2)}}} \le a_1 \le \hat{a}_1 + z_p\frac{\sqrt{\sigma_e^{(2)}}}{\sqrt{nS_{xx}^{(2)}}}
\tag{89}
$$

$$
\hat{a}_0 - z_p\sqrt{\left(\frac{1}{n}+\frac{\overline{x}^2}{nS_{xx}^{(2)}}\right)\sigma_e^{(2)}} \le a_0 \le \hat{a}_0 + z_p\sqrt{\left(\frac{1}{n}+\frac{\overline{x}^2}{nS_{xx}^{(2)}}\right)\sigma_e^{(2)}}
\tag{90}
$$

In the practical cases, we do not know $\sigma_e^{(2)}$, and hence use $s_e^{(2)}$ instead. Then, the normalized variables are assumed to follow a t distribution function with a freedom of $n-2$. Therefore, the predictive range is given by

$$\hat{a}_1 - t_p\left(n-2\right)\frac{\sqrt{s_e^{(2)}}}{\sqrt{nS_{xx}^{(2)}}} \le a_1 \le \hat{a}_1 + t_p\left(n-2\right)\frac{\sqrt{s_e^{(2)}}}{\sqrt{nS_{xx}^{(2)}}} \tag{91}$$

$$\hat{a}_0 - t_p\sqrt{\left(\frac{1}{n}+\frac{\overline{x}^2}{nS_{xx}^{(2)}}\right)s_e^{(2)}} \le a_0 \le \hat{a}_0 + t_p\sqrt{\left(\frac{1}{n}+\frac{\overline{x}^2}{nS_{xx}^{(2)}}\right)s_e^{(2)}} \tag{92}$$

10. Estimation of Error Range of Regression Line

We can estimate the error range of value of a regression line. The expected regression value for value of x_0 is denoted as Y_0.

We assume that we obtain n set of data $\left(x_i, y_i\right)$. The corresponding regression value is given by

$$\hat{Y}_0 = \hat{a}_0 + \hat{a}_1 x_0 \tag{93}$$

The expected value is given by

$$\begin{aligned} E\left[\hat{Y}_0\right] &= E\left[\hat{a}_0\right] + E\left[\hat{a}_1\right]x_0 \\ &= a_0 + a_1 x_0 \equiv Y_0 \end{aligned} \tag{94}$$

The variance of $\hat{Y}_0$ is given by

$$\begin{aligned} V\left[\hat{Y}_0\right] &= V\left[\hat{a}_0 + \hat{a}_1 x_0\right] \\ &= V\left[\hat{a}_0\right] + 2Cov\left[\hat{a}_0, \hat{a}_1\right]x_0 + V\left[\hat{a}_1\right]x_0^2 \\ &= \left(\frac{1}{n}+\frac{\overline{x}^2}{nS_{xx}^{(2)}}\right)\sigma_e^{(2)} - \frac{2x_0\overline{x}}{nS_{xx}^{(2)}}\sigma_e^{(2)} + \frac{x_0^2}{nS_{xx}^{(2)}}\sigma_e^{(2)} \\ &= \left[\frac{1}{n}+\frac{\left(x_0-\overline{x}\right)^2}{nS_{xx}^{(2)}}\right]\sigma_e^{(2)} \end{aligned} \tag{95}$$

We assume that it follows a normal distribution and hence the predicted value of Y_0 is given by

$$\hat{Y}_0 - z_p \sqrt{\left[\frac{1}{n} + \frac{(x_0 - \bar{x})^2}{nS_{xx}^{(2)}}\right]\sigma_e^{(2)}} \le Y_o \le \hat{Y}_0 + z_p \sqrt{\left[\frac{1}{n} + \frac{(x_0 - \bar{x})^2}{nS_{xx}^{(2)}}\right]\sigma_e^{(2)}} \tag{96}$$

If we use $s_e^{(2)}$ instead of $\sigma_e^{(2)}$ for the variance, we need to use a t distribution with a freedom of $n-2$, and the predictive range is given by

$$\hat{Y}_0 - t_p(n-2)\sqrt{\left[\frac{1}{n} + \frac{(x_0 - \bar{x})^2}{nS_{xx}^{(2)}}\right]s_e^{(2)}} \le Y_0 \le \hat{Y}_0 + t_p(n-2)\sqrt{\left[\frac{1}{n} + \frac{(x_0 - \bar{x})^2}{nS_{xx}^{(2)}}\right]s_e^{(2)}} \tag{97}$$

When an explanatory variable x changes from x_0 to x_1, the corresponding variation of objective variable $\hat{Y}$ from $\hat{Y}_0$ to $\hat{Y}_1$. The variation of the objective variable is then given by

$$\Delta\hat{Y} = \hat{Y}_1 - \hat{Y}_0 \tag{98}$$

Considering the error range, the total variation D is given by

$$\begin{aligned}
D &= \left\{\hat{Y}_1 - t_p(n-2)\sqrt{\left[\frac{1}{n} + \frac{(x_1 - \bar{x})^2}{nS_{xx}^{(2)}}\right]s_e^{(2)}}\right\} - \left\{\hat{Y}_0 + t_p(n-2)\sqrt{\left[\frac{1}{n} + \frac{(x_0 - \bar{x})^2}{nS_{xx}^{(2)}}\right]s_e^{(2)}}\right\} \\
&= \hat{Y}_1 - \hat{Y}_0 - t_p(n-2)\left\{\sqrt{\left[\frac{1}{n} + \frac{(x_1 - \bar{x})^2}{nS_{xx}^{(2)}}\right]s_e^{(2)}} + \sqrt{\left[\frac{1}{n} + \frac{(x_0 - \bar{x})^2}{nS_{xx}^{(2)}}\right]s_e^{(2)}}\right\} \\
&= \Delta\hat{Y} - t_p(n-2)\left\{\sqrt{\left[\frac{1}{n} + \frac{(x_1 - \bar{x})^2}{nS_{xx}^{(2)}}\right]s_e^{(2)}} + \sqrt{\left[\frac{1}{n} + \frac{(x_0 - \bar{x})^2}{nS_{xx}^{(2)}}\right]s_e^{(2)}}\right\}
\end{aligned} \tag{99}$$

If this is the positive, we can judge that the objective variable certainly changes. Therefore, we can evaluate this as below.

$$\Delta\hat{Y} > t_p(n-2)\left\{\sqrt{\left[\frac{1}{n} + \frac{(x_1 - \bar{x})^2}{nS_{xx}^{(2)}}\right]s_e^{(2)}} + \sqrt{\left[\frac{1}{n} + \frac{(x_0 - \bar{x})^2}{nS_{xx}^{(2)}}\right]s_e^{(2)}}\right\} \tag{100}$$

Next, we consider that the predictive value of y instead of the regression value of Y for a given x of x_0. We predict $y_i = a_o + a_1 x_i + e_i$ instead of $y_i = a_o + a_1 x_i$. Therefore, the variance is given by

$$\left[1 + \frac{1}{n} + \frac{\left(x_0 - \bar{x}\right)^2}{nS_{xx}^{(2)}}\right]\sigma^2 \tag{101}$$

The predictive range for y_0 is given by

$$\hat{Y}_0 - z_p \sqrt{\left[1 + \frac{1}{n} + \frac{\left(x_0 - \bar{x}\right)^2}{nS_{xx}^{(2)}}\right]\sigma_e^{(2)}} \le y_o \le \hat{Y}_0 + z_p \sqrt{\left[1 + \frac{1}{n} + \frac{\left(x_0 - \bar{x}\right)^2}{nS_{xx}^{(2)}}\right]\sigma_e^{(2)}} \tag{102}$$

If we use $s_e^{(2)}$ instead of $\sigma_e^{(2)}$ for the variance, we need to use a t distribution with a freedom of $n-2$, and the predictive range is given by

$$\hat{Y}_0 - t_p\left(n-2\right) \sqrt{\left[1 + \frac{1}{n} + \frac{\left(x_0 - \bar{x}\right)^2}{nS_{xx}^{(2)}}\right]s_e^{(2)}} \le y_o \le \hat{Y}_0 + t_p\left(n-2\right) \sqrt{\left[1 + \frac{1}{n} + \frac{\left(x_0 - \bar{x}\right)^2}{nS_{xx}^{(2)}}\right]s_e^{(2)}} \tag{103}$$

When we obtain the data set (x, y), we can evaluate corresponding expected objective value as well as its error range for the x. We can compare these values with given y, and can judge whether it is reasonable or not.

Let us consider an example. Table 2 shows scores of mathematics and physics of 40 members, where the score of mathematics is denoted as x and that of physics as y.

The corresponding average and sample variances are given by

$$\bar{x} = 52.38 \tag{104}$$

$$\bar{y} = 72.78 \tag{105}$$

$$S_{xx}^{(2)} = 213.38 \tag{106}$$

$$S_{yy}^{(2)} = 176.52 \tag{107}$$

$$S_{xy}^{(2)} = 170.26 \tag{108}$$

The sample correlation factor is given by

$$r_{xy} = 0.88 \tag{109}$$

The center value for the correlation factor is given by

$$\rho_{cent} = 0.87 \tag{110}$$

which is a little smaller than r_{xy}. The correlation factor range is predicted as

$$0.78 \leq \rho \leq 0.93 \tag{111}$$

The regression line is given by

$$\hat{Y} = 30.98 + 0.80x \tag{112}$$

The determinant factor and one with considering the freedom are given by

$$R^2 = 0.77 \tag{113}$$

$$R^{*2} = 0.76 \tag{114}$$

The predicted regression and objective values are shown in Figure 3. The error range increases with the distance from the average location to the x increases. The original data are almost within the range of predicted objected values.

Table 2. Score of mathematics and physics of 40 members

ID	Mathematics	Physics	ID	Mathematics	Physics
1	34	57	21	49	68
2	56	78	22	35	49
3	64	76	23	59	87
4	34	64	24	27	56
5	62	78	25	77	96
6	52	82	26	24	49
7	79	91	27	58	76
8	50	67	28	47	63
9	46	64	29	56	78
10	49	58	30	75	96
11	42	62	31	68	100
12	60	72	32	70	76
13	39	63	33	51	72
14	45	65	34	69	98
15	40	68	35	30	57
16	42	73	36	38	68
17	65	82	37	85	95
18	41	67	38	65	74
19	50	66	39	46	56
20	54	80	40	62	84

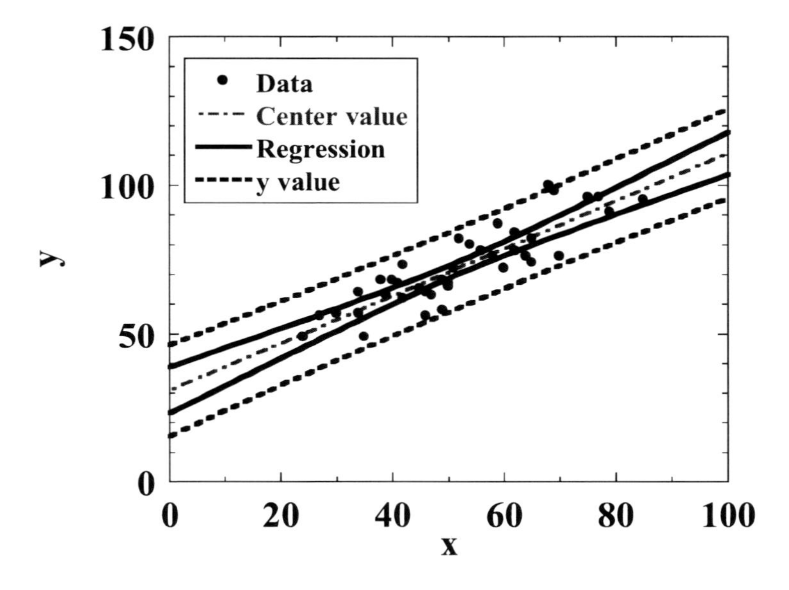

Figure 3. Predicted regression and objective values. The original data is also shown.

11. LOGARITHMIC REGRESSION

We assume the linear regression up to here. However, there are various cases where an appropriate fitted function is not a linear line.

When the target function is an exponential one, given by

$$y = be^{ax} \tag{115}$$

Taking logarithms for both sides of Eq. (115), we obtain

$$\ln y = ax + \ln b \tag{116}$$

Converting variables as

$$\begin{cases} u = x \\ v = \ln y \end{cases} \tag{117}$$

We can apply the linear regression.

When y is proportional to power law of x, that is, it is given by

$$y = bx^{a} \tag{118}$$

Taking logarithms for both sides of Eq.(118), we obtain

$$\ln y = a \ln x + \ln b \tag{119}$$

Converting variables as

$$\begin{cases} u = \ln x \\ v = \ln y \end{cases} \tag{120}$$

We can apply the linear regression.

12. Power Law Regression

We assume the data are expressed with a power law equation given by

$$y = a_0 + a_1 x + a_2 x^2 + \cdots + a_m x^m \tag{121}$$

The error is given by

$$\begin{aligned}\sigma_e^{(2)} &= \frac{1}{N}\left(e_1^{\ 2} + e_2^{\ 2} + \cdots + e_N^{\ 2}\right) \\ &= \frac{1}{N}\sum_{i=1}^{N}\left[y_i - \left(a_0 + a_1 x_i + a_2 x_i^2 + \cdots + a_m x_i^{\ m}\right)\right]^2\end{aligned} \tag{122}$$

We then evaluate

$$\frac{\partial \sigma_e^{(2)}}{\partial a_0} = 0, \frac{\partial \sigma_e^{(2)}}{\partial a_1} = 0, \frac{\partial \sigma_e^{(2)}}{\partial a_2} = 0, \cdots, \frac{\partial \sigma_e^{(2)}}{\partial a_m} = 0 \tag{123}$$

This gives us $m+1$ equations, and we can determine $\left(a_0, a_1, a_2, \cdots, a_m\right)$ as follows.

$$\frac{\partial \sigma_e^{(2)}}{\partial a_0} = -\frac{2}{N}\sum_{i=1}^{N}\left[y_i - \left(a_0 + a_1 x_i + a_2 x_i^2 + \cdots + a_m x_i^{\ m}\right)\right] = 0 \tag{124}$$

This gives

$$a_0 = \bar{y} - \left(a_1 \bar{x} + a_1 \overline{x^2} + \cdots + a_m \overline{x^k}\right) \tag{125}$$

We also have

$$\begin{aligned}\frac{\partial \sigma_e^{(2)}}{\partial a_k} &= -\frac{2}{N}\sum_{i=1}^{N}\left[y_i - \left(a_0 + a_1 x_i + a_2 x_i^2 + \cdots + a_m x_i^{\ m}\right)\right] x_i^k \\ &= -\frac{2}{N}\sum_{i=1}^{N}\left[\left(y_i - \bar{y}\right) - \left[a_1\left(x_i - \bar{x}\right) + a_2\left(x_i^2 - \overline{x^2}\right) + \cdots + a_m\left(x_i^m - x^m\right)\right]\right] x_i^k \\ &= -\frac{2}{N}\sum_{i=1}^{N}\left[\left(y_i - \bar{y}\right) - \left[a_1\left(x_i - \bar{x}\right) + a_2\left(x_i^2 - \overline{x^2}\right) + \cdots + a_m\left(x_i^m - x^m\right)\right]\right]\left(x_i^k - \overline{x^k}\right) \\ &= 0\end{aligned} \tag{126}$$

We then obtain

$$\begin{pmatrix} K_{x11} & K_{x12} & \cdots & K_{x1m} \\ K_{x21} & K_{x22} & \cdots & K_{x22} \\ \vdots & \vdots & \ddots & \vdots \\ K_{xm1} & K_{xm2} & \cdots & K_{xmm} \end{pmatrix} \begin{pmatrix} a_1 \\ a_2 \\ \vdots \\ a_m \end{pmatrix} = \begin{pmatrix} K_{y1} \\ K_{y2} \\ \vdots \\ K_{ym} \end{pmatrix} \tag{127}$$

where

$$K_{xkl} = \frac{1}{N}\sum_{i=1}^{N}\left(x_i^k - \overline{x^k}\right)\left(x_i^l - \overline{x^l}\right) \tag{128}$$

$$K_{yk} = \frac{1}{N}\sum_{i=1}^{N}\left(y_i - \overline{y}\right)\left(x_i^k - \overline{x^k}\right) \tag{129}$$

13. Regression for an Arbitrarily Function

We further treat an arbitrarily function with m parameters denoted as $f\left(a_1, a_2, \cdots, a_m, x\right)$.

The error $\sigma_e^{(2)}$ is expressed by

$$\sigma_e^{(2)} = \frac{1}{N}\sum_{i=1}^{N}\left[y_i - f\left(a_1, a_2, \cdots, a_m, x_i\right)\right]^2 \tag{130}$$

Partial differentiating this with respect to a_1, we obtain

$$\frac{\partial \sigma_e^{(2)}}{\partial a_1} = \frac{1}{N}\sum_{i=1}^{N} -2\left[y_i - f\left(a_1, a_2, \cdots, a_m, x_i\right)\right]\frac{\partial f\left(a_1, a_2, \cdots, a_m, x_i\right)}{\partial a_1} = 0 \tag{131}$$

Arranging this equation, we obtain

$$\frac{1}{N}\sum_{i=1}^{N}\left[y_i - f\left(a_1, a_2, \cdots, a_m, x_i\right)\right]\frac{\partial f\left(a_1, a_2, \cdots, a_m, x_i\right)}{\partial a_1} = 0 \tag{132}$$

We set the left side of Eq. (132) as $g_1\left(a_1, a_2, \cdots, a_m\right)$, that is,

$$g_1\left(a_1,a_2,\cdots,a_m\right)=\frac{1}{N}\sum_{i=1}^{N}\left[y_i-f\left(a_1,a_2,\cdots,a_m,x_i\right)\right]\frac{\partial f\left(a_1,a_2,\cdots,a_m,x_i\right)}{\partial a_1} \tag{133}$$

Performing similar analysis for the other parameters, we obtain

$$\begin{cases} g_1\left(a_1,a_2,\cdots,a_m\right)=0 \\ g_2\left(a_1,a_2,\cdots,a_m\right)=0 \\ \cdots \\ g_m\left(a_1,a_2,\cdots,a_m\right)=0 \end{cases} \tag{134}$$

where

$$g_k\left(a_1,a_2,\cdots,a_m\right)=\frac{1}{N}\sum_{i=1}^{N}\left[y_i-f\left(a_1,a_2,\cdots,a_m,x_i\right)\right]\frac{\partial f\left(a_1,a_2,\cdots,a_m,x_i\right)}{\partial a_k} \tag{135}$$

The parameters $\left(a_1,a_2,\cdots,a_m\right)$ can be solved numerically as follows.

We set initial value set as $\left(a_1^{(0)},a_2^{(0)},\cdots,a_m^{(0)}\right)$.

The deviation from the updating value $\left(a_1^{(1)},a_2^{(1)},\cdots,a_m^{(1)}\right)$ is given by

$$\left(\Delta a_1,\Delta a_2,\cdots,\Delta a_m\right)=\left(a_1^{(1)}-a_1^{(0)},a_2^{(1)}-a_2^{(0)},\cdots,a_m^{(1)}-a_m^{(0)}\right) \tag{136}$$

Let us consider the equation of $g_k\left(a_1,a_2,\cdots,a_m\right)=0$, which we want to obtain the corresponding set of $\left(a_1,a_2,\cdots,a_m\right)$. However, we do not know the set. We assume that the tentative set $\left(a_1^{(1)},a_2^{(1)},\cdots,a_m^{(1)}\right)$ is close to the one, then

$$\begin{aligned} 0 &= g_k\left(a_1,a_2,\cdots,a_m\right) \\ &\approx g_k\left(a_1^{(0)},a_2^{(0)},\cdots,a_m^{(0)}\right)+\frac{\partial g_k}{\partial a_1}\Delta a_1+\frac{\partial g_k}{\partial a_2}\Delta a_2+\cdots+\frac{\partial g_k}{\partial a_m}\Delta a_m \end{aligned} \tag{137}$$

where

$$\frac{\partial g_k\left(a_1,a_2,\cdots,a_m\right)}{a_l}$$
$$=\frac{1}{N}\sum_{i=1}^{N}\left\{-\frac{\partial f\left(a_1,a_2,\cdots,a_m,x_i\right)}{\partial a_l}\frac{\partial f\left(a_1,a_2,\cdots,a_m,x_i\right)}{\partial a_k}+\left[y_i-f\left(a_1,a_2,\cdots,a_m,x_i\right)\right]\frac{\partial^2 f\left(a_1,a_2,\cdots,a_m,x_i\right)}{\partial a_l \partial a_k}\right\} \tag{138}$$

Therefore, we obtain

$$\frac{\partial g_k}{\partial a_1}\Delta a_1+\frac{\partial g_k}{\partial a_2}\Delta a_2+\cdots+\frac{\partial g_k}{\partial a_m}\Delta a_m \approx -g_k\left(a_1^{(0)},a_2^{(0)},\cdots,a_m^{(0)}\right) \tag{139}$$

We then obtain

$$\begin{cases} \frac{\partial g_1}{\partial a_1}\Delta a_1+\frac{\partial g_1}{\partial a_2}\Delta a_2+\cdots+\frac{\partial g_1}{\partial a_m}\Delta a_m = -g_1\left(a_1^{(0)},a_2^{(0)},\cdots,a_m^{(0)}\right) \\ \frac{\partial g_2}{\partial a_1}\Delta a_1+\frac{\partial g_2}{\partial a_2}\Delta a_2+\cdots+\frac{\partial g_2}{\partial a_m}\Delta a_m = -g_2\left(a_1^{(0)},a_2^{(0)},\cdots,a_m^{(0)}\right) \\ \cdots \\ \frac{\partial g_m}{\partial a_1}\Delta a_1+\frac{\partial g_m}{\partial a_2}\Delta a_2+\cdots+\frac{\partial g_m}{\partial a_m}\Delta a_m = -g_m\left(a_1^{(0)},a_2^{(0)},\cdots,a_m^{(0)}\right) \end{cases} \tag{140}$$

We then obtain the below matrix.

$$\begin{pmatrix} \frac{\partial g_1}{\partial a_1} & \frac{\partial g_1}{\partial a_2} & \cdots & \frac{\partial g_1}{\partial a_m} \\ \frac{\partial g_2}{\partial a_1} & \frac{\partial g_2}{\partial a_2} & \cdots & \frac{\partial g_2}{\partial a_m} \\ \vdots & \vdots & \ddots & \vdots \\ \frac{\partial g_m}{\partial a_1} & \frac{\partial g_m}{\partial a_2} & \cdots & \frac{\partial g_m}{\partial a_m} \end{pmatrix} \begin{pmatrix} \Delta a_1 \\ \Delta a_2 \\ \vdots \\ \Delta a_m \end{pmatrix} = \begin{pmatrix} -g_1\left(a_1^{(0)},a_2^{(0)},\cdots,a_m^{(0)}\right) \\ -g_2\left(a_1^{(0)},a_2^{(0)},\cdots,a_m^{(0)}\right) \\ \vdots \\ -g_m\left(a_1^{(0)},a_2^{(0)},\cdots,a_m^{(0)}\right) \end{pmatrix} \tag{141}$$

We then have

$$\begin{pmatrix} \Delta a_1 \\ \Delta a_2 \\ \vdots \\ \Delta a_m \end{pmatrix} = \begin{pmatrix} \frac{\partial g_1}{\partial a_1} & \frac{\partial g_1}{\partial a_2} & \cdots & \frac{\partial g_1}{\partial a_m} \\ \frac{\partial g_2}{\partial a_1} & \frac{\partial g_2}{\partial a_2} & \cdots & \frac{\partial g_2}{\partial a_m} \\ \vdots & \vdots & \ddots & \vdots \\ \frac{\partial g_m}{\partial a_1} & \frac{\partial g_m}{\partial a_2} & \cdots & \frac{\partial g_m}{\partial a_m} \end{pmatrix}^{-1} \begin{pmatrix} -g_1\left(a_1^{(0)},a_2^{(0)},\cdots,a_m^{(0)}\right) \\ -g_2\left(a_1^{(0)},a_2^{(0)},\cdots,a_m^{(0)}\right) \\ \vdots \\ -g_m\left(a_1^{(0)},a_2^{(0)},\cdots,a_m^{(0)}\right) \end{pmatrix} \tag{142}$$

The updating value is then given by

$$\begin{cases} a_1^{(1)} = a_1^{(0)} + \Delta a_1 \\ a_2^{(1)} = a_2^{(0)} + \Delta a_2 \\ \cdots \\ a_m^{(1)} = a_m^{(0)} + \Delta a_m \end{cases} \tag{143}$$

If the following hold

$$\begin{cases} abs(\Delta a_1) < \delta_c \\ abs(\Delta a_2) < \delta_c \\ \cdots \\ abs(\Delta a_m) < \delta_c \end{cases} \tag{144}$$

the updating value is the ones we need. If they do not hold, we set

$$\begin{cases} a_1^{(0)} = a_1^{(1)} \\ a_2^{(0)} = a_2^{(1)} \\ \cdots \\ a_m^{(0)} = a_m^{(1)} \end{cases} \tag{145}$$

We perform this cycle while Eq. (144) becomes valid.

14. Logistic Regression

We apply the above procedure to a logistic curve, which is frequently used.
The linear regression line is expressed as

$$y = ax + b \tag{146}$$

The objective value y increases monotonically with increasing an explanatory value of x. We want to consider a two levels categorical objective variable such as yes or no, good or bad, or so on. Then we regard y as the probability for the categorical data and modify the left hand of Eq. (146), and express it as

$$\ln\left(\frac{y}{1-y}\right) = ax + b \tag{147}$$

When we consider the case of yes or no, y corresponds to the probability of yes, and 1-y corresponds to the one of no. Eq. (147) can be solved with respect to y as

$$\begin{aligned} y &= \frac{1}{1+\exp(-ax-b)} \\ &= \frac{1}{1+k\exp(-ax)} \end{aligned} \tag{148}$$

where

$$k = e^{-b} \tag{149}$$

Y has a value range between 0 and 1, and saturates when it approaches to 1. We want to extend this form to have any positive value, and add one more parameter as

$$f(a_1, a_2, a_3; x) = \frac{a_1}{1 + a_2 \exp(-a_3 x)} \tag{150}$$

This is the logistic curve. In many cases, the dependence of the objective variable on explanatory variable is not simple, that is, even when the dependence is clear, it saturate for certain large explanatory variable values. A logistic curve is frequently used to express the saturation of phenomenon as mentioned above.

We can easily extend this to include various explanatory variables given by

$$f(a_1, a_2, a_{3i}; x_i) = \frac{a_1}{1 + a_2 \exp\left[-\left(a_{31}x_1 + a_{32}x_2 + \cdots + a_{3m}x_m\right)\right]} \tag{151}$$

In this analysis, we focus on one variable, and use Eq. (150).

The error $\sigma_e^{(2)}$ for the logistic curve is expressed by

$$\sigma_e^{(2)} = \frac{1}{N}\sum_{i=1}^{N}\left[y_i - f(a_1, a_2, a_3; x)\right]^2 \tag{152}$$

Partial differentiating this with respect to a, we obtain

$$\begin{aligned}\frac{\partial \sigma_e^{(2)}}{\partial a_1} &= \frac{1}{N}\sum_{i=1}^{N} -2\left[y_i - f\left(a_1,a_2,a_3;x_i\right)\right]\frac{\partial f\left(a_1,a_2,a_3;x_i\right)}{\partial a_1} \\ &= \frac{1}{N}\sum_{i=1}^{N} -2\left[y_i - f\left(a_1,a_2,a_3;x_i\right)\right]\frac{1}{1+a_2\exp\left(-a_3x_i\right)} = 0\end{aligned} \tag{153}$$

Arranging this equation, we obtain

$$\frac{1}{N}\sum_{i=1}^{N}\left[y_i - f\left(a_1,a_2,a_3;x_i\right)\right]\frac{1}{1+a_2\exp\left(-a_3x_i\right)} = 0 \tag{154}$$

We set the left side of Eq. (132) as $g_{a_1}\left(a_1,a_2,a_3\right)$, that is,

$$g_{a_1}\left(a_1,a_2,a_3\right) = \frac{1}{N}\sum_{i=1}^{N}\left[y_i - \frac{a_1}{1+a_2\exp\left(-a_3x_i\right)}\right]\frac{1}{1+a_2\exp\left(-a_3x_i\right)} = 0 \tag{155}$$

Performing a similar analysis for the other parameters, we obtain

$$g_{a_2}\left(a_1,a_2,a_3\right) = \frac{1}{N}\sum_{i=1}^{N}\left[y_i - \frac{a_1}{1+a_2\exp\left(-a_3x_i\right)}\right]\frac{a_1\exp\left(-a_3x_i\right)}{\left[1+a_2\exp\left(-a_3x_i\right)\right]^2} \tag{156}$$

$$g_{a_3}\left(a_1,a_2,a_3\right) = \frac{1}{N}\sum_{i=1}^{N}\left[y_i - \frac{a_1}{1+a_2\exp\left(-a_3x_i\right)}\right]\frac{a_1a_2x_i\exp\left(-a_3x_i\right)}{\left[1+a_2\exp\left(-a_3x_i\right)\right]^2} \tag{157}$$

We then have

$$\begin{aligned}&\frac{\partial g_{a_1}\left(a_1,a_2,a_3\right)}{\partial a_1} \\ &= \frac{\partial}{\partial a_1}\left\{\frac{1}{N}\sum_{i=1}^{N}\left[y_i\frac{1}{1+a_2\exp\left(-a_3x_i\right)}\right] - \frac{a_1}{N}\sum_{i=1}^{N}\left[\frac{1}{\left[1+a_2\exp\left(-a_3x_i\right)\right]^2}\right]\right\} \\ &= -\frac{1}{N}\sum_{i=1}^{N}\frac{1}{\left[1+a_2\exp\left(-a_3x_i\right)\right]^2}\end{aligned} \tag{158}$$

$$\begin{aligned}
&\frac{\partial g_{a_1}(a_1,a_2,a_3)}{\partial a_2} \\
&= \frac{\partial}{\partial a_2}\left\{\frac{1}{N}\sum_{i=1}^{N}\left[y_i\frac{1}{1+a_2\exp(-a_3x_i)}\right]-\frac{a_1}{N}\sum_{i=1}^{N}\left[\frac{1}{\left[1+a_2\exp(-a_3x_i)\right]^2}\right]\right\} \\
&= \frac{1}{N}\sum_{i=1}^{N}\left[y_i\frac{-\exp(-a_3x_i)}{\left[1+a_2\exp(-a_3x_i)\right]^2}\right]-\frac{a_1}{N}\sum_{i=1}^{N}\left[\frac{-2\left[1+a_2\exp(-a_3x_i)\right]\exp(-a_3x_i)}{\left[1+a_2\exp(-a_3x_i)\right]^4}\right] \\
&= -\frac{1}{N}\sum_{i=1}^{N}\left[y_i\frac{\exp(-a_3x_i)}{\left[1+a_2\exp(-a_3x_i)\right]^2}\right]+\frac{2a_1}{N}\sum_{i=1}^{N}\left[\frac{\exp(-a_3x_i)}{\left[1+a_2\exp(-a_3x_i)\right]^3}\right]
\end{aligned} \tag{159}$$

$$\begin{aligned}
&\frac{\partial g_{a_1}(a_1,a_2,a_3)}{\partial a_3} \\
&= \frac{\partial}{\partial a_3}\left\{\frac{1}{N}\sum_{i=1}^{N}\left[y_i\frac{1}{1+a_2\exp(-a_3x_i)}\right]-\frac{a_1}{N}\sum_{i=1}^{N}\left[\frac{1}{\left[1+a_2\exp(-a_3x_i)\right]^2}\right]\right\} \\
&= \frac{1}{N}\sum_{i=1}^{N}\left[y_i\frac{a_2x_i\exp(-a_3x_i)}{\left[1+a_2\exp(-a_3x_i)\right]^2}\right]-\frac{a_1}{N}\sum_{i=1}^{N}\left[\frac{2a_2x_i\left[1+a_2\exp(-a_3x_i)\right]\exp(-a_3x_i)}{\left[1+a_2\exp(-a_3x_i)\right]^4}\right] \\
&= \frac{a_2}{N}\sum_{i=1}^{N}\left[y_i\frac{x_i\exp(-a_3x_i)}{\left[1+a_2\exp(-a_3x_i)\right]^2}\right]-\frac{2a_1a_2}{N}\sum_{i=1}^{N}\left[\frac{x_i\exp(-a_3x_i)}{\left[1+a_2\exp(-a_3x_i)\right]^3}\right]
\end{aligned} \tag{160}$$

$$\begin{aligned}
&\frac{\partial g_{a_2}(a_1,a_2,a_3)}{\partial a_1} \\
&= \frac{\partial}{\partial a_1}\left\{\frac{1}{N}\sum_{i=1}^{N}\left[y_i\frac{a_1\exp(-a_3x_i)}{\left[1+a_2\exp(-a_3x_i)\right]^2}\right]-\frac{1}{N}\sum_{i=1}^{N}\left[\frac{a_1^2\exp(-a_3x_i)}{\left[1+a_2\exp(-a_3x_i)\right]^3}\right]\right\} \\
&= \frac{1}{N}\sum_{i=1}^{N}\left[y_i\frac{\exp(-a_3x_i)}{\left[1+a_2\exp(-a_3x_i)\right]^2}\right]-\frac{1}{N}\sum_{i=1}^{N}\left[\frac{2a_1\exp(-a_3x_i)}{\left[1+a_2\exp(-a_3x_i)\right]^3}\right] \\
&= \frac{1}{N}\sum_{i=1}^{N}\left[y_i\frac{\exp(-a_3x_i)}{\left[1+a_2\exp(-a_3x_i)\right]^2}\right]-\frac{2a_1}{N}\sum_{i=1}^{N}\left[\frac{\exp(-a_3x_i)}{\left[1+a_2\exp(-a_3x_i)\right]^3}\right]
\end{aligned} \tag{161}$$

$$
\begin{aligned}
&\frac{\partial g_{a_2}(a_1,a_2,a_3)}{\partial a_2} \\
&= \frac{\partial}{\partial a_2}\left\{ \frac{1}{N}\sum_{i=1}^{N}\left[y_i \frac{a_1\exp(-a_3x_i)}{\left[1+a_2\exp(-a_3x_i)\right]^2}\right] - \frac{1}{N}\sum_{i=1}^{N}\left[\frac{a_1^2\exp(-a_3x_i)}{\left[1+a_2\exp(-a_3x_i)\right]^3}\right]\right\} \\
&= \frac{1}{N}\sum_{i=1}^{N}\left[y_i \frac{-2a_1\exp(-2a_3x_i)\left[1+a_2\exp(-a_3x_i)\right]}{\left[1+a_2\exp(-a_3x_i)\right]^4}\right] \\
&\quad - \frac{1}{N}\sum_{i=1}^{N}\left[\frac{-3a_1^2\exp(-2a_3x_i)\left[1+a_2\exp(-a_3x_i)\right]^2}{\left[1+a_2\exp(-a_3x_i)\right]^6}\right] \\
&= -\frac{2a_1}{N}\sum_{i=1}^{N}\left[y_i \frac{\exp(-2a_3x_i)}{\left[1+a_2\exp(-a_3x_i)\right]^3}\right] + \frac{3a_1^2}{N}\sum_{i=1}^{N}\left[\frac{\exp(-2a_3x_i)}{\left[1+a_2\exp(-a_3x_i)\right]^4}\right]
\end{aligned} \tag{162}
$$

$$
\begin{aligned}
&\frac{\partial g_{a_2}(a_1,a_2,a_3)}{\partial a_3} \\
&= \frac{\partial}{\partial a_3}\left\{ \frac{1}{N}\sum_{i=1}^{N}\left[y_i \frac{a_1\exp(-a_3x_i)}{\left[1+a_2\exp(-a_3x_i)\right]^2}\right] - \frac{1}{N}\sum_{i=1}^{N}\left[\frac{a_1^2\exp(-a_3x_i)}{\left[1+a_2\exp(-a_3x_i)\right]^3}\right]\right\} \\
&= \frac{1}{N}\sum_{i=1}^{N}\left[y_i \frac{-a_1x_i\exp(-a_3x_i)\left[1+a_2\exp(-a_3x_i)\right]^2 + 2a_1a_2x_i\exp(-2a_3x_i)\left[1+a_2\exp(-a_3x_i)\right]}{\left[1+a_2\exp(-a_3x_i)\right]^4}\right] \\
&\quad - \frac{1}{N}\sum_{i=1}^{N}\left[\frac{-a_1^2x_i\exp(-a_3x_i)\left[1+a_2\exp(-a_3x_i)\right]^3 + 3a_1^2a_2x_i\exp(-2a_3x_i)\left[1+a_2\exp(-a_3x_i)\right]^2}{\left[1+a_2\exp(-a_3x_i)\right]^6}\right] \\
&= -\frac{a_1}{N}\sum_{i=1}^{N}\left[y_i \frac{x_i\exp(-a_3x_i)}{\left[1+a_2\exp(-a_3x_i)\right]^2}\right] + \frac{2a_1a_2}{N}\sum_{i=1}^{N}\left[y_i \frac{x_i\exp(-2a_3x_i)}{\left[1+a_2\exp(-a_3x_i)\right]^3}\right] \\
&\quad + \frac{a_1^2}{N}\sum_{i=1}^{N}\left[\frac{x_i\exp(-a_3x_i)}{\left[1+a_2\exp(-a_3x_i)\right]^3}\right] - \frac{3a_1^2a_2}{N}\sum_{i=1}^{N}\left[\frac{x_i\exp(-2a_3x_i)}{\left[1+a_2\exp(-a_3x_i)\right]^4}\right]
\end{aligned} \tag{163}
$$

$$
\begin{aligned}
&\frac{\partial g_{a_3}(a_1,a_2,a_3)}{\partial a_1} \\
&= \frac{\partial}{\partial a_1}\left\{ \frac{1}{N}\sum_{i=1}^{N}\left[y_i \frac{a_1a_2x_i\exp(-a_3x_i)}{\left[1+a_2\exp(-a_3x_i)\right]^2}\right] - \frac{1}{N}\sum_{i=1}^{N}\left[\frac{a_1^2a_2x_i\exp(-a_3x_i)}{\left[1+a_2\exp(-a_3x_i)\right]^3}\right]\right\} \\
&= \frac{1}{N}\sum_{i=1}^{N}\left[y_i \frac{a_2x_i\exp(-a_3x_i)}{\left[1+a_2\exp(-a_3x_i)\right]^2}\right] - \frac{1}{N}\sum_{i=1}^{N}\left[\frac{2a_1a_2x_i\exp(-a_3x_i)}{\left[1+a_2\exp(-a_3x_i)\right]^3}\right] \\
&= \frac{a_2}{N}\sum_{i=1}^{N}\left[y_i \frac{x_i\exp(-a_3x_i)}{\left[1+a_2\exp(-a_3x_i)\right]^2}\right] - \frac{2a_1a_2}{N}\sum_{i=1}^{N}\left[\frac{x_i\exp(-a_3x_i)}{\left[1+a_2\exp(-a_3x_i)\right]^3}\right]
\end{aligned} \tag{164}
$$

$$\begin{aligned}
&\frac{\partial g_{a_3}\left(a_1,a_2,a_3\right)}{\partial a_2} \\
&= \frac{\partial}{\partial a_2}\left\{\frac{1}{N}\sum_{i=1}^{N}\left[y_i \frac{a_1 a_2 x_i \exp\left(-a_3 x_i\right)}{\left[1+a_2\exp\left(-a_3 x_i\right)\right]^2}\right] - \frac{1}{N}\sum_{i=1}^{N}\left[\frac{a_1^2 a_2 x_i \exp\left(-a_3 x_i\right)}{\left[1+a_2\exp\left(-a_3 x_i\right)\right]^3}\right]\right\} \\
&= \frac{1}{N}\sum_{i=1}^{N}\left[y_i \frac{a_1 x_i \exp\left(-a_3 x_i\right)\left[1+a_2\exp\left(-a_3 x_i\right)\right]^2 - 2a_1 a_2 x_i \exp\left(-2a_3 x_i\right)\left[1+a_2\exp\left(-a_3 x_i\right)\right]}{\left[1+a_2\exp\left(-a_3 x_i\right)\right]^4}\right] \\
&-\frac{1}{N}\sum_{i=1}^{N}\left[\frac{a_1^2 x_i \exp\left(-a_3 x_i\right)\left[1+a_2\exp\left(-a_3 x_i\right)\right]^3 - 3a_1^2 a_2 x_i \exp\left(-2a_3 x_i\right)\left[1+a_2\exp\left(-a_3 x_i\right)\right]^2}{\left[1+a_2\exp\left(-a_3 x_i\right)\right]^6}\right] \\
&= \frac{a_1}{N}\sum_{i=1}^{N}\left[y_i \frac{x_i \exp\left(-a_3 x_i\right)}{\left[1+a_2\exp\left(-a_3 x_i\right)\right]^2}\right] - \frac{2a_1 a_2}{N}\sum_{i=1}^{N}\left[y_i \frac{x_i \exp\left(-2a_3 x_i\right)}{\left[1+a_2\exp\left(-a_3 x_i\right)\right]^3}\right] \\
&-\frac{a_1^2}{N}\sum_{i=1}^{N}\left[\frac{x_i \exp\left(-a_3 x_i\right)}{\left[1+a_2\exp\left(-a_3 x_i\right)\right]^3}\right] + \frac{3a_1^2 a_2}{N}\sum_{i=1}^{N}\left[\frac{x_i \exp\left(-2a_3 x_i\right)}{\left[1+a_2\exp\left(-a_3 x_i\right)\right]^4}\right]
\end{aligned} \tag{165}$$

$$\begin{aligned}
&\frac{\partial g_{a_3}\left(a_1,a_2,a_3\right)}{\partial a_3} \\
&= \frac{\partial}{\partial a_3}\left\{\frac{1}{N}\sum_{i=1}^{N}\left[y_i \frac{a_1 a_2 x_i \exp\left(-a_3 x_i\right)}{\left[1+a_2\exp\left(-a_3 x_i\right)\right]^2}\right] - \frac{1}{N}\sum_{i=1}^{N}\left[\frac{a_1^2 a_2 x_i \exp\left(-a_3 x_i\right)}{\left[1+a_2\exp\left(-a_3 x_i\right)\right]^3}\right]\right\} \\
&= \frac{1}{N}\sum_{i=1}^{N}\left[y_i \frac{-a_1 a_2 x_i^2 \exp\left(-a_3 x_i\right)\left[1+a_2\exp\left(-a_3 x_i\right)\right]^2 + 2a_1 a_2^2 x_i^2 \operatorname{xp}\left(-2a_3 x_i\right)\left[1+a_2\exp\left(-a_3 x_i\right)\right]}{\left[1+a_2\exp\left(-a_3 x_i\right)\right]^4}\right] \\
&-\frac{1}{N}\sum_{i=1}^{N}\left[\frac{-a_1^2 a_2 x_i^2 \exp\left(-a_3 x_i\right)\left[1+a_2\exp\left(-a_3 x_i\right)\right]^3 + 3a_1^2 a_2^2 x_i^2 \exp\left(-2a_3 x_i\right)\left[1+a_2\exp\left(-a_3 x_i\right)\right]^2}{\left[1+a_2\exp\left(-a_3 x_i\right)\right]^6}\right] \\
&= -\frac{a_1 a_2}{N}\sum_{i=1}^{N}\left[y_i \frac{x_i^2 \exp\left(-a_3 x_i\right)}{\left[1+a_2\exp\left(-a_3 x_i\right)\right]^2}\right] + \frac{2a_1 a_2^2}{N}\sum_{i=1}^{N}\left[y_i \frac{x_i^2 \operatorname{xp}\left(-2a_3 x_i\right)}{\left[1+a_2\exp\left(-a_3 x_i\right)\right]^3}\right] \\
&+\frac{a_1^2 a_2}{N}\sum_{i=1}^{N}\left[\frac{x_i^2 \exp\left(-a_3 x_i\right)}{\left[1+a_2\exp\left(-a_3 x_i\right)\right]^3}\right] - \frac{3a_1^2 a_2^2}{N}\sum_{i=1}^{N}\left[\frac{x_i^2 \exp\left(-2a_3 x_i\right)}{\left[1+a_2\exp\left(-a_3 x_i\right)\right]^4}\right]
\end{aligned} \tag{166}$$

The incremental variation is given by

$$\begin{pmatrix}\Delta a_1\\ \Delta a_2\\ \Delta a_3\end{pmatrix} = \begin{pmatrix}\frac{\partial g_1}{\partial a_1} & \frac{\partial g_1}{\partial a_2} & \frac{\partial g_1}{\partial a_3}\\ \frac{\partial g_2}{\partial a_1} & \frac{\partial g_2}{\partial a_2} & \frac{\partial g_2}{\partial a_3}\\ \frac{\partial g_3}{\partial a_1} & \frac{\partial g_3}{\partial a_2} & \frac{\partial g_3}{\partial a_3}\end{pmatrix}^{-1} \begin{pmatrix}-g_1\left(a_1^{(0)},a_2^{(0)},a_3^{(0)}\right)\\ -g_2\left(a_1^{(0)},a_2^{(0)},a_3^{(0)}\right)\\ -g_3\left(a_1^{(0)},a_2^{(0)},a_3^{(0)}\right)\end{pmatrix} \tag{167}$$

The updating value is then given by

$$\begin{cases} a_1^{(1)} = a_1^{(0)} + \Delta a_1 \\ a_2^{(1)} = a_2^{(0)} + \Delta a_2 \\ \cdots \\ a_m^{(1)} = a_m^{(0)} + \Delta a_m \end{cases} \tag{168}$$

If the following hold

$$\begin{cases} abs(\Delta a_1) < \delta_c \\ abs(\Delta a_2) < \delta_c \\ \cdots \\ abs(\Delta a_m) < \delta_c \end{cases} \tag{169}$$

the updating value is the ones we need. If they do not hold, we set

$$\begin{cases} a_1^{(0)} = a_1^{(1)} \\ a_2^{(0)} = a_2^{(1)} \\ \cdots \\ a_m^{(0)} = a_m^{(1)} \end{cases} \tag{170}$$

We perform this cycle while Eq. (144) becomes valid.
We can set initial conditions roughly as follows.
We can see the saturated value inspecting the data and set it as a_{10}.
Selecting two points of (x_1, y_1), and (x_2, y_2), we obtain

$$y_1 = \frac{a_{10}}{1 + a_{20}\exp(-a_{30}x_1)} \tag{171}$$

$$y_2 = \frac{a_{10}}{1 + a_{20}\exp(-a_{30}x_2)} \tag{172}$$

We then modify them as follows.

$$\frac{a_{10}}{y_1} - 1 = a_{20}\exp(-a_{30}x_1) \tag{173}$$

$$\frac{a_{10}}{y_2} - 1 = a_{20} \exp\left(-a_{30} x_2\right) \tag{174}$$

Therefore, we obtain

$$\ln\left(\frac{a_{10}}{y_1} - 1\right) = \ln a_{20} - a_{30} x_1 \tag{175}$$

$$\ln\left(\frac{a_{10}}{y_2} - 1\right) = \ln a_{20} - a_{30} x_2 \tag{176}$$

We then obtain

$$a_{30} = \frac{1}{x_2 - x_1} \ln \frac{\left(\frac{a_{10}}{y_1} - 1\right)}{\left(\frac{a_{10}}{y_2} - 1\right)} \tag{177}$$

$$\begin{aligned} \ln a_{20} &= \ln\left(\frac{a_{10}}{y_1} - 1\right) + a_{30} x_1 \\ &= \frac{x_2}{x_2 - x_1} \ln\left(\frac{a_{10}}{y_1} - 1\right) - \frac{x_1}{x_2 - x_1} \ln\left(\frac{a_{10}}{y_2} - 1\right) \\ &= \ln \frac{\left(\frac{a_{10}}{y_1} - 1\right)^{\frac{x_2}{x_2 - x_1}}}{\left(\frac{a_{10}}{y_2} - 1\right)^{\frac{x_1}{x_2 - x_1}}} \end{aligned} \tag{178}$$

Therefore, we obtain

$$a_{20} = \frac{\left(\frac{a_{10}}{y_1} - 1\right)^{\frac{x_2}{x_2 - x_1}}}{\left(\frac{a_{10}}{y_2} - 1\right)^{\frac{x_1}{x_2 - x_1}}} \tag{179}$$

Let us consider the data given in Table 3. The value of y saturate with increasing x as shown in Figure 4.

Table 3. Data for logistic curve

x	y
1	3
2	5
3	9
4	14
5	23
6	35
7	55
8	65
9	80
10	90
11	91
12	94
13	97
14	99
15	100

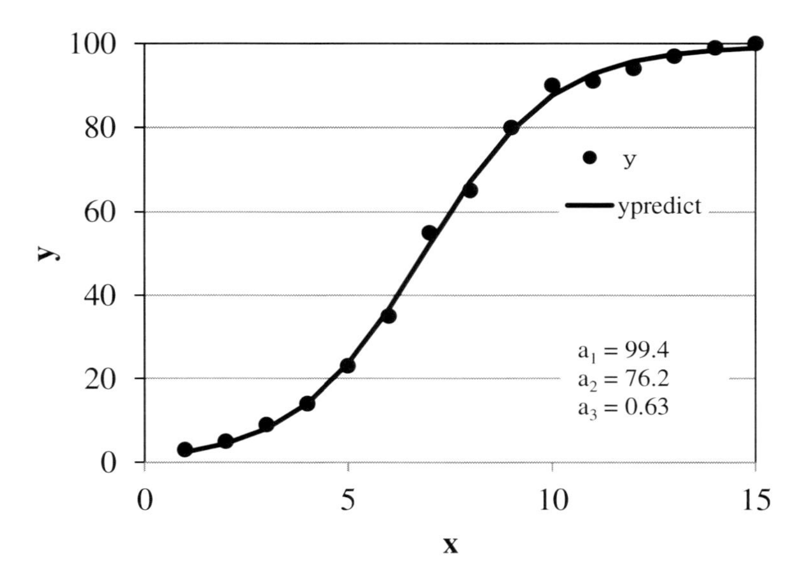

Figure 4. Logistic cured with regression one.

We set a_{10} as 100, and use the data set of $(5,23)$ and $(10,90)$. We then have

$$\begin{aligned} a_{30} &= \frac{1}{x_2 - x_1} \ln \frac{\left(\frac{a_{10}}{y_1} - 1\right)}{\left(\frac{a_{10}}{y_2} - 1\right)} \\ &= \frac{1}{10-5} \ln \frac{\left(\frac{100}{23} - 1\right)}{\left(\frac{100}{90} - 1\right)} \\ &= 0.68 \end{aligned} \tag{180}$$

$$\begin{aligned} a_{20} &= \frac{\left(\frac{a_{10}}{y_1} - 1\right)^{\frac{x_2}{x_2 - x_1}}}{\left(\frac{a_{10}}{y_2} - 1\right)^{\frac{x_1}{x_2 - x_1}}} \\ &= \frac{\left(\frac{100}{23} - 1\right)^{\frac{10}{10-5}}}{\left(\frac{100}{90} - 1\right)^{\frac{5}{10-5}}} \\ &= 100.9 \end{aligned} \tag{181}$$

Using the initial conditions above, we perform a regression procedure shown in this section, and obtain the regression curve with factors of

$$\begin{cases} a_1 = 99.4 \\ a_2 = 76.2 \\ a_3 = 0.63 \end{cases} \tag{182}$$

The data and a regression cure agree well.

15. Matrix Expression for Regression

The regression can be expressed with a matrix form. This gives no new information. However, the expression is simple and one can use the method alternatively.

The objective and an explanatory variables are related as follows.

$$y_i = a_0 + a_1 \left(x_i - \overline{x} \right) + e_i \tag{183}$$

where we have n samples. Therefore, all data are related as

$$\begin{cases} y_1 = a_0 + a_1 \left(x_1 - \overline{x} \right) + e_1 \\ y_2 = a_0 + a_1 \left(x_2 - \overline{x} \right) + e_2 \\ \cdots \\ y_n = a_0 + a_1 \left(x_n - \overline{x} \right) + e_1 \end{cases} \tag{184}$$

Therefore, the regression analysis is to obtain the factors a_0, a_1.
We introduce matrixes and vectors as

$$Y = \begin{pmatrix} y_1 \\ y_2 \\ \vdots \\ y_n \end{pmatrix} \tag{185}$$

$$X = \begin{pmatrix} 1 & x_1 - \overline{x} \\ 1 & x_2 - \overline{x} \\ \vdots & \vdots \\ 1 & x_n - \overline{x} \end{pmatrix} \tag{186}$$

$$\boldsymbol{\beta} = \begin{pmatrix} a_0 \\ a_1 \end{pmatrix} \tag{187}$$

$$\mathbf{e} = \begin{pmatrix} e_1 \\ e_2 \\ \vdots \\ e_n \end{pmatrix} \tag{188}$$

Using the matrixes and vectors, the corresponding equation is expressed by

$$Y = X\boldsymbol{\beta} + \mathbf{e} \tag{189}$$

The total sum of the square of error Q is then expressed by

$$\begin{aligned} Q &= \mathbf{e}^T \mathbf{e} \\ &= (Y - X\boldsymbol{\beta})^T (Y - X\boldsymbol{\beta}) \\ &= (Y^T - \boldsymbol{\beta}^T X^T)(Y - X\boldsymbol{\beta}) \\ &= Y^T Y - Y^T X\boldsymbol{\beta} - \boldsymbol{\beta}^T X^T Y + \boldsymbol{\beta}^T X^T X\boldsymbol{\beta} \\ &= Y^T Y - 2\boldsymbol{\beta}^T X^T Y + \boldsymbol{\beta}^T X^T X\boldsymbol{\beta} \end{aligned} \tag{190}$$

We obtain the optimum value by minimizing Q. This can be done by vector differentiating (see Appendix 1-10 of volume 3) with respect to $\boldsymbol{\beta}$ it given by

$$\begin{aligned} \frac{\partial Q}{\partial \boldsymbol{\beta}} &= -2\frac{\partial}{\partial \boldsymbol{\beta}}\left(\boldsymbol{\beta}^T X^T Y\right) + \frac{\partial}{\partial \boldsymbol{\beta}}\left(\boldsymbol{\beta}^T X^T X\boldsymbol{\beta}\right) \\ &= -2X^T Y + 2X^T X\boldsymbol{\beta} \\ &= \mathbf{0} \end{aligned} \tag{191}$$

Therefore, we obtain

$$\boldsymbol{\beta} = \left(X^T X\right)^{-1} X^T Y \tag{192}$$

SUMMARY

To summarize the results in this chapter—
The regression line is given by

$$\frac{y - \mu_y}{\sqrt{\sigma_{yy}^{(2)}}} = \rho_{xy}\left(\frac{x - \mu_x}{\sqrt{\sigma_{xx}^{(2)}}}\right)$$

If we want to force the regression to start from $(0, a_0)$, the regression line is given by

$$y = \frac{\rho_{xy}\sqrt{\sigma_{xx}^{(2)}}\sqrt{\sigma_{yy}^{(2)}} + \mu_x\mu_y - a_0\mu_x}{\left(\sigma_{xx}^{(2)} + \mu_x^2\right)} x + a_0$$

The accuracy of regression is evaluated by

$$R^2 = \frac{\sigma_r^{(2)}}{\sigma_{yy}^{(2)}} = \rho_{xy}^2$$

where $\sigma_r^{(2)}$ is the variance of Y given by

$$\sigma_r^{(2)} = \frac{1}{N}\sum_{i=1}^{N}\left(Y_i - \mu_y\right)^2$$

The error range for each factors are given by

$$\hat{a}_1 - t_p\left(n-2\right)\frac{\sqrt{s_e^{(2)}}}{\sqrt{nS_{xx}^{(2)}}} \le a_1 \le \hat{a}_1 + t_p\left(n-2\right)\frac{\sqrt{s_e^{(2)}}}{\sqrt{nS_{xx}^{(2)}}}$$

$$\hat{a}_0 - t_p\sqrt{\left(\frac{1}{n} + \frac{\overline{x}^2}{nS_{xx}^{(2)}}\right)s_e^{(2)}} \le a_0 \le \hat{a}_0 + t_p\sqrt{\left(\frac{1}{n} + \frac{\overline{x}^2}{nS_{xx}^{(2)}}\right)s_e^{(2)}}$$

The predictive value for the regression Y_0 is given by

$$\hat{Y}_0 - t_p\left(n-2\right)\sqrt{\left[\frac{1}{n} + \frac{\left(x_0 - \overline{x}\right)^2}{nS_{xx}^{(2)}}\right]s_e^{(2)}} \le Y_o \le \hat{Y}_0 + t_p\left(n-2\right)\sqrt{\left[\frac{1}{n} + \frac{\left(x_0 - \overline{x}\right)^2}{nS_{xx}^{(2)}}\right]s_e^{(2)}}$$

The predictive y value y_0 is given by

$$\hat{Y}_0 - t_p\left(n-2\right)\sqrt{\left[1 + \frac{1}{n} + \frac{\left(x_0 - \overline{x}\right)^2}{nS_{xx}^{(2)}}\right]s_e^{(2)}} \le y_o \le \hat{Y}_0 + t_p\left(n-2\right)\sqrt{\left[1 + \frac{1}{n} + \frac{\left(x_0 - \overline{x}\right)^2}{nS_{xx}^{(2)}}\right]s_e^{(2)}}$$

where

$$\hat{Y}_0 = \hat{a}_0 + \hat{a}_1 x_0$$

When the target function is an exponential one, given by

$$y = be^{ax}$$

We converting variables as

$$\begin{cases} u = x \\ v = \ln y \end{cases}$$

We can apply the linear regression to the variables.

When y is proportional to power law of x, that is, it is given by

$$y = bx^a$$

We converting variables as

$$\begin{cases} u = \ln x \\ v = \ln y \end{cases}$$

We can apply the linear regression to the variables.

We assume the data are expressed with a power law equation given by

$$y = a_0 + a_1 x + a_2 x^2 + \cdots + a_m x^m$$

The factors are given by

$$\begin{pmatrix} a_1 \\ a_2 \\ \vdots \\ a_m \end{pmatrix} = \begin{pmatrix} K_{x11} & K_{x12} & \cdots & K_{x1m} \\ K_{x21} & K_{x22} & \cdots & K_{x22} \\ \vdots & \vdots & \ddots & \vdots \\ K_{xm1} & K_{xm2} & \cdots & K_{xmm} \end{pmatrix}^{-1} \begin{pmatrix} K_{y1} \\ K_{y2} \\ \vdots \\ K_{ym} \end{pmatrix}$$

where

$$K_{xkl} = \frac{1}{N}\sum_{i=1}^{N}\left(x_i^k - \overline{x^k}\right)\left(x_i^l - \overline{x^l}\right)$$

$$K_{yk} = \frac{1}{N}\sum_{i=1}^{N}\left(y_i - \overline{y}\right)\left(x_i^k - \overline{x^k}\right)$$

We assume an arbitrarily function having many factors is given by $f\left(a_1, a_2, \cdots, a_m, x\right)$

We define the following function

$$g_1(a_1,a_2,\cdots,a_m)=\frac{1}{N}\sum_{i=1}^{N}\left[y_i-f(a_1,a_2,\cdots,a_m,x_i)\right]\frac{\partial f(a_1,a_2,\cdots,a_m,x_i)}{\partial a_1}$$

$$\begin{pmatrix}\Delta a_1\\ \Delta a_2\\ \vdots\\ \Delta a_m\end{pmatrix}=\begin{pmatrix}\frac{\partial g_1}{\partial a_1} & \frac{\partial g_1}{\partial a_2} & \cdots & \frac{\partial g_1}{\partial a_m}\\ \frac{\partial g_2}{\partial a_1} & \frac{\partial g_2}{\partial a_2} & \cdots & \frac{\partial g_2}{\partial a_m}\\ \vdots & \vdots & \ddots & \vdots\\ \frac{\partial g_m}{\partial a_1} & \frac{\partial g_m}{\partial a_2} & \cdots & \frac{\partial g_m}{\partial a_m}\end{pmatrix}^{-1}\begin{pmatrix}-g_1\left(a_1^{(0)},a_2^{(0)},\cdots,a_m^{(0)}\right)\\ -g_2\left(a_1^{(0)},a_2^{(0)},\cdots,a_m^{(0)}\right)\\ \vdots\\ -g_m\left(a_1^{(0)},a_2^{(0)},\cdots,a_m^{(0)}\right)\end{pmatrix}$$

The updating value is then given by

$$\begin{cases}a_1^{(1)}=a_1^{(0)}+\Delta a_1\\ a_2^{(1)}=a_2^{(0)}+\Delta a_2\\ \cdots\\ a_m^{(1)}=a_m^{(0)}+\Delta a_m\end{cases}$$

We repeat this process until the followings hold.

$$\begin{cases}abs(\Delta a_1)<\delta_c\\ abs(\Delta a_2)<\delta_c\\ \cdots\\ abs(\Delta a_m)<\delta_c\end{cases}$$

We apply this procedure to the logistic curve.
The residual error is evaluated by

$$t=\frac{z_{ek}}{\sqrt{1-h_{kk}}}$$

where h_{kk} is the leverage ratio given by

$$h_{kk}=\frac{1}{n}+\frac{(x_k-\bar{x})^2}{nS_{xx}^{(2)}}$$

The regression can be expressed using a matrix form as

$$\boldsymbol{\beta} = \left(X^T X\right)^{-1} X^T Y$$

where

$$y_i = a_0 + a_1\left(x_i - \bar{x}\right) + e_i$$

The matrixes and vectors are defined as

$$Y = \begin{pmatrix} y_1 \\ y_2 \\ \vdots \\ y_n \end{pmatrix}$$

$$X = \begin{pmatrix} 1 & x_1 - \bar{x} \\ 1 & x_2 - \bar{x} \\ \vdots & \vdots \\ 1 & x_n - \bar{x} \end{pmatrix}$$

$$\boldsymbol{\beta} = \begin{pmatrix} a_0 \\ a_1 \end{pmatrix}$$

$$\mathbf{e} = \begin{pmatrix} e_1 \\ e_2 \\ \vdots \\ e_n \end{pmatrix}$$

Chapter 4

MULTIPLE REGRESSION ANALYSIS

ABSTRACT

The objective variable is influenced by many kinds of variables in general. We therefore extend the analysis of a regression with one explanatory variable to multiple ones. We predict a theoretical prediction for given multi parameters with their error range. We also discuss the reality of the correlation relationship where the correlation factor is influenced by the other variables. The pure correlation without the influenced factor is also discussed. We also evaluate the effectiveness of the explanatory variables.

Keywords: regression, multiple regression, objective variable, explanatory variable, multicollinearity, partial correlation factor

1. INTRODUCTION

We treated one explanatory variable in the previous chapter, and studied the relationship between an explanatory variable and one objective variable. However, the objective variable is influenced by various parameters. For example, the price of a house is influenced by location, house age, site area, etc. Therefore, we need to treat multi explanatory variables to predict the objective variable robustly. We first treat two explanatory variables, and then treat ones of more than two.

We treat a sample regression in this chapter. The regression for population can be easily derived by changing sample averages and sample variances to the population's ones as shown in the previous chapter.

We perform matrix operation in this chapter. The basics of matrix operation are described in Chapter 15 of volume 3.

2. Regression with Two Explanatory Variables

We treat one objective variable y, and two explanatory variables x_1 and x_2 in this section for simplicity. We derive concrete forms for the two variables and extend it to the one for multiple ones later.

2.1. Regression Line

The objective variable y_i and two explanatory variables x_{i1} and x_{i2} in population can be related to

$$y_i = a_0 + a_1 x_{i1} + a_2 x_{i2} + e_i \tag{1}$$

where a_0, a_1, and a_2 are constant factors.

We pick up n set data from the population, and want to predict the characteristics of the population.

We can decide the sample factors a_0, a_1, and a_2 from the sample data, which is not the same as the ones for population, and hence we denote them as $\hat{a}_0$, $\hat{a}_1$, and $\hat{a}_2$. The sample data can be expressed by

$$\begin{aligned} y_1 &= \hat{a}_0 + \hat{a}_1 x_{11} + \hat{a}_2 x_{12} + e_1 \\ y_2 &= \hat{a}_0 + \hat{a}_1 x_{21} + \hat{a}_2 x_{22} + e_2 \\ &\cdots \\ y_n &= \hat{a}_0 + \hat{a}_1 x_{n1} + \hat{a}_2 x_{n2} + e_n \end{aligned} \tag{2}$$

e_i is called as residual error. A sample regression value Y_i is expressed by

$$Y_i = \hat{a}_0 + \hat{a}_1 x_{i1} + \hat{a}_2 x_{i2} \tag{3}$$

e_i can then be expressed by

$$e_i = y_i - Y_i = y_i - \left(\hat{a}_0 + \hat{a}_1 x_{i1} + \hat{a}_2 x_{i2}\right) \tag{4}$$

The variance of e_i is denoted as $S_e^{(2)}$, and is given by

$$S_e^{(2)} = \frac{1}{n}\sum_{i=1}^{n}\left[y_i - \left(\hat{a}_0 + \hat{a}_1 x_{i1} + \hat{a}_2 x_{i2}\right)\right]^2 \tag{5}$$

We decide $\hat{a}_0$, $\hat{a}_1$, and $\hat{a}_2$ so that $S_e^{(2)}$ has the minimum value.

Partial differentiating $S_e^{(2)}$ with respect to $\hat{a}_0$ and set 0 for them, we obtain

$$\frac{\partial S_e^{(2)}}{\partial \hat{a}_0} = -2\frac{1}{n}\sum_{i=1}^{n}\left[y_i - \left(\hat{a}_0 + \hat{a}_1 x_{i1} + \hat{a}_2 x_{i2}\right)\right] = 0 \tag{6}$$

We then obtain

$$\begin{aligned}\frac{1}{n}\sum_{i=1}^{n}\left[y_i - \left(\hat{a}_0 + \hat{a}_1 x_{i1} + \hat{a}_2 x_{i2}\right)\right] &= \frac{1}{n}\sum_{i=1}^{n} y_i - \hat{a}_0 - \hat{a}_1\frac{1}{n}\sum_{i=1}^{n} x_{i1} - \hat{a}_2\frac{1}{n}\sum_{i=1}^{r} x_{i2} \\ &= \bar{y} - \hat{a}_0 - \hat{a}_1\bar{x}_1 - \hat{a}_2\bar{x}_2 \\ &= 0\end{aligned} \tag{7}$$

Partial differentiating $S_e^{(2)}$ with respect to $\hat{a}_1$ and $\hat{a}_2$, and set 0 for them, we obtain

$$\frac{\partial S_e^{(2)}}{\partial \hat{a}_1} = -2\frac{1}{n}\sum_{i=1}^{n} x_{i1}\left[y_i - \left(\hat{a}_0 + \hat{a}_1 x_{i1} + \hat{a}_2 x_{i2}\right)\right] = 0 \tag{8}$$

$$\frac{\partial S_e^{(2)}}{\partial \hat{a}_2} = -2\frac{1}{n}\sum_{i=1}^{n} x_{i2}\left[y_i - \left(\hat{a}_0 + \hat{a}_1 x_{i1} + \hat{a}_2 x_{i2}\right)\right] = 0 \tag{9}$$

Substituting Eq. (7) into Eqs. (8) and (9), we obtain

$$\frac{1}{n}\sum_{i=1}^{n} x_{i1}\left[\left(y_i - \bar{y}\right) - \hat{a}_1\left(x_{i1} - \bar{x}_1\right) - \hat{a}_2\left(x_{i2} - \bar{x}_2\right)\right] = 0 \tag{10}$$

$$\frac{1}{n}\sum_{i=1}^{n} x_{i2}\left[\left(y_i - \bar{y}\right) - \hat{a}_1\left(x_{i1} - \bar{x}_1\right) - \hat{a}_2\left(x_{i2} - \bar{x}_2\right)\right] = 0 \tag{11}$$

Note that the followings are valid.

$$\frac{1}{n}\sum_{i=1}^{n}\bar{x}_1\left[\left(y_i-\bar{y}\right)-\hat{a}_1\left(x_{i1}-\bar{x}_1\right)-\hat{a}_2\left(x_{i2}-\bar{x}_2\right)\right]=0 \tag{12}$$

$$\frac{1}{n}\sum_{i=1}^{n}\bar{x}_2\left[\left(y_i-\bar{y}\right)-\hat{a}_1\left(x_{i1}-\bar{x}_1\right)-\hat{a}_2\left(x_{i2}-\bar{x}_2\right)\right]=0 \tag{13}$$

Therefore, we obtain

$$\begin{aligned}\frac{1}{n}\sum_{i=1}^{n}\left(x_{i1}-\bar{x}_1\right)\left[\left(y_i-\bar{y}\right)-\hat{a}_1\left(x_{i1}-\bar{x}_1\right)-\hat{a}_2\left(x_{i2}-\bar{x}_2\right)\right]&=S_{1y}^{(2)}-\hat{a}_1S_{11}^{(2)}-\hat{a}_2S_{12}^{(2)}\\&=0\end{aligned} \tag{14}$$

$$\begin{aligned}\frac{1}{n}\sum_{i=1}^{n}\left(x_{i2}-\bar{x}_2\right)\left[\left(y_i-\bar{y}\right)-\hat{a}_1\left(x_{i1}-\bar{x}_1\right)-\hat{a}_2\left(x_{i2}-\bar{x}_2\right)\right]&=S_{2y}^{(2)}-\hat{a}_1S_{21}^{(2)}-\hat{a}_2S_{22}^{(2)}\\&=0\end{aligned} \tag{15}$$

where

$$S_{11}^{(2)}\equiv\frac{1}{n}\sum_{i=1}^{n}\left(x_{i1}-\bar{x}_1\right)^2 \tag{16}$$

$$S_{22}^{(2)}\equiv\frac{1}{n}\sum_{i=1}^{n}\left(x_{i2}-\bar{x}_2\right)^2 \tag{17}$$

$$\begin{aligned}S_{12}^{(2)}&=S_{21}^{(2)}\\&=\frac{1}{n}\sum_{i=1}^{n}\left(x_{i1}-\bar{x}_1\right)\left(x_{i2}-\bar{x}_2\right)\end{aligned} \tag{18}$$

$$S_{yy}^{(2)}\equiv\frac{1}{n}\sum_{i=1}^{n}\left(y_i-\bar{y}\right)^2 \tag{19}$$

$$S_{1y}^{(2)}\equiv\frac{1}{n}\sum_{i=1}^{n}\left(x_{i1}-\bar{x}_1\right)\left(y_i-\bar{y}\right) \tag{20}$$

$$S_{2y}^{(2)}=\frac{1}{n}\sum_{i=1}^{n}\left(x_{i2}-\bar{x}_2\right)\left(y_i-\bar{y}\right) \tag{21}$$

Eqs. (14) and (15) are expressed with a matrix form as

$$\begin{pmatrix} S_{11}^{(2)} & S_{12}^{(2)} \\ S_{21}^{(2)} & S_{22}^{(2)} \end{pmatrix} \begin{pmatrix} \hat{a}_1 \\ \hat{a}_2 \end{pmatrix} = \begin{pmatrix} S_{1y}^{(2)} \\ S_{2y}^{(2)} \end{pmatrix} \tag{22}$$

Therefore, we can obtain $\hat{a}_1$ and $\hat{a}_2$ as

$$\begin{aligned} \begin{pmatrix} \hat{a}_1 \\ \hat{a}_2 \end{pmatrix} &= \begin{pmatrix} S_{11}^{(2)} & S_{12}^{(2)} \\ S_{12}^{(2)} & S_{22}^{(2)} \end{pmatrix}^{-1} \begin{pmatrix} S_{1y}^{(2)} \\ S_{2y}^{(2)} \end{pmatrix} \\ &= \begin{pmatrix} S^{11(2)} & S^{12(2)} \\ S^{21(2)} & S^{22(2)} \end{pmatrix} \begin{pmatrix} S_{1y}^{(2)} \\ S_{2y}^{(2)} \end{pmatrix} \end{aligned} \tag{23}$$

where

$$\begin{pmatrix} S^{11(2)} & S^{12(2)} \\ S^{21(2)} & S^{22(2)} \end{pmatrix} = \begin{pmatrix} S_{11}^{(2)} & S_{12}^{(2)} \\ S_{12}^{(2)} & S_{22}^{(2)} \end{pmatrix}^{-1} \tag{24}$$

Once we obtain the coefficients $\hat{a}_1$ and $\hat{a}_2$ from Eq. (23), we can obtain $\hat{a}_0$ from Eq. (7) as

$$\hat{a}_0 = \bar{y} - \hat{a}_1\bar{x}_1 - \hat{a}_2\bar{x}_2 \tag{25}$$

2.2. Accuracy of the Regression

The accuracy of the regression can be evaluated similarly to the case of one explanatory variable by a determinant factor given by

$$R^2 = \frac{S_r^{(2)}}{S_{yy}^{(2)}} \tag{26}$$

where $\sigma_r^{(2)}$ is the variance of Y given by

$$S_r^{(2)} = \frac{1}{n}\sum_{i=1}^{n}\left(Y_i - \mu_y\right)^2 \tag{27}$$

This is the ratio of the variance of the regression line to the variance of y. If whole data are on the regression line, that is, the regression is perfect, $S_{yy}^{(2)}$ is equal to $S_r^{(2)}$, and $R^2 = 1$.

On the other hand, when the deviation of data from the regression line is significant, the determinant ratio is expected to become small.

We further modify the variance of y can also be modified in the previous chapter as

$$\begin{aligned} S_{yy}^{(2)} &= \frac{1}{n}\sum_{i=1}^{n}\left(y_i - \bar{y}\right)^2 \\ &= \frac{1}{n}\sum_{i=1}^{n}\left(y_i - Y_i + Y_i - \bar{y}\right)^2 \\ &= \frac{1}{n}\sum_{i=1}^{n}\left(y_i - Y_i\right)^2 + \frac{1}{n}\sum_{i=1}^{n}\left(Y_i - \bar{y}\right)^2 + 2\frac{1}{n}\sum_{i=1}^{n}\left(y_i - Y_i\right)\left(Y_i - \bar{y}\right) \\ &= \frac{1}{n}\sum_{i=1}^{n} e_i^2 + \frac{1}{n}\sum_{i=1}^{n}\left(Y_i - \bar{y}\right)^2 + 2\frac{1}{n}\sum_{i=1}^{n} e_i\left(a_0 + a_1 x_{i1} + a_2 x_{i2}\right) \\ &= S_e^{(2)} + S_r^{(2)} \end{aligned} \tag{28}$$

where we utilize

$$Cov\left[e, x_1\right] = Cov\left[e, x_2\right] = 0 \tag{29}$$

Therefore, the determination coefficient is reduced to

$$\begin{aligned} R^2 &= \frac{S_r^{(2)}}{S_e^{(2)} + S_r^{(2)}} \\ &= 1 - \frac{S_e^{(2)}}{S_{yy}^{(2)}} \end{aligned} \tag{30}$$

Consequently, the determination coefficient is the square of the correlation factor, and has a value range between 0 and 1.

The multiple correlation factor r_{mult} is defined as the correlation between y_i and Y_i, and is given by

$$r_{mult} = \frac{\sum_{i=1}^{n}\left(y_i - \bar{y}\right)\left(Y_i - \bar{y}\right)}{\sqrt{\sum_{i=1}^{n}\left(y_i - \bar{y}\right)^2}\sqrt{\sum_{i=1}^{n}\left(Y_i - \bar{y}\right)^2}} \tag{31}$$

We can modify the denominator of (31) as

$$\begin{aligned} S_e^{(2)} &= \frac{1}{n}\sum_{i=1}^{n}\left(y_i - Y_i\right)^2 \\ &= \frac{1}{n}\sum_{i=1}^{n}\left[\left(y_i - \bar{y}\right) - \left(Y_i - \bar{y}\right)\right]^2 \\ &= \frac{1}{n}\sum_{i=1}^{n}\left(y_i - \bar{y}\right)^2 + \frac{1}{n}\sum_{i=1}^{n}\left(Y_i - \bar{y}\right)^2 - 2\frac{1}{n}\sum_{i=1}^{n}\left(y_i - \bar{y}\right)\left(Y_i - \bar{y}\right) \\ &= S_{yy}^{(2)} + S_R^{(2)} - 2\frac{1}{n}\sum_{i=1}^{n}\left(y_i - \bar{y}\right)\left(Y_i - \bar{y}\right) \end{aligned} \tag{32}$$

The cross term in Eq. (32) is performed as

$$\begin{aligned} \frac{1}{n}\sum_{i=1}^{n}\left(y_i - \bar{y}\right)\left(Y_i - \bar{y}\right) &= \frac{1}{2}\left(S_{yy}^{(2)} + S_R^{(2)} - S_e^{(2)}\right) \\ &= \frac{1}{2}\left[S_{yy}^{(2)} + S_R^{(2)} - \left(S_{yy}^{(2)} - S_R^{(2)}\right)\right] \\ &= S_R^{(2)} \end{aligned} \tag{33}$$

Then, the multiple correlation factor is given by

$$\begin{aligned} r_{mult} &= \frac{S_R^{(2)}}{\sqrt{S_{yy}^{(2)}}\sqrt{S_R^{(2)}}} \\ &= \sqrt{\frac{S_R^{(2)}}{S_{yy}^{(2)}}} \end{aligned} \tag{34}$$

The square of r_{mult} is the determinant of the multi regression, and is denoted as R^2 and is given by

$$R^2 = \frac{S_R^{(2)}}{S_{yy}^{(2)}} = \frac{S_{yy}^{(2)} - S_e^{(2)}}{S_{yy}^{(2)}} = 1 - \frac{S_e^{(2)}}{S_{yy}^{(2)}} \tag{35}$$

Substituting Eq. (34) into Eq. (35), we can express the determinant factor with a multiple correlation factor as

$$R^2 = r_{mult}^2 \tag{36}$$

The variances used in Eq. (35) are related to the samples, and hence the determinant with unbiased variances are also defined as

$$R^{*2} = 1 - \frac{\frac{n}{\phi_e} S_e^{(2)}}{\frac{n}{\phi_T} S_{yy}^{(2)}} \tag{37}$$

Three variables a_0, a_1, and a_2 are used in $S_e^{(2)}$, and hence the freedom of it ϕ_e is given by

$$\phi_e = n - 3 \tag{38}$$

$\bar{y}$ is used in $S_{yy}^{(2)}$, and hence the freedom of it ϕ_T is given by

$$\phi_T = n - 1 \tag{39}$$

R^{*2} is called as a degree of freedom adjusted coefficient of determination for a multiple regression.

R^{*2} can also be related to R^2 as

$$R^{*2} = 1 - \frac{\phi_T}{\phi_e}\left(1 - R^2\right) \tag{40}$$

2.3. Residual Error and Leverage Ratio

We evaluate the accuracy of the regression.

The residual error can be evaluated as

$$e_k = y_k - Y_k \tag{41}$$

This is then normalized by the biased variance and is given by

$$z_{ek} = \frac{e_k}{\sqrt{s_e^{(2)}}} \tag{42}$$

where $s_e^{(2)}$ is given by

$$s_e^{(2)} = \frac{n}{\phi_e} S_e^{(2)} \tag{43}$$

This follows a standard normal distribution with $N\left(0,1^2\right)$.

Further, it is modified using leverage ratio h_{kk} as

$$t_k = \frac{z_{ek}}{\sqrt{1-h_{kk}}} \tag{44}$$

This can be used for judging the outlier or not.

We should consider the leverage ratio for a multiple regression, which is shown below.

The regression data can be expressed by a leverage ratio as

$$Y_k = h_{k1}y_1 + h_{k2}y_2 + \cdots + h_{kk}y_k + \cdots + h_{kn}y_n \tag{45}$$

On the other hand, Y_k is given by

$$\begin{aligned} Y_k &= \hat{a}_0 + \hat{a}_1 x_{k1} + \hat{a}_2 x_{k2} \\ &= \bar{y} - \left(\hat{a}_1\bar{x}_1 + \hat{a}_2\bar{x}_2\right) + \hat{a}_1 x_{k1} + \hat{a}_2 x_{k2} \\ &= \bar{y} - \hat{a}_1\left(x_{k1} - \bar{x}_1\right) + \hat{a}_2\left(x_{k2} - \bar{x}_2\right) \\ &= \bar{y} - \left(x_{k1} - \bar{x}_1, x_{k2} - \bar{x}_2\right)\begin{pmatrix}\hat{a}_1 \\ \hat{a}_2\end{pmatrix} \end{aligned} \tag{46}$$

Each term is evaluated below.

$$\bar{y} = \frac{1}{n}\sum_{i=1}^{n} y_k \tag{47}$$

$$\begin{aligned}
\left(x_{k1}-\overline{x}_1, x_{k2}-\overline{x}_2\right)\begin{pmatrix}\hat{a}_1\\ \hat{a}_2\end{pmatrix} &= \left(x_{k1}-\overline{x}_1, x_{k2}-\overline{x}_2\right)\begin{pmatrix}S^{11(2)} & S^{12(2)}\\ S^{21(2)} & S^{22(2)}\end{pmatrix}\begin{pmatrix}S_{1y}^{(2)}\\ S_{2y}^{(2)}\end{pmatrix}\\
&= \left(x_{k1}-\overline{x}_1, x_{k2}-\overline{x}_2\right)\begin{pmatrix}S^{11(2)} & S^{12(2)}\\ S^{21(2)} & S^{22(2)}\end{pmatrix}\begin{pmatrix}\frac{1}{n}\sum_{i=1}^{n}\left(x_{i1}-\overline{x}_1\right)\left(y_i-\overline{y}\right)\\ \frac{1}{n}\sum_{i=1}^{n}\left(x_{i2}-\overline{x}_2\right)\left(y_i-\overline{y}\right)\end{pmatrix}\\
&= \left(x_{k1}-\overline{x}_1, x_{k2}-\overline{x}_2\right)\begin{pmatrix}S^{11(2)} & S^{12(2)}\\ S^{21(2)} & S^{22(2)}\end{pmatrix}\begin{pmatrix}\frac{1}{n}\sum_{i=1}^{n}\left(x_{i1}-\overline{x}_1\right)y_i\\ \frac{1}{n}\sum_{i=1}^{n}\left(x_{i2}-\overline{x}_2\right)y_i\end{pmatrix}
\end{aligned} \tag{48}$$

Focusing on the k -th term, we obtain

$$h_{kk} = \frac{1}{n} + \frac{D_k^2}{n} \tag{49}$$

where

$$\begin{aligned}
D_k^2 &= \begin{pmatrix}x_{k1}-\overline{x}_1 & x_{k2}-\overline{x}_2\end{pmatrix}\begin{pmatrix}S^{11(2)} & S^{12(2)}\\ S^{12(2)} & S^{22(2)}\end{pmatrix}\begin{pmatrix}x_{k1}-\overline{x}_1\\ x_{k2}-\overline{x}_2\end{pmatrix}\\
&= \begin{pmatrix}x_{k1}-\overline{x}_1 & x_{k1}-\overline{x}\end{pmatrix}\begin{pmatrix}\left(x_{k1}-\overline{x}_1\right)S^{11(2)}+\left(x_{k2}-\overline{x}_2\right)S^{12(2)}\\ \left(x_{k1}-\overline{x}_1\right)S^{12(2)}+\left(x_{k2}-\overline{x}_2\right)S^{22(2)}\end{pmatrix}\\
&= \left(x_{k1}-\overline{x}_1\right)^2 S^{11(2)} + 2\left(x_{k1}-\overline{x}_1\right)\left(x_{k2}-\overline{x}_2\right)S^{12(2)} + \left(x_{k2}-\overline{x}_2\right)^2 S^{22(2)}
\end{aligned} \tag{50}$$

This D_k is called as Mahalanobis' distance.
Sum of the leverage ratio is given by

$$\begin{aligned}
\sum_{k=1}^{n} h_{kk} &= \sum_{k=1}^{n}\frac{1}{n} + \sum_{k=1}^{n}\frac{D_k^2}{n}\\
&= 1 + \frac{1}{n}\sum_{k=1}^{n}\sum_{i=1}^{2}\sum_{j=1}^{2}\left(x_{ki}-\overline{x}_i\right)\left(x_{kj}-\overline{x}_j\right)S^{ij(2)}\\
&= 1 + \sum_{i=1}^{2}\sum_{i=1}^{2}\left[\frac{\sum_{k=1}^{n}\left(x_{ki}-\overline{x}_i\right)\left(x_{kj}-\overline{x}_j\right)}{n}\right]S^{ij(2)}\\
&= 1 + \sum_{i=1}^{2}\sum_{i=1}^{2}S_{ij}^{(2)}S^{ij(2)}\\
&= 1+2
\end{aligned} \tag{51}$$

The average leverage ratio is then given by

$$\mu_{hkk} = \frac{3}{n} \tag{52}$$

Therefore, we use a critical value for h_{kk} as

$$h_{kk} \leq \kappa \times \mu_{hkk} \tag{53}$$

$\kappa = 2$ is frequently used, and we evaluate whether it is outlier or not using the value.

When x_1 and x_2 are independent on each other, the inverse matrix is reduced to

$$\begin{aligned} \begin{pmatrix} S^{11(2)} & S^{12(2)} \\ S^{12(2)} & S^{22(2)} \end{pmatrix} &= \begin{pmatrix} S_{11}^{(2)} & S_{12}^{(2)} \\ S_{12}^{(2)} & S_{22}^{(2)} \end{pmatrix}^{-1} \\ &= \begin{pmatrix} \frac{1}{S_{11}^{(2)}} & 0 \\ 0 & \frac{1}{S_{22}^{(2)}} \end{pmatrix} \end{aligned} \tag{54}$$

and we obtain

$$D_k^2 = \frac{\left(x_{k1} - \bar{x}_1\right)^2}{S_{11}^{(2)}} + \frac{\left(x_{k2} - \bar{x}_2\right)^2}{S_{22}^{(2)}} \tag{55}$$

The corresponding leverage ratio is given by

$$h_{kk} = \frac{1}{n} + \frac{\left(x_{k1} - \bar{x}_1\right)^2}{nS_{11}^{(2)}} + \frac{\left(x_{k2} - \bar{x}_2\right)^2}{nS_{22}^{(2)}} \tag{56}$$

Therefore, we can simply add the terms in the single regression. There are interactions between the explanatory variables, and the expression becomes more complicated in general, and we need to use the corresponding models.

2.4. Prediction of Coefficients of a Regression Line

When we extract samples, the corresponding regression line varies, that is, the factors $\hat{a}_0$, $\hat{a}_1$, and $\hat{a}_2$ vary. We study averages and variances of $\hat{a}_0$, $\hat{a}_1$, and $\hat{a}_2$ in this section.

We assume the pollution data are given by

$$y_i = a_0 + a_1 x_{i1} + a_2 x_{i2} + e_i \quad for\ i = 1, 2, \cdots, n \tag{57}$$

where (x_{i1}, x_{i2}) is the given variable, a_0, a_1, and a_2 are established values for population. e_i is independent on each other and follows a normal distribution with $\left(0, \sigma_e^{(2)}\right)$. The probability variable in Eq. (57) is then only e_i.

When we get sample data (x_{i1}, x_{i2}, y_i) for $i = 1, 2, \cdots, n$. The coefficients are evaluated as

$$\begin{aligned}
\begin{pmatrix} \hat{a}_1 \\ \hat{a}_2 \end{pmatrix} &= \begin{pmatrix} S^{11(2)} & S^{12(2)} \\ S^{21(2)} & S^{22(2)} \end{pmatrix} \begin{pmatrix} S_{1y}^{(2)} \\ S_{2y}^{(2)} \end{pmatrix} \\
&= \begin{pmatrix} S^{11(2)} S_{1y}^{(2)} + S^{12(2)} S_{2y}^{(2)} \\ S^{21(2)} S_{1y}^{(2)} + S^{22(2)} S_{2y}^{(2)} \end{pmatrix}
\end{aligned} \tag{58}$$

Each coefficient is expanded as

$$\begin{aligned}
\hat{a}_1 &= S^{11(2)} S_{1y}^{(2)} + S^{12(2)} S_{2y}^{(2)} \\
&= \sum_{p=1}^{2} S^{1p(2)} S_{py}^{(2)} \\
&= \frac{1}{n} \sum_{i=1}^{n} \left[\sum_{p=1}^{2} S^{1p(2)} \left(x_{ip} - \bar{x}_p \right) \right] y_i \\
&= \frac{1}{n} \sum_{i=1}^{n} \left[\sum_{p=1}^{2} S^{1p(2)} \left(x_{ip} - \bar{x}_p \right) \right] \left(a_0 + a_1 x_{i1} + a_2 x_{i2} + e_i \right)
\end{aligned} \tag{59}$$

$$\begin{aligned}
\hat{a}_2 &= S^{21(2)} S_{1y}^{(2)} + S^{22(2)} S_{2y}^{(2)} \\
&= \sum_{p=1}^{2} S^{2p(2)} S_{py}^{(2)} \\
&= \frac{1}{n} \sum_{i=1}^{n} \left[\sum_{p=1}^{2} S^{2p(2)} \left(x_{ip} - \bar{x}_p \right) \right] y_i \\
&= \frac{1}{n} \sum_{i=1}^{n} \left[\sum_{p=1}^{2} S^{2p(2)} \left(x_{ip} - \bar{x}_p \right) \right] \left(a_0 + a_1 x_{i1} + a_2 x_{i2} + e_i \right)
\end{aligned} \tag{60}$$

$$\begin{aligned}
\hat{a}_0 &= \bar{y} - \left(\hat{a}_1 \bar{x}_1 + \hat{a}_2 \bar{x}_2 \right) \\
&= \frac{1}{n} \sum_{i=1}^{n} y_i - \frac{1}{n} \sum_{i=1}^{n} \left[\bar{x}_1 \sum_{p=1}^{2} S^{1p(2)} \left(x_{ip} - \bar{x}_p \right) \right] y_i - \frac{1}{n} \sum_{i=1}^{n} \left[\bar{x}_2 \sum_{p=1}^{2} S^{2p(2)} \left(x_{ip} - \bar{x}_p \right) \right] y_i \\
&= \frac{1}{n} \sum_{i=1}^{n} \left\{ 1 - \bar{x}_1 \sum_{p=1}^{2} S^{1p(2)} \left(x_{ip} - \bar{x}_p \right) - \bar{x}_2 \sum_{p=1}^{2} S^{2p(2)} \left(x_{ip} - \bar{x}_p \right) \right\} y_i \\
&= \frac{1}{n} \sum_{i=1}^{n} \left\{ 1 - \sum_{j=1}^{2} \bar{x}_j \sum_{p=1}^{2} S^{jp(2)} \left(x_{ip} - \bar{x}_p \right) \right\} y_i \\
&= \frac{1}{n} \sum_{i=1}^{n} \left\{ 1 - \sum_{j=1}^{2} \bar{x}_j \sum_{p=1}^{2} S^{jp(2)} \left(x_{ip} - \bar{x}_p \right) \right\} \left(a_0 + a_1 x_{i1} + a_2 x_{i2} + e_i \right)
\end{aligned} \tag{61}$$

We notice that only e_i is a probability variable, and evaluate the expected value as

$$\begin{aligned}
E\left[\hat{a}_1 \right] &= \frac{1}{n} \sum_{i=1}^{n} \left[\sum_{p=1}^{2} S^{1p(2)} \left(x_{ip} - \bar{x}_p \right) \right] \left(a_0 + a_1 x_{i1} + a_2 x_{i2} + e_i \right) \\
&= a_0 \frac{1}{n} \sum_{i=1}^{n} \left[\sum_{p=1}^{2} S^{1p(2)} \left(x_{ip} - \bar{x}_p \right) \right] \\
&\quad + a_1 \frac{1}{n} \sum_{i=1}^{n} \left[\sum_{p=1}^{2} S^{1p(2)} \left(x_{ip} - \bar{x}_p \right) \right] x_{i1} \\
&\quad + a_2 \frac{1}{n} \sum_{i=1}^{n} \left[\sum_{p=1}^{2} S^{1p(2)} \left(x_{ip} - \bar{x}_p \right) \right] x_{i2} \\
&\quad + \frac{1}{n} \sum_{i=1}^{n} \left[\sum_{p=1}^{2} S^{1p(2)} \left(x_{ip} - \bar{x}_p \right) \right] e_i \\
&= a_1 \sum_{p=1}^{2} S^{1p(2)} S_{p1}^{(2)} + a_2 \sum_{p=1}^{2} S^{1p(2)} S_{p2}^{(2)} \\
&= a_1
\end{aligned} \tag{62}$$

where

$$\sum_{p=1}^{2} S^{ip(2)} S_{pj}^{(2)} = \delta_{ij} \tag{63}$$

$$\begin{aligned} E\left[\hat{a}_2\right] &= \frac{1}{n}\sum_{i=1}^{n}\left[\sum_{p=1}^{2} S^{2p(2)}\left(x_{ip}-\bar{x}_p\right)\right]\left(a_0+a_1x_{i1}+a_2x_{i2}+e_i\right) \\ &= a_0\frac{1}{n}\sum_{i=1}^{n}\left[\sum_{p=1}^{2} S^{2p(2)}\left(x_{ip}-\bar{x}_p\right)\right] \\ &\quad +a_1\frac{1}{n}\sum_{i=1}^{n}\left[\sum_{p=1}^{2} S^{2p(2)}\left(x_{ip}-\bar{x}_p\right)\right]x_{i1} \\ &\quad +a_2\frac{1}{n}\sum_{i=1}^{n}\left[\sum_{p=1}^{2} S^{2p(2)}\left(x_{ip}-\bar{x}_p\right)\right]x_{i2} \\ &\quad +\frac{1}{n}\sum_{i=1}^{n}\left[\sum_{p=1}^{2} S^{2p(2)}\left(x_{ip}-\bar{x}_p\right)\right]e_i \\ &= a_1\sum_{p=1}^{2} S^{2p(2)}S_{p1}^{(2)} + a_2\sum_{p=1}^{2} S^{2p(2)}S_{p2}^{(2)} \\ &= a_2 \end{aligned} \tag{64}$$

$$\begin{aligned} E\left[\hat{a}_0\right] &= \frac{1}{n}\sum_{i=1}^{n}\left\{1-\sum_{j=1}^{2}\bar{x}_j\sum_{p=1}^{2} S^{jp(2)}\left(x_{ip}-\bar{x}_p\right)\right\}\left(a_0+a_1x_{i1}+a_2x_{i2}+e_i\right) \\ &= a_0\frac{1}{n}\sum_{i=1}^{n}\left[1-\sum_{j=1}^{2}\bar{x}_j\sum_{p=1}^{2} S^{jp(2)}\left(x_{ip}-\bar{x}_p\right)\right] \\ &\quad +a_1\frac{1}{n}\sum_{i=1}^{n}\left[1-\sum_{j=1}^{2}\bar{x}_j\sum_{p=1}^{2} S^{jp(2)}\left(x_{ip}-\bar{x}_p\right)\right]x_{i1} \\ &\quad +a_2\frac{1}{n}\sum_{i=1}^{n}\left[1-\sum_{j=1}^{2}\bar{x}_j\sum_{p=1}^{2} S^{jp(2)}\left(x_{ip}-\bar{x}_p\right)\right]x_{i2} \\ &\quad +\frac{1}{n}\sum_{i=1}^{n}\left[1-\sum_{j=1}^{2}\bar{x}_j\sum_{p=1}^{2} S^{jp(2)}\left(x_{ip}-\bar{x}_p\right)\right]e_i \end{aligned} \tag{65}$$

$$\begin{aligned} & a_0\frac{1}{n}\sum_{i=1}^{n}\left[1-\sum_{j=1}^{2}\bar{x}_j\sum_{p=1}^{2} S^{jp(2)}\left(x_{ip}-\bar{x}_p\right)\right] \\ &= a_0 - a_0\sum_{i=1}^{n}\left[\bar{x}_1\sum_{p=1}^{2} S^{1p(2)}\left(x_{ip}-\bar{x}_p\right)\right] - a_0\sum_{i=1}^{n}\left[\bar{x}_2\sum_{p=1}^{2} S^{2p(2)}\left(x_{ip}-\bar{x}_p\right)\right] \\ &= a_0 - a_0\left[\bar{x}_1\sum_{p=1}^{2} S^{1p(2)}\sum_{i=1}^{n}\left(x_{ip}-\bar{x}_p\right)\right] - a_0\left[\bar{x}_2\sum_{p=1}^{2} S^{2p(2)}\sum_{i=1}^{n}\left(x_{ip}-\bar{x}_p\right)\right] \\ &= a_0 \end{aligned} \tag{66}$$

$$
\begin{aligned}
& a_1 \frac{1}{n}\sum_{i=1}^{n}\left[1-\sum_{j=1}^{2}\bar{x}_j\sum_{p=1}^{2}S^{jp(2)}\left(x_{ip}-\bar{x}_p\right)\right]x_{i1} \\
&= a_1\bar{x}_1 - a_1\frac{1}{n}\sum_{i=1}^{n}\left[\bar{x}_1\sum_{p=1}^{2}S^{1p(2)}\left(x_{ip}-\bar{x}_p\right)\right]x_{i1} - a_1\frac{1}{n}\sum_{i=1}^{n}\left[\bar{x}_2\sum_{p=1}^{2}S^{2p(2)}\left(x_{ip}-\bar{x}_p\right)\right]x_{i1} \\
&= a_1\bar{x}_1 - a_1\bar{x}_1\sum_{p=1}^{2}S^{1p(2)}S_{p1}^{(2)} - a_1\bar{x}_2\sum_{p=1}^{2}S^{2p(2)}S_{p1}^{(2)} \\
&= a_1\bar{x}_1 - a_1\bar{x}_1 S^{11(2)}S_{11}^{(2)} \\
&= 0
\end{aligned} \tag{67}
$$

$$
\begin{aligned}
& a_2 \frac{1}{n}\sum_{i=1}^{n}\left[1-\sum_{j=1}^{2}\bar{x}_j\sum_{p=1}^{2}S^{jp(2)}\left(x_{ip}-\bar{x}_p\right)\right]x_{i2} \\
&= a_2\bar{x}_2 - a_2\bar{x}_1\sum_{p=1}^{2}S^{1p(2)}\sum_{i=1}^{n}\left(x_{ip}-\bar{x}_p\right)x_{i2} - a_2\bar{x}_2\sum_{p=1}^{2}S^{2p(2)}\sum_{i=1}^{n}\left(x_{ip}-\bar{x}_p\right)x_{i2} \\
&= a_2\bar{x}_2 - a_2\bar{x}_1\sum_{p=1}^{2}S^{1p(2)}S_{p2}^{(2)} - a_2\bar{x}_2\sum_{p=1}^{2}S^{2p(2)}S_{p2}^{(2)} \\
&= a_2\bar{x}_2 - a_2\bar{x}_2 S^{22(2)}S_{22}^{(2)} \\
&= a_2\bar{x}_2 - a_2\bar{x}_2 \\
&= 0
\end{aligned} \tag{68}
$$

$$
\frac{1}{n}\sum_{i=1}^{n}\left[1-\sum_{j=1}^{2}\bar{x}_j\sum_{p=1}^{2}S^{jp(2)}\left(x_{ip}-\bar{x}_p\right)\right]e_i = 0 \tag{69}
$$

Therefore, we obtain

$$
E\left[\hat{a}_0\right] = a_0 \tag{70}
$$

We then evaluate variances of factors.

$$
\begin{aligned}
V\left[\hat{a}_1\right] &= \sum_{i=1}^{n}\left[\frac{1}{n}\sum_{p=1}^{2}S^{1p(2)}\left(x_{ip}-\bar{x}_p\right)\right]^2 V\left[e_i\right] \\
&= \frac{1}{n}\frac{1}{n}\sum_{i=1}^{n}\left[S^{11(2)}\left(x_{i1}-\bar{x}_1\right)+S^{12(2)}\left(x_{i2}-\bar{x}_2\right)\right]^2 V\left[e_i\right] \\
&= \frac{1}{n}\frac{1}{n}\sum_{i=1}^{n}\left[\left(S^{11(2)}\right)^2\left(x_{i1}-\bar{x}_1\right)^2 + 2S^{12(2)}S^{11(2)}\left(x_{i1}-\bar{x}_1\right)\left(x_{i2}-\bar{x}_2\right)+\left(S^{12(2)}\right)^2\left(x_{i2}-\bar{x}_2\right)^2\right]s_e^{(2)} \\
&= \frac{1}{n}\left[\left(S^{11(2)}\right)^2 S_{11}^{(2)} + 2S^{12(2)}S^{11(2)}S_{12}^{(2)} + \left(S^{12(2)}\right)^2 S_{22}^{(2)}\right]s_e^{(2)} \\
&= \frac{1}{n}\left[S^{11(2)}\left[S^{11(2)}S_{11}^{(2)}+S^{12(2)}S_{21}^{(2)}\right]+S^{12(2)}\left[S^{11(2)}S^{12(2)}+S^{12(2)}S_{22}^{(2)}\right]\right]s_e^{(2)} \\
&= \frac{S^{11(2)}}{n}s_e^{(2)}
\end{aligned} \tag{71}
$$

$$
\begin{aligned}
V\left[\hat{a}_2\right] &= \sum_{i=1}^{n}\left[\frac{1}{n}\sum_{p=1}^{2} S^{2p(2)}\left(x_{ip}-\bar{x}_p\right)\right]^2 V\left[e\right] \\
&= \frac{1}{n}\frac{1}{n}\sum_{i=1}^{n}\left[S^{21(2)}\left(x_{i1}-\bar{x}_1\right)+S^{22(2)}\left(x_{i2}-\bar{x}_2\right)\right]^2 s_e^{(2)} \\
&= \frac{1}{n}\frac{1}{n}\sum_{i=1}^{n}\left[\left(S^{21(2)}\right)^2\left(x_{i2}-\bar{x}_2\right)^2+2S^{21(2)}S^{22(2)}\left(x_{i1}-\bar{x}_1\right)\left(x_{i2}-\bar{x}_2\right)+\left(S^{22(2)}\right)^2\left(x_{i2}-\bar{x}_2\right)^2\right]^2 s_e^{(2)} \\
&= \frac{1}{n}\left[\left(S^{21(2)}\right)^2 S_{22}^{(2)}+2S^{21(2)}S^{22(2)}S_{12}^{(2)}+\left(S^{22(2)}\right)^2 S_{22}^{(2)}\right]s_e^{(2)} \\
&= \frac{1}{n}\left[S^{21(2)}\left[S^{21(2)}S_{22}^{(2)}+S^{22(2)}S_{12}^{(2)}\right]+S^{22(2)}\left[S^{21(2)}S_{12}^{(2)}+S^{22(2)}S_{22}^{(2)}\right]\right]s_e^{(2)} \\
&= \frac{S^{22(2)}}{n}s_e^{(2)}
\end{aligned}
\tag{72}
$$

$$
\begin{aligned}
V\left[\hat{a}_0\right] &= V\left[\frac{1}{n}\sum_{i=1}^{n}\left\{1-\sum_{j=1}^{2}\bar{x}_j\sum_{p=1}^{2}S^{jp(2)}\left(x_{ip}-\bar{x}_p\right)\right\}\left(a_0+a_1x_{i1}+a_2x_{i2}+e_i\right)\right] \\
&= \left[\sum_{i=1}^{n}\left[\frac{1}{n}\left\{1-\sum_{j=1}^{2}\bar{x}_j\sum_{p=1}^{2}S^{jp(2)}\left(x_{ip}-\bar{x}_p\right)\right\}\right]^2\right]V\left[e\right] \\
&= \frac{1}{n}\frac{1}{n}\left[\sum_{i=1}^{n}\left[\left\{1-\sum_{j=1}^{2}\bar{x}_j\left[S^{j1(2)}\left(x_{i1}-\bar{x}_1\right)+S^{j2(2)}\left(x_{i2}-\bar{x}_2\right)\right]\right\}\right]^2\right]V\left[e\right] \\
&= \frac{1}{n}\frac{1}{n}\left[\sum_{i=1}^{n}\left[1-\left\{\bar{x}_1\left[S^{11(2)}\left(x_{i1}-\bar{x}_1\right)+S^{12(2)}\left(x_{i2}-\bar{x}_2\right)\right]+\bar{x}_2\left[S^{21(2)}\left(x_{i1}-\bar{x}_1\right)+S^{22(2)}\left(x_{i2}-\bar{x}_2\right)\right]\right\}\right]^2\right]s_e^{(2)} \\
&= \frac{1}{n}\frac{1}{n}\left[\sum_{i=1}^{n}\left[\begin{array}{l}1+\left\{\bar{x}_1\left[S^{11(2)}\left(x_{i1}-\bar{x}_1\right)+S^{12(2)}\left(x_{i2}-\bar{x}_2\right)\right]+\bar{x}_2\left[S^{21(2)}\left(x_{i1}-\bar{x}_1\right)+S^{22(2)}\left(x_{i2}-\bar{x}_2\right)\right]\right\}^2 \\ -2\left\{\bar{x}_1\left[S^{11(2)}\left(x_{i1}-\bar{x}_1\right)+S^{12(2)}\left(x_{i2}-\bar{x}_2\right)\right]+\bar{x}_2\left[S^{21(2)}\left(x_{i1}-\bar{x}_1\right)+S^{22(2)}\left(x_{i2}-\bar{x}_2\right)\right]\right\}\end{array}\right]\right]s_e^{(2)} \\
&= \frac{1}{n}\frac{1}{n}\left[\sum_{i=1}^{n}\left[1+\left\{\begin{array}{l}\bar{x}_1^{\,2}\left[S^{11(2)}\left(x_{i1}-\bar{x}_1\right)+S^{12(2)}\left(x_{i2}-\bar{x}_2\right)\right]^2+\bar{x}_2^{\,2}\left[S^{21(2)}\left(x_{i1}-\bar{x}_1\right)+S^{22(2)}\left(x_{i2}-\bar{x}_2\right)\right]^2 \\ +2\bar{x}_1\bar{x}_2\left[S^{11(2)}\left(x_{i1}-\bar{x}_1\right)+S^{12(2)}\left(x_{i2}-\bar{x}_2\right)\right]\left[S^{21(2)}\left(x_{i1}-\bar{x}_1\right)+S^{22(2)}\left(x_{i2}-\bar{x}_2\right)\right]\end{array}\right\}\right]\right]s_e^{(2)} \\
&= \frac{1}{n}+\frac{1}{n}\frac{1}{n}\left[\sum_{i=1}^{n}\left[\begin{array}{l}\bar{x}_1^{\,2}\left[\left(S^{11(2)}\right)^2\left(x_{i1}-\bar{x}_1\right)^2+\left(S^{12(2)}\right)^2\left(x_{i2}-\bar{x}_2\right)^2+2S^{11(2)}S^{12(2)}\left(x_{i1}-\bar{x}_1\right)\left(x_{i2}-\bar{x}_2\right)\right] \\ +\bar{x}_2^{\,2}\left[\left(S^{21(2)}\right)^2\left(x_{i1}-\bar{x}_1\right)^2+\left(S^{22(2)}\right)^2\left(x_{i2}-\bar{x}_2\right)^2+2S^{21(2)}S^{22(2)}\left(x_{i1}-\bar{x}_1\right)\left(x_{i2}-\bar{x}_2\right)\right] \\ +2\bar{x}_1\bar{x}_2\left[\begin{array}{l}S^{11(2)}S^{21(2)}\left(x_{i1}-\bar{x}_1\right)^2+S^{11(2)}S^{22(2)}\left(x_{i1}-\bar{x}_1\right)\left(x_{i2}-\bar{x}_2\right) \\ +S^{12(2)}S^{21(2)}\left(x_{i2}-\bar{x}_2\right)\left(x_{i1}-\bar{x}_1\right)+S^{12(2)}S^{22(2)}\left(x_{i2}-\bar{x}_2\right)\left(x_{i2}-\bar{x}_2\right)\end{array}\right]\end{array}\right]\right]s_e^{(2)} \\
&= \left\{\frac{1}{n}+\frac{1}{n}\left[\begin{array}{l}\bar{x}_1^{\,2}\left[\left(S^{11(2)}\right)^2S_{11}^{(2)}+\left(S^{12(2)}\right)^2S_{22}^{(2)}+2S^{11(2)}S^{12(2)}S_{12}^{(2)}\right] \\ +\bar{x}_2^{\,2}\left[\left(S^{21(2)}\right)^2S_{11}^{(2)}+\left(S^{22(2)}\right)^2S_{22}^{(2)2}+2S^{21(2)}S^{22(2)}S_{12}^{(2)}\right] \\ +2\bar{x}_1\bar{x}_2\left[\begin{array}{l}S^{11(2)}\left[S^{21(2)}S_{11}^{(2)}+S^{22(2)}S_{12}^{(2)}\right] \\ +S^{12(2)}\left[S^{21(2)}S_{21}^{(2)}+S^{22(2)}S_{22}^{(2)}\right]\end{array}\right]\end{array}\right]\right\}s_e^{(2)} \\
&= \left\{\frac{1}{n}+\frac{1}{n}\left[\bar{x}_1^{\,2}\left[S^{11(2)}\right]+\bar{x}_2^{\,2}\left[S^{22(2)}\right]+2\bar{x}_1\bar{x}_2S^{12(2)}\right]\right\}s_e^{(2)} \\
&= \left\{\frac{1}{n}+\frac{1}{n}\left[\bar{x}_1^{\,2}\left[S^{11(2)}\right]+\bar{x}_2^{\,2}\left[S^{22(2)}\right]+\bar{x}_1\bar{x}_2\left[S^{12(2)}+S^{21(2)}\right]\right]\right\}s_e^{(2)} \\
&= \frac{1}{n}\left[1+\sum_{j=1}^{2}\sum_{l=1}^{2}\bar{x}_j\bar{x}_lS^{jl(2)}\right]s_e^{(2)} \\
&= \frac{1}{n}\left\{1+\begin{pmatrix}\bar{x}_1 & \bar{x}_2\end{pmatrix}\begin{pmatrix}S^{11(2)} & S^{12(2)} \\ S^{21(2)} & S^{22(2)}\end{pmatrix}\begin{pmatrix}\bar{x}_1 \\ \bar{x}_2\end{pmatrix}\right\}s_e^{(2)}
\end{aligned}
\tag{73}
$$

where we utilize the following relationship in the last derivation step.

$$S^{12(2)} = S^{21(2)} \tag{74}$$

The covariance between $\hat{a}_1$ and $\hat{a}_2$ is given by

$$\begin{aligned}
Cov\left[\hat{a}_1, \hat{a}_2\right] &= \frac{1}{n}\frac{1}{n}\left\{\sum_{i=1}^{n}\left[\sum_{p=1}^{2} S^{1p(2)}\left(x_{ip} - \bar{x}_p\right)\right]\left[\sum_{p'=1}^{2} S^{2p'(2)}\left(x_{ip'} - \bar{x}_{p'}\right)\right]\right\} V[e] \\
&= \frac{1}{n}\frac{1}{n}\left\{\sum_{i=1}^{n}\left[S^{11(2)}\left(x_{i1} - \bar{x}_1\right) + S^{12(2)}\left(x_{i2} - \bar{x}_2\right)\right]\left[S^{21(2)}\left(x_{i1} - \bar{x}_1\right) + S^{22(2)}\left(x_{i2} - \bar{x}_2\right)\right]\right\} s_e^{(2)} \\
&= \frac{1}{n}\frac{1}{n}\sum_{i=1}^{n}\left[\begin{array}{l} S^{11(2)}S^{21(2)}\left(x_{i1} - \bar{x}_1\right)^2 + S^{11(2)}S^{22(2)}\left(x_{i1} - \bar{x}_1\right)\left(x_{i2} - \bar{x}_2\right) \\ +S^{12(2)}S^{21(2)}\left(x_{i2} - \bar{x}_2\right)\left(x_{i1} - \bar{x}_1\right) + S^{12(2)}S^{22(2)}\left(x_{i2} - \bar{x}_2\right)^2 \end{array}\right] s_e^{(2)} \\
&= \frac{1}{n}\left[\begin{array}{l} S^{11(2)}\left[S^{21(2)}S_{11}^{(2)} + S^{22(2)}S_{12}^{(2)}\right] \\ +S^{12(2)}\left[S^{21(2)}S_{21}^{(2)} + S^{22(2)}S_{22}^{(2)}\right] \end{array}\right] s_e^{(2)} \\
&= \frac{1}{n} S^{12(2)} s_e^{(2)}
\end{aligned} \tag{75}$$

Similarly, we obtain

$$Cov\left[\hat{a}_2, \hat{a}_1\right] = \frac{1}{n} S^{21(2)} s_e^{(2)} \tag{76}$$

Consequently, the $V\left[\hat{a}_j\right]$ and $V\left[\hat{a}_j, \hat{a}_l\right]$ is generally expressed by

$$Cov\left[\hat{a}_j, \hat{a}_l\right] = \frac{1}{n} S^{jl(2)} s_e^{(2)} \tag{77}$$

The covariance between $\hat{a}_0$ and $\hat{a}_1$ is given by

$$
\begin{aligned}
Cov\left[\hat{a}_0,\hat{a}_1\right] &= \frac{1}{n}\frac{1}{n}\sum_{i=1}^{n}\left\{1-\sum_{j=1}^{2}\bar{x}_j\sum_{p=1}^{2}S^{jp(2)}\left(x_{ip}-\bar{x}_p\right)\right\}\left[\sum_{p=1}^{2}S^{1p(2)}\left(x_{ip}-\bar{x}_p\right)\right]V\left[e\right] \\
&= \frac{1}{n}\frac{1}{n}\sum_{i=1}^{n}\left\{1-\sum_{j=1}^{2}\bar{x}_j\left[S^{j1(2)}\left(x_{i1}-\bar{x}_1\right)+S^{j2(2)}\left(x_{i2}-\bar{x}_2\right)\right]\right\}\left[S^{11(2)}\left(x_{i1}-\bar{x}_1\right)+S^{12(2)}\left(x_{i2}-\bar{x}_2\right)\right]s_e^{(2)} \\
&= \frac{1}{n}\frac{1}{n}\sum_{i=1}^{n}\left\{\begin{array}{l}1-\bar{x}_1\left[S^{11(2)}\left(x_{i1}-\bar{x}_1\right)+S^{12(2)}\left(x_{i2}-\bar{x}_2\right)\right] \\ -\bar{x}_2\left[S^{21(2)}\left(x_{i1}-\bar{x}_1\right)+S^{22(2)}\left(x_{i2}-\bar{x}_2\right)\right]\end{array}\right\}\left[S^{11(2)}\left(x_{i1}-\bar{x}_1\right)+S^{12(2)}\left(x_{i2}-\bar{x}_2\right)\right]s_e^{(2)} \\
&= -\frac{1}{n}\frac{1}{n}\sum_{i=1}^{n}\left\{\begin{array}{l}\bar{x}_1\left[S^{11(2)}\left(x_{i1}-\bar{x}_1\right)+S^{12(2)}\left(x_{i2}-\bar{x}_2\right)\right] \\ +\bar{x}_2\left[S^{21(2)}\left(x_{i1}-\bar{x}_1\right)+S^{22(2)}\left(x_{i2}-\bar{x}_2\right)\right]\end{array}\right\}\left[S^{11(2)}\left(x_{i1}-\bar{x}_1\right)+S^{12(2)}\left(x_{i2}-\bar{x}_2\right)\right]s_e^{(2)} \\
&= -\frac{1}{n}\frac{1}{n}\sum_{i=1}^{n}\left\{\begin{array}{l}\bar{x}_1\left[S^{11(2)}S^{11(2)}\left(x_{i1}-\bar{x}_1\right)^2+S^{12(2)}S^{11(2)}\left(x_{i2}-\bar{x}_2\right)\left(x_{i1}-\bar{x}_1\right)\right] \\ +\bar{x}_2\left[S^{21(2)}S^{11(2)}\left(x_{i1}-\bar{x}_1\right)^2+S^{22(2)}S^{11(2)}\left(x_{i2}-\bar{x}_2\right)\left(x_{i1}-\bar{x}_1\right)\right] \\ \bar{x}_1\left[S^{11(2)}S^{12(2)}\left(x_{i1}-\bar{x}_1\right)\left(x_{i2}-\bar{x}_2\right)+S^{12(2)}S^{12(2)}\left(x_{i2}-\bar{x}_2\right)^2\right] \\ +\bar{x}_2\left[S^{21(2)}S^{12(2)}\left(x_{i1}-\bar{x}_1\right)\left(x_{i2}-\bar{x}_2\right)+S^{22(2)}S^{12(2)}\left(x_{i2}-\bar{x}_2\right)^2\right]\end{array}\right\}s_e^{(2)} \\
&= -\frac{1}{n}\left\{\begin{array}{l}\bar{x}_1\left[S^{11(2)}S^{11(2)}S_{11}^{(2)}+S^{12(2)}S^{11(2)}S_{21}^{(2)}\right] \\ +\bar{x}_2\left[S^{21(2)}S^{11(2)}S_{11}^{(2)}+S^{22(2)}S^{11(2)}S_{21}^{(2)}\right] \\ \bar{x}_1\left[S^{11(2)}S^{12(2)}S_{12}^{(2)}+S^{12(2)}S^{12(2)}S_{22}^{(2)}\right] \\ +\bar{x}_2\left[S^{21(2)}S^{12(2)}S_{12}^{(2)}+S^{22(2)}S^{12(2)}S_{22}^{(2)}\right]\end{array}\right\}s_e^{(2)} \\
&= -\frac{1}{n}\left\{\bar{x}_1S^{11(2)}+\bar{x}_2S^{12(2)}\right\}s_e^{(2)} \\
&= -\frac{1}{n}\begin{pmatrix}S^{11(2)} & S^{12(2)}\end{pmatrix}\begin{pmatrix}\bar{x}_1 \\ \bar{x}_2\end{pmatrix}
\end{aligned}
\tag{78}
$$

Similarly, we obtain

$$
\begin{aligned}
Cov\left[\hat{a}_0,\hat{a}_2\right] &= -\frac{1}{n}\left\{\bar{x}_1S^{21(2)}+\bar{x}_2S^{22(2)}\right\}s_e^{(2)} \\
&= -\frac{1}{n}\begin{pmatrix}S^{21(2)} & S^{22(2)}\end{pmatrix}\begin{pmatrix}\bar{x}_1 \\ \bar{x}_2\end{pmatrix}s_e^{(2)}
\end{aligned}
\tag{79}
$$

Consequently, the $Cov\left[\hat{a}_0,\hat{a}_j\right]$ is generally expressed by

$$
Cov\left[\hat{a}_0,\hat{a}_j\right] = -\frac{1}{n}\sum_{l=1}^{2}\bar{x}_lS^{jl(2)}s_e^{(2)} \tag{80}
$$

We assume that $\hat{a}_1$ and $\hat{a}_0$ follow normal distributions, and hence assume the normalized variables.

The parameter given by

$$t = \frac{\hat{a}_i - a_i}{\sqrt{V[a_i]}} \quad for\ i = 0,1,2 \tag{81}$$

follows a t distribution function with a freedom of ϕ_e, where

$$\phi_e = n - 3 \tag{82}$$

Therefore, the predictive range is given by

$$\hat{a}_i - t_p(\phi_e; P)V[\hat{a}_i] \leq a_i \leq \hat{a}_i + t_p(\phi_e; P)V[\hat{a}_i] \tag{83}$$

2.5. Estimation of a Regression Line

We assume that we obtain n set of data (x_{i1}, x_{i2}, y_i). The corresponding regression value is given by

$$Y = \hat{a}_0 + \hat{a}_1 x_1 + \hat{a}_2 x_2 \tag{84}$$

The expected value is given by

$$\begin{aligned} E[Y] &= E[\hat{a}_0] + E[\hat{a}_1]x_1 + E[\hat{a}_2]x_2 \\ &= a_0 + a_1 x_1 + a_2 x_2 \end{aligned} \tag{85}$$

The variance of Y is given by

$$\begin{aligned}V[Y] &= V[\hat{a}_0 + \hat{a}_1 x_1 + \hat{a}_2 x_2] \\ &= V[\hat{a}_0] + 2\sum_{j=1}^{2} x_j Cov[\hat{a}_0, \hat{a}_j] + \sum_{j=1}^{2}\sum_{l=1}^{2} x_j x_l Cov[\hat{a}_j, \hat{a}_l] \\ &= \frac{1}{n}\left[1 + \sum_{j=1}^{2}\sum_{l=1}^{2} \bar{x}_j \bar{x}_l S^{jl(2)}\right] s_e^{(2)} - \frac{2}{n}\left[\sum_{j=1}^{2} x_j \sum_{l=1}^{2} \bar{x}_l S^{jl(2)}\right] s_e^{(2)} + \frac{1}{n}\left[\sum_{j=1}^{2}\sum_{l=1}^{2} x_j x_l S^{jl(2)}\right] s_e^{(2)} \\ &= \frac{1}{n}\left[1 + \sum_{j=1}^{2}\sum_{l=1}^{2} \left(\bar{x}_j \bar{x}_l - 2x_j \bar{x}_l + x_j x_l\right) S^{jl(2)}\right] s_e^{(2)}\end{aligned} \tag{86}$$

Let us consider the term below.

$$\begin{aligned}&\sum_{j=1}^{2}\sum_{l=1}^{2} 2x_j \bar{x}_l S^{jl(2)} \\ &= \sum_{j=1}^{2}\sum_{l=1}^{2} x_j \bar{x}_l S^{jl(2)} + \sum_{j=1}^{2}\sum_{l=1}^{2} x_j \bar{x}_l S^{jl(2)}\end{aligned} \tag{87}$$

The second term is modified as

$$\begin{aligned}\sum_{j=1}^{2}\sum_{l=1}^{2} x_j \bar{x}_l S^{jl(2)} &= \sum_{j=1}^{2}\sum_{l=1}^{2} x_l \bar{x}_j S^{lj(2)} \\ &= \sum_{j=1}^{2}\sum_{l=1}^{2} x_l \bar{x}_j S^{jl(2)}\end{aligned} \tag{88}$$

where we utilize the relationship of

$$S^{lj(2)} = S^{jl(2)} \tag{89}$$

Therefore, we obtain

$$\sum_{j=1}^{2}\sum_{l=1}^{2} 2x_j \bar{x}_l S^{jl(2)} = \sum_{j=1}^{2}\sum_{l=1}^{2} \left(x_j \bar{x}_l + x_l \bar{x}_j\right) S^{jl(2)} \tag{90}$$

Substituting Eq. (90) into Eq. (86), we obtain

$$
\begin{aligned}
V[Y] &= \frac{1}{n}\left[1+\sum_{j=1}^{2}\sum_{l=1}^{2}\left(\bar{x}_j\bar{x}_l - 2x_j\bar{x}_l + x_j x_l\right)S^{jl(2)}\right]s_e^{(2)} \\
&= \frac{1}{n}\left[1+\sum_{j=1}^{2}\sum_{l=1}^{2}\left(\bar{x}_j\bar{x}_l - x_j\bar{x}_l - x_l\bar{x}_j + x_j x_l\right)S^{jl(2)}\right]s_e^{(2)} \\
&= \frac{1}{n}\left[1+\sum_{j=1}^{2}\sum_{l=1}^{2}\left(x_j - \bar{x}_j\right)\left(x_l - \bar{x}_l\right)S^{jl(2)}\right]s_e^{(2)} \\
&= \frac{1}{n}\left[1+\begin{pmatrix} x_1-\bar{x}_1 & x_2-\bar{x}_2 \end{pmatrix}\begin{pmatrix} S^{11(2)} & S^{12(2)} \\ S^{21(2)} & S^{22(2)} \end{pmatrix}\begin{pmatrix} x_1-\bar{x}_1 \\ x_2-\bar{x}_2 \end{pmatrix}\right]s_e^{(2)} \\
&= \frac{1}{n}\left(1+D^2\right)s_e^{(2)}
\end{aligned}
\tag{91}
$$

where

$$
D^2 = \begin{pmatrix} x_1-\bar{x}_1 & x_2-\bar{x}_2 \end{pmatrix}\begin{pmatrix} S^{11(2)} & S^{12(2)} \\ S^{21(2)} & S^{22(2)} \end{pmatrix}\begin{pmatrix} x_1-\bar{x}_1 \\ x_2-\bar{x}_2 \end{pmatrix}
\tag{92}
$$

Therefore, the variable given by

$$
\begin{aligned}
t &= \frac{Y-E[Y]}{V[Y]} \\
&= \frac{\left(\hat{a}_0+\hat{a}_1x_1+\hat{a}_2x_2\right)-\left(a_0+a_1x_1+a_2x_2\right)}{\sqrt{\left(\frac{1}{n}+\frac{D^2}{n}\right)s_e^{(2)}}}
\end{aligned}
\tag{93}
$$

follows a t distribution. We can then estimate the value range of the regression given by

$$
\hat{a}_0+\hat{a}_1x_1+\hat{a}_2x_2 - t_p\left(\phi_e;P\right)\sqrt{\left(\frac{1}{n}+\frac{D^2}{n}\right)s_e^{(2)}} \le Y \le \hat{a}_0+\hat{a}_1x_1+\hat{a}_2x_2 + t_p\left(\phi_e;P\right)\sqrt{\left(\frac{1}{n}+\frac{D^2}{n}\right)s_e^{(2)}}
\tag{94}
$$

The objective variable range is given by

$$
y = \hat{a}_0+\hat{a}_1x_1+\hat{a}_2x_2+e
\tag{95}
$$

and the corresponding range is given by

$$\hat{a}_0 + \hat{a}_1 x_1 + \hat{a}_2 x_2 - t(\phi_e, P)\sqrt{\left(1 + \frac{1}{n} + \frac{D^2}{n}\right) s_e^{(2)}} \le y \le \hat{a}_0 + \hat{a}_1 x_1 + \hat{a}_2 x_2 + t(\phi_e, P)\sqrt{\left(1 + \frac{1}{n} + \frac{D^2}{n}\right) s_e^{(2)}} \tag{96}$$

2.6. Selection of Explanatory Variables

We used two explanatory variables. However, we want to know how each variable contributes to express the subject variable. We want to know the validity of each explanatory variable. One difficulty exists that with increasing number of explanatory variable, the determinant factor increases. Therefore, we cannot use the determinant factor to evaluate the validity.

We start with regression without explanatory variable, and denote it as model 0. The regression is given by

$$Model0: Y_i = \bar{y} \tag{97}$$

The corresponding variance $S_{e(M0)}^{(2)}$ is given by

$$S_{e(M0)}^{(2)} = \frac{1}{n}\sum_{i=1}^{n}(y_i - Y_i)^2 = \frac{1}{n}\sum_{i=1}^{n}(y_i - \bar{y})^2 = S_{yy}^{(2)} \tag{98}$$

In the next step, we evaluate the validity of x_1 and x_2, and denote the model as model 1.

The regression using explanatory variable is given by x_l

$$Y_i = a_0 + a_1 x_{il_1} \tag{99}$$

The corresponding variance $S_{e(M1)}^{(2)}$ is given by

$$S_{e(M1)}^{(2)} = \frac{1}{n}\sum_{i=1}^{n}(y_i - Y_i)^2 = \frac{1}{n}\sum_{i=1}^{n}\left[y_i - \left(a_0 + a_1 x_{il_1}\right)\right]^2 \tag{100}$$

Then the variable

$$F_1 = \frac{\left(nS_{e(M0)}^{(2)} - nS_{e(M1)}^{(2)}\right) / \left(\phi_{e(M0)} - \phi_{e(M1)}\right)}{nS_{e(M1)}^{(2)} / \phi_{e(M1)}} \tag{101}$$

follows F distribution with $F\left(\phi_{e(M0)} - \phi_{e(M1)}, \phi_{e(M1)}\right)$ where $\phi_{e(M0)}$ and $\phi_{e(M1)}$ are the freedom and are given by

$$\phi_{e(M0)} = n - 1 \tag{102}$$

$$\phi_{e(M1)} = n - 2 \tag{103}$$

We can judge the validity of the explanatory variable as

$$\begin{cases} F_1 \geq F\left(\phi_{e(M0)} - \phi_{e(M1)}, \phi_{e(M1)}\right) & valid \\ F_1 < F\left(\phi_{e(M0)} - \phi_{e(M1)}, \phi_{e(M1)}\right) & invalid \end{cases} \tag{104}$$

We evaluate F_1 for x_1 and x_2, that is, $l_1 = 1, 2$, and evaluate the corresponding F_1.

If both F_1 for $l_1 = 1, 2$ are invalid, we use the model 0 and the process is end.

If at least one of the F_1 is valid, we select the larger one, and the model 1 regression is determined as

$$Y_i = a_0 + a_1 x_{il_1} \tag{105}$$

where

$$l_1 = 1 \; or \; 2 \tag{106}$$

depending on the value of the corresponding F_1.

We then evaluate the second variable and denote it as model 2. The corresponding regression is given by

$$Y_i = a_0 + a_1 x_{il_1} + a_2 x_{il_2} \tag{107}$$

The corresponding variance is given by

$$S_{e(M2)}^{(2)} = \frac{1}{n}\sum_{i=1}^{n}\left(y_i - Y_i\right)^2 = \frac{1}{n}\sum_{i=1}^{n}\left[y_i - \left(a_0 + a_1 x_{il_1} + a_2 x_{il_2}\right)\right]^2 \tag{108}$$

Then the variable

$$F_2 = \frac{\left(nS_{e(M1)}^{(2)} - nS_{e(M2)}^{(2)}\right)\Big/\left(\phi_{e(M1)} - \phi_{e(M2)}\right)}{nS_{e(M2)}^{(2)}\Big/\phi_{e(M2)}} \tag{109}$$

follows F distribution with $F\left(\phi_{e(M1)} - \phi_{e(M2)}, \phi_{e(M2)}\right)$ where $\phi_{e(M1)}$ and $\phi_{e(M2)}$ are the freedom and are given by

$$\phi_{e(M1)} = n - 2 \tag{110}$$

$$\phi_{e(M2)} = n - 3 \tag{111}$$

We can judge the validity of the explanatory variable as

$$\begin{cases} F_2 \geq F\left(\phi_{e(M1)} - \phi_{e(M2)}, \phi_{e(M2)}\right) & valid \\ F_2 < F\left(\phi_{e(M1)} - \phi_{e(M2)}, \phi_{e(M2)}\right) & invalid \end{cases} \tag{112}$$

If both F_2 are invalid, we use the model 1 and the process is end.
If F_2 is valid, we use two variables as the explanatory ones.

3. REGRESSION WITH MORE THAN TWO EXPLANATORY VARIABLES

We move to an analysis where the explanatory variable number is bigger than two and we denote the number as m. The forms for the multiple regressions are basically derived in the previous section. We then neglect some parts of the derivation in this section.

3.1. Derivation of Regression Line

We pick up n set data from the population, and want to predict the characteristics of the population.

We can decide the sample factors a_0 , a_1, a_2 , $\cdots$, a_m from the sample data, which is not the same as the ones for the population, and hence we denote them as $\hat{a}_0$, $\hat{a}_1$, $\cdots$, and $\hat{a}_m$ a. The sample data can be expressed by

$$\begin{aligned} y_1 &= \hat{a}_0 + \hat{a}_1 x_{11} + \hat{a}_2 x_{12} + \cdots + \hat{a}_m x_{1m} + e_1 \\ y_2 &= \hat{a}_0 + \hat{a}_1 x_{21} + \hat{a}_2 x_{22} + \cdots + \hat{a}_m x_{2m} + e_2 \\ &\cdots \\ y_n &= \hat{a}_\mathbf{0} + \hat{a}_1 x_{n1} + \hat{a}_2 x_{n2} + \cdots + \hat{a}_m x_{np} + e_n \end{aligned} \tag{113}$$

where e_i is a residual error. A sample regression value Y_i is expressed by

$$Y_i = \hat{a}_0 + \hat{a}_1 x_{i1} + \hat{a}_2 x_{i2} + \cdots + \hat{a}_m x_{im} \tag{114}$$

e_i can then be expressed by

$$e_i = y_i - Y_i = y_i - \left(\hat{a}_0 + \hat{a}_1 x_{i1} + \hat{a}_2 x_{i2} + \cdots + \hat{a}_m x_{im} \right) \tag{115}$$

The variance of e_i is denoted as $S_e^{(2)}$, and is given by

$$S_e^{(2)} = \frac{1}{n} \sum_{i=1}^{n} \left[y_i - \left(\hat{a}_0 + \hat{a}_1 x_{i1} + \hat{a}_2 x_{i2} + \cdots + \hat{a}_m x_{im} \right) \right]^2 \tag{116}$$

We decide $\hat{a}_0$, $\hat{a}_1$, $\cdots$, and $\hat{a}_m$ so that $S_e^{(2)}$ has the minimum value.

Partial differentiating $S_e^{(2)}$ with respect to $\hat{a}_0$, $\hat{a}_1$, $\cdots$, and $\hat{a}_m$ and setting 0 for them, we obtain

$$\begin{pmatrix} S_{11}^{(2)} & S_{12}^{(2)} & \cdots & S_{1m}^{(2)} \\ S_{21}^{(2)} & S_{22}^{(2)} & \cdots & S_{2m}^{(2)} \\ \cdots & \cdots & \ddots & \cdots \\ S_{m1}^{(2)} & S_{m2}^{(2)} & \cdots & S_{mm}^{(2)} \end{pmatrix} \begin{pmatrix} \hat{a}_1 \\ \hat{a}_2 \\ \cdots \\ \hat{a}_p \end{pmatrix} = \begin{pmatrix} S_{1y}^{(2)} \\ S_{2y}^{(2)} \\ \cdots \\ S_{py}^{(2)} \end{pmatrix} \tag{117}$$

Therefore, we can obtain the coefficients as

$$\begin{aligned} \begin{pmatrix} \hat{a}_1 \\ \hat{a}_2 \\ \cdots \\ \hat{a}_p \end{pmatrix} &= \begin{pmatrix} S_{11}^{(2)} & S_{12}^{(2)} & \cdots & S_{1m}^{(2)} \\ S_{21}^{(2)} & S_{22}^{(2)} & \cdots & S_{2m}^{(2)} \\ \cdots & \cdots & \ddots & \cdots \\ S_{m1}^{(2)} & S_{m2}^{(2)} & \cdots & S_{mm}^{(2)} \end{pmatrix}^{-1} \begin{pmatrix} S_{1y}^{(2)} \\ S_{2y}^{(2)} \\ \cdots \\ S_{py}^{(2)} \end{pmatrix} \\ &= \begin{pmatrix} S^{11(2)} & S^{12(2)} & \cdots & S^{1m(2)} \\ S^{21(2)} & S^{22(2)} & \cdots & S^{2m(2)} \\ \cdots & \cdots & \ddots & \cdots \\ S^{m1(2)} & S^{m2(2)} & \cdots & S^{mm(2)} \end{pmatrix} \begin{pmatrix} S_{1y}^{(2)} \\ S_{2y}^{(2)} \\ \cdots \\ S_{py}^{(2)} \end{pmatrix} \end{aligned} \tag{118}$$

where

$$\begin{pmatrix} S^{11(2)} & S^{12(2)} & \cdots & S^{1m(2)} \\ S^{21(2)} & S^{22(2)} & \cdots & S^{2m(2)} \\ \cdots & \cdots & \ddots & \cdots \\ S^{m1(2)} & S^{m2(2)} & \cdots & S^{mm(2)} \end{pmatrix} = \begin{pmatrix} S_{11}^{(2)} & S_{12}^{(2)} & \cdots & S_{1m}^{(2)} \\ S_{21}^{(2)} & S_{22}^{(2)} & \cdots & S_{2m}^{(2)} \\ \cdots & \cdots & \ddots & \cdots \\ S_{m1}^{(2)} & S_{m2}^{(2)} & \cdots & S_{mm}^{(2)} \end{pmatrix}^{-1} \tag{119}$$

After obtaining $\hat{a}_1$, $\cdots$, $\hat{a}_m$, we can decide $\hat{a}_0$ as

$$\hat{a}_0 = \bar{y} - \hat{a}_1\bar{x}_1 - \hat{a}_2\bar{x}_2 - \cdots - \hat{a}_m\bar{x}_m \tag{120}$$

3.2. Accuracy of the Regression

The accuracy of the regression can be evaluated similarly to the case of two explanatory variables by determinant factor given by

$$R^2 = \frac{S_r^{(2)}}{S_{yy}^{(2)}} \tag{121}$$

where $S_r^{(2)}$ is the variance of Y given by

$$S_r^{(2)} = \frac{1}{n}\sum_{i=1}^{n}\left(Y_i - \bar{y}\right)^2 \tag{122}$$

The variance of y has the same expression as the one with two explanatory variables given by

$$S_{yy}^{(2)} = S_e^{(2)} + S_r^{(2)} \tag{123}$$

Therefore, the determination coefficient is reduced to

$$\begin{aligned} R^2 &= \frac{S_r^{(2)}}{S_e^{(2)} + S_r^{(2)}} \\ &= 1 - \frac{S_e^{(2)}}{S_{yy}^{(2)}} \end{aligned} \tag{124}$$

Consequently, the determination coefficient is the square of the correlation factor, and has a value range between 0 and 1.

The multiple correlation factor r_{mult} is defined as the correlation between y_i and Y_i, and is given by

$$\begin{aligned} r_{mult} &= \frac{\sum_{i=1}^{n}\left(y_i - \bar{y}\right)\left(Y_i - \bar{y}\right)}{\sqrt{\sum_{i=1}^{n}\left(y_i - \bar{y}\right)^2}\sqrt{\sum_{i=1}^{n}\left(Y_i - \bar{y}\right)^2}} \\ &= \sqrt{\frac{S_R^{(2)}}{S_{yy}^{(2)}}} \end{aligned} \tag{125}$$

We can modify the denominator of (31) as

$$\begin{aligned}S_e^{(2)} &= \frac{1}{n}\sum_{i=1}^{n}(y_i - Y_i)^2 \\ &= \frac{1}{n}\sum_{i=1}^{n}\left[(y_i - \bar{y}) - (Y_i - \bar{y})\right]^2 \\ &= \frac{1}{n}\sum_{i=1}^{n}(y_i - \bar{y})^2 + \frac{1}{n}\sum_{i=1}^{n}(Y_i - \bar{y})^2 - 2\frac{1}{n}\sum_{i=1}^{n}(y_i - \bar{y})(Y_i - \bar{y}) \\ &= S_{yy}^{(2)} + S_R^{(2)} - 2\frac{1}{n}\sum_{i=1}^{n}(y_i - \bar{y})(Y_i - \bar{y})\end{aligned} \tag{126}$$

The cross term in Eq. (32) is performed as

$$\begin{aligned}\frac{1}{n}\sum_{i=1}^{n}(y_i - \bar{y})(Y_i - \bar{y}) &= \frac{1}{2}\left(S_{yy}^{(2)} + S_R^{(2)} - S_e^{(2)}\right) \\ &= \frac{1}{2}\left[S_{yy}^{(2)} + S_R^{(2)} - \left(S_{yy}^{(2)} - S_R^{(2)}\right)\right] \\ &= S_R^{(2)}\end{aligned} \tag{127}$$

Then, the multiple correlation factor is also the same form with two explanatory variables given by

$$\begin{aligned}r_{mult} &= \frac{S_R^{(2)}}{\sqrt{S_{yy}^{(2)}}\sqrt{S_R^{(2)}}} \\ &= \sqrt{\frac{S_R^{(2)}}{S_{yy}^{(2)}}}\end{aligned} \tag{128}$$

The square of r_{mult} is the determinant of the multi regression, and is denoted as R^2 and is given by

$$R^2 = \frac{S_R^{(2)}}{S_{yy}^{(2)}} = \frac{S_{yy}^{(2)} - S_e^{(2)}}{S_{yy}^{(2)}} = 1 - \frac{S_e^{(2)}}{S_{yy}^{(2)}} \tag{129}$$

Substituting Eq. (34) into Eq. (35), we can express the determinant factor with multiple correlation factor as

$$R^2 = r_{mult}^2 \tag{130}$$

The variances used in Eq. (35) are related to the samples, and hence the determinant with unbiased variances are also defined as

$$R^{*2} = 1 - \frac{\frac{n}{\phi_e} S_e^{(2)}}{\frac{n}{\phi_T} S_{yy}^{(2)}} \tag{131}$$

where

$$\phi_e = n - (m+1) \tag{132}$$

and

$$\phi_T = n - 1 \tag{133}$$

The validity of the regression can be evaluated with a parameter as

$$F = \frac{s_r^{(2)}}{s_e^{(2)}} \tag{134}$$

where

$$s_r^{(2)} = \frac{nS_r^{(2)}}{m} \tag{135}$$

$$s_e^{(2)} = \frac{nS_e^{(2)}}{n-(m+1)} \tag{136}$$

We also evaluate the critical F value of $F_p(m, n-(m+1))$, and judge as

$$\begin{cases} F \le F_p(m, n-(m+1)) \Rightarrow invalid \\ F > F_p(m, n-(m+1)) \Rightarrow valid \end{cases} \tag{137}$$

3.3. Residual Error and Leverage Ratio

We evaluate the accuracy of the regression.
The residual error can be evaluated as

$$e_k = y_k - Y_k \tag{138}$$

This is then normalized by the biased variance and is given by

$$z_{ek} = \frac{e_k}{\sqrt{s_e^{(2)}}} \tag{139}$$

where $s_e^{(2)}$ is given by

$$s_e^{(2)} = \frac{n}{\phi_e} S_e^{(2)} \tag{140}$$

This follows a standard normal distribution with $N\left(0,1^2\right)$.
Further, it is modified as

$$t_k = \frac{z_{ek}}{\sqrt{1-h_{kk}}} \tag{141}$$

This can be used for judging the outlier or not.
We should consider the leverage ratio h_{kk} for a multiple regression.
The regression data can be expressed by leverage ratio as

$$Y_k = h_{k1}y_1 + h_{k2}y_2 + \cdots + h_{kk}y_k + \cdots + h_{kn}y_n \tag{142}$$

On the other hand, Y_k is given by

$$\begin{aligned}
Y_k &= \hat{a}_0 + \hat{a}_1 x_{k1} + \hat{a}_2 x_{k2} + \cdots + \hat{a}_m x_{km} \\
&= \overline{y} - \left(\hat{a}_1 \overline{x}_1 + \hat{a}_2 \overline{x}_2 + \cdots + \hat{a}_m \overline{x}_m\right) + \hat{a}_1 x_{k1} + \hat{a}_2 x_{k2} + \cdots + \hat{a}_m x_{km} \\
&= \overline{y} - \hat{a}_1 \left(x_{k1} - \overline{x}_1\right) + \hat{a}_2 \left(x_{k2} - \overline{x}_2\right) + \cdots + \hat{a}_m \left(x_{km} - \overline{x}_m\right) \\
&= \overline{y} - \begin{pmatrix} x_{k1} - \overline{x}_1 & x_{k2} - \overline{x}_2 & \cdots & x_{km} - \overline{x}_m \end{pmatrix} \begin{pmatrix} \hat{a}_1 \\ \hat{a}_2 \\ \vdots \\ \hat{a}_m \end{pmatrix}
\end{aligned} \tag{143}$$

Each term is evaluated below.

$$\overline{y} = \frac{1}{n} \sum_{i=1}^{n} y_k \tag{144}$$

$$\begin{aligned}
&\begin{pmatrix} x_{k1} - \overline{x}_1 & x_{k2} - \overline{x}_2 & \cdots & x_{km} - \overline{x}_m \end{pmatrix} \begin{pmatrix} \hat{a}_1 \\ \hat{a}_2 \\ \vdots \\ \hat{a}_m \end{pmatrix} \\
&= \begin{pmatrix} x_{k1} - \overline{x}_1 & x_{k2} - \overline{x}_2 & \cdots & x_{km} - \overline{x}_m \end{pmatrix} \begin{pmatrix} S_{11}^{(2)} & S_{12}^{(2)} & \cdots & S_{1m}^{(2)} \\ S_{21}^{(2)} & S_{22}^{(2)} & \cdots & S_{2m}^{(2)} \\ \cdots & \cdots & \ddots & \cdots \\ S_{m1}^{(2)} & S_{m2}^{(2)} & \cdots & S_{mm}^{(2)} \end{pmatrix}^{-1} \begin{pmatrix} S_{1y}^{(2)} \\ S_{2y}^{(2)} \\ \cdots \\ S_{py}^{(2)} \end{pmatrix} \\
&= \begin{pmatrix} x_{k1} - \overline{x}_1 & x_{k2} - \overline{x}_2 & \cdots & x_{km} - \overline{x}_m \end{pmatrix} \begin{pmatrix} S^{11(2)} & S^{12(2)} & \cdots & S^{1m(2)} \\ S^{21(2)} & S^{22(2)} & \cdots & S^{2m(2)} \\ \cdots & \cdots & \ddots & \cdots \\ S^{m1(2)} & S^{m2(2)} & \cdots & S^{mm(2)} \end{pmatrix} \begin{pmatrix} S_{1y}^{(2)} \\ S_{2y}^{(2)} \\ \cdots \\ S_{py}^{(2)} \end{pmatrix} \\
&= \begin{pmatrix} x_{k1} - \overline{x}_1 & x_{k2} - \overline{x}_2 & \cdots & x_{km} - \overline{x}_m \end{pmatrix} \begin{pmatrix} S^{11(2)} & S^{12(2)} & \cdots & S^{1m(2)} \\ S^{21(2)} & S^{22(2)} & \cdots & S^{2m(2)} \\ \cdots & \cdots & \ddots & \cdots \\ S^{m1(2)} & S^{m2(2)} & \cdots & S^{mm(2)} \end{pmatrix} \begin{pmatrix} \frac{1}{n}\sum_{i=1}^{n}\left(x_{i1} - \overline{x}_1\right)\left(y_i - \overline{y}\right) \\ \frac{1}{n}\sum_{i=1}^{n}\left(x_{i2} - \overline{x}_2\right)\left(y_i - \overline{y}\right) \\ \cdots \\ \frac{1}{n}\sum_{i=1}^{n}\left(x_{im} - \overline{x}_m\right)\left(y_i - \overline{y}\right) \end{pmatrix} \\
&= \begin{pmatrix} x_{k1} - \overline{x}_1 & x_{k2} - \overline{x}_2 & \cdots & x_{km} - \overline{x}_m \end{pmatrix} \begin{pmatrix} S^{11(2)} & S^{12(2)} & \cdots & S^{1m(2)} \\ S^{21(2)} & S^{22(2)} & \cdots & S^{2m(2)} \\ \cdots & \cdots & \ddots & \cdots \\ S^{m1(2)} & S^{m2(2)} & \cdots & S^{mm(2)} \end{pmatrix} \begin{pmatrix} \frac{1}{n}\sum_{i=1}^{n}\left(x_{i1} - \overline{x}_1\right)y_i \\ \frac{1}{n}\sum_{i=1}^{n}\left(x_{i2} - \overline{x}_2\right)y_i \\ \cdots \\ \frac{1}{n}\sum_{i=1}^{n}\left(x_{im} - \overline{x}_m\right)y_i \end{pmatrix}
\end{aligned} \tag{145}$$

Focusing on the k -th term, we obtain

$$h_{kk} = \frac{1}{n} + \frac{D_k^2}{n} \tag{146}$$

where

$$D_k^2 = \begin{pmatrix} x_{k1} - \bar{x}_1 & x_{k2} - \bar{x}_2 & \cdots & x_{km} - \bar{x}_m \end{pmatrix} \begin{pmatrix} S^{11(2)} & S^{12(2)} & \cdots & S^{1m(2)} \\ S^{21(2)} & S^{22(2)} & \cdots & S^{2m(2)} \\ \cdots & \cdots & \ddots & \cdots \\ S^{m1(2)} & S^{m2(2)} & \cdots & S^{mm(2)} \end{pmatrix} \begin{pmatrix} x_{k1} - \bar{x}_1 \\ x_{k2} - \bar{x}_2 \\ \cdots \\ x_{km} - \bar{x}_m \end{pmatrix} \tag{147}$$

This D_k^2 is called as square of Mahalanobis' distance.

Sum of the leverage ratio is given by

$$\begin{aligned} \sum_{k=1}^{n} h_{kk} &= \sum_{k=1}^{n} \frac{1}{n} + \sum_{k=1}^{n} \frac{D_k^2}{n} \\ &= 1 + \frac{1}{n} \sum_{k=1}^{n} \sum_{i=1}^{p} \sum_{j=1}^{p} \left(x_{ki} - \bar{x}_i \right) \left(x_{kj} - \bar{x}_j \right) S^{ij(2)} \\ &= 1 + \sum_{i=1}^{p} \sum_{i=1}^{p} \left[\frac{\sum_{k=1}^{n} \left(x_{ki} - \bar{x}_i \right) \left(x_{kj} - \bar{x}_j \right)}{n} \right] S^{ij(2)} \\ &= 1 + \sum_{i=1}^{p} \sum_{i=1}^{p} S_{ij}^{(2)} S^{ij(2)} \end{aligned} \tag{148}$$

Since we obtain

$$\left(S_{ij}^{(2)} \right) \left(S^{ij(2)} \right) = \begin{pmatrix} 1 & 0 & \cdots & 0 \\ 0 & \ddots & \ddots & \vdots \\ \vdots & \ddots & \ddots & 0 \\ 0 & \cdots & 0 & 1 \end{pmatrix} \tag{149}$$

Therefore, we obtain

$$\sum_{k=1}^{n} h_{kk} = 1 + \sum_{i=1}^{m}\sum_{j=1}^{m} S_{ij}^{(2)} S^{ij(2)} = 1 + m \tag{150}$$

The average leverage ratio is then given by

$$\mu_{hkk} = \frac{m+1}{n} \tag{151}$$

Therefore, we use a critical value for h_{kk} as

$$h_{kk} \leq \kappa \times \mu_{hkk} \tag{152}$$

$\kappa = 2$ is frequently used, and evaluate whether it is outlier or not.

3.4. Prediction of Coefficients of a Regression Line for Multi Variables

We can simply extend the range of the variables of $\hat{a}_0$, $\hat{a}_1$, $\hat{a}_2$, $\cdots$, $\hat{a}_m$ from the two variables forms.

We assume the pollution set has a regression line is given by

$$y_i = a_0 + a_1 x_{i1} + a_2 x_{i2} + \cdots + a_m x_{im} + e_i \quad for\ i = 1,2,\cdots,n \tag{153}$$

where $(x_{i1}, x_{i2}, \cdots, x_{im})$ are the given variables, a_0, a_1, a_2, $\cdots$, a_m are established values for population. e_i is independent on each other and follows a normal distribution with $\left(0, \sigma^{(2)}\right)$. The probability variable in Eq. (57) is then only e_i.

Extending the forms for two variables, we can evaluate the averages and variables of the factors as the followings.

The averages of the factors are given by

$$E\left[\hat{a}_k\right] = a_k \quad for\ k = 0,1,2,\cdots,m \tag{154}$$

The variances of the factors are given by

$$V\left[\hat{a}_k\right]=\frac{S^{kk(2)}}{n}s_e^{(2)} \quad for\ k=1,2,\cdots,m \tag{155}$$

$$V\left[\hat{a}_0\right]=\frac{1}{n}\left[1+\begin{pmatrix}\bar{x}_1 & \bar{x}_2 & \cdots & \bar{x}_m\end{pmatrix}\begin{pmatrix}S^{11(2)} & S^{12(2)} & \cdots & S^{1m(2)}\\ S^{21(2)} & S^{22(2)} & \cdots & S^{2m(2)}\\ \cdots & \cdots & \ddots & \cdots\\ S^{m1(2)} & S^{m2(2)} & \cdots & S^{mm(2)}\end{pmatrix}\begin{pmatrix}\bar{x}_1\\ \bar{x}_2\\ \vdots\\ \bar{x}_m\end{pmatrix}\right]s_e^{(2)} \tag{156}$$

The covariance between $\hat{a}_j$ and $\hat{a}_l$ are given by

$$Cov\left[\hat{a}_j,\hat{a}_l\right]=\frac{1}{n}S^{jl(2)}s_e^{(2)} \quad for\ j,l=1,2,\cdots,m \tag{157}$$

The covariance between $\hat{a}_0$ and $\hat{a}_j$ are given by

$$Cov\left[\hat{a}_0,\hat{a}_j\right]=-\frac{1}{n}\begin{pmatrix}S^{j1(2)} & S^{j2(2)} & \cdots & S^{jm(2)}\end{pmatrix}\begin{pmatrix}\bar{x}_1\\ \bar{x}_2\\ \vdots\\ \bar{x}_m\end{pmatrix}s_e^{(2)} \tag{158}$$

We assume that a_0, a_1, a_2, $\cdots$, a_m follow normal distributions, and hence assume that the normalized variables

$$t=\frac{\hat{a}_i-a_i}{\sqrt{V\left[a_i\right]}} \quad for\ i=0,1,2 \tag{159}$$

follows a t distribution function with a freedom of ϕ_e, where

$$\phi_e=n-\left(m+1\right) \tag{160}$$

Therefore, the predictive range is given by

$$\hat{a}_i-t_p\left(\phi_e;P\right)V\left[\hat{a}_i\right]\le a_i\le \hat{a}_i+t_p\left(\phi_e;P\right)V\left[\hat{a}_i\right] \tag{161}$$

3.5. Estimation of a Regression Line

We assume that we obtain n set of data $(x_{i1}, x_{i2}, \cdots, x_{im}, y_i)$. The corresponding regression value is given by

$$Y = \hat{a}_0 + \hat{a}_1 x_1 + \hat{a}_2 x_2 + \cdots + \hat{a}_m x_m \tag{162}$$

The expected value is given by

$$\begin{aligned} E[Y] &= E[\hat{a}_0] + E[\hat{a}_1] x_1 + E[\hat{a}_2] x_2 + \cdots + E[\hat{a}_m] x_m \\ &= a_0 + a_1 x_1 + a_2 x_2 + \cdots + a_m x_m \end{aligned} \tag{163}$$

The variance of Y is given by

$$\begin{aligned} V[Y] &= V[\hat{a}_0 + \hat{a}_1 x_1 + \hat{a}_2 x_2 + \cdots + \hat{a}_m x_m] \\ &= \frac{1}{n}\left[1 + \begin{pmatrix} x_1 - \bar{x}_1 & x_2 - \bar{x}_2 & \cdots & x_m - \bar{x}_m \end{pmatrix} \begin{pmatrix} S^{11(2)} & S^{12(2)} & \cdots & S^{1m(2)} \\ S^{21(2)} & S^{22(2)} & \cdots & S^{2m(2)} \\ \cdots & \cdots & \ddots & \cdots \\ S^{m1(2)} & S^{m2(2)} & \cdots & S^{mm(2)} \end{pmatrix} \begin{pmatrix} x_1 - \bar{x}_1 \\ x_2 - \bar{x}_2 \\ \cdots \\ x_m - \bar{x}_m \end{pmatrix}\right] s_e^{(2)} \\ &= \frac{1}{n}\left(1 + D^2\right) s_e^{(2)} \end{aligned} \tag{164}$$

where

$$D^2 = \begin{pmatrix} x_1 - \bar{x}_1 & x_2 - \bar{x}_2 & \cdots & x_m - \bar{x}_m \end{pmatrix} \begin{pmatrix} S^{11(2)} & S^{12(2)} & \cdots & S^{1m(2)} \\ S^{21(2)} & S^{22(2)} & \cdots & S^{2m(2)} \\ \cdots & \cdots & \ddots & \cdots \\ S^{m1(2)} & S^{m2(2)} & \cdots & S^{mm(2)} \end{pmatrix} \begin{pmatrix} x_1 - \bar{x}_1 \\ x_2 - \bar{x}_2 \\ \cdots \\ x_m - \bar{x}_m \end{pmatrix} \tag{165}$$

We can then estimate the value range of the regression.
The regression range is given by

$$\begin{aligned} &\hat{a}_0 + \hat{a}_1 x_1 + \hat{a}_2 x_2 + \cdots + \hat{a}_m x_m - t_p(\phi_e; P)\sqrt{\left(\frac{1}{n} + \frac{D^2}{n}\right) s_e^{(2)}} \\ &\le Y \le \\ &\hat{a}_0 + \hat{a}_1 x_1 + \hat{a}_2 x_2 + \cdots + \hat{a}_m x_m + t_p(\phi_e; P)\sqrt{\left(\frac{1}{n} + \frac{D^2}{n}\right) s_e^{(2)}} \end{aligned} \tag{166}$$

The objective variable range is given by

$$y = \hat{a}_0 + \hat{a}_1 x_1 + \hat{a}_2 x_2 + \cdots + \hat{a}_m x_m + e \tag{167}$$

and the corresponding range is given by

$$\begin{aligned} &\hat{a}_0 + \hat{a}_1 x_1 + \hat{a}_2 x_2 + \cdots + \hat{a}_m x_m - t(\phi_e, P)\sqrt{\left(1 + \frac{1}{n} + \frac{D^2}{n}\right) s_e^{(2)}} \\ &\leq y \leq \\ &\hat{a}_0 + \hat{a}_1 x_1 + \hat{a}_2 x_2 + \cdots + \hat{a}_m x_m + t(\phi_e, P)\sqrt{\left(1 + \frac{1}{n} + \frac{D^2}{n}\right) s_e^{(2)}} \end{aligned} \tag{168}$$

3.6. The Relationship between Multiple Coefficient and Correlation Factors

We study the relationship between multiple factors and correlation factors. We focus on the two variables x_1 and x_2 for simplicity, and also assume that the variables are normalized.

The factors a_1, a_2 are related to the correlation factors as

$$\begin{pmatrix} 1 & r_{12} \\ r_{12} & 1 \end{pmatrix} \begin{pmatrix} a_1 \\ a_2 \end{pmatrix} = \begin{pmatrix} r_1 \\ r_2 \end{pmatrix} \tag{169}$$

Therefore, the factors are solved as

$$\begin{aligned} \begin{pmatrix} a_1 \\ a_2 \end{pmatrix} &= \begin{pmatrix} 1 & r_{12} \\ r_{12} & 1 \end{pmatrix}^{-1} \begin{pmatrix} r_1 \\ r_2 \end{pmatrix} \\ &= \begin{pmatrix} \dfrac{1}{1 - r_{12}^{\ 2}} & -\dfrac{r_{12}}{1 - r_{12}^{\ 2}} \\ -\dfrac{r_{12}}{1 - r_{12}^{\ 2}} & \dfrac{1}{1 - r_{12}^{\ 2}} \end{pmatrix} \begin{pmatrix} r_1 \\ r_2 \end{pmatrix} \\ &= \begin{pmatrix} \dfrac{r_1 - r_{12} r_2}{1 - r_{12}^{\ 2}} \\ \dfrac{r_2 - r_{12} r_1}{1 - r_{12}^{\ 2}} \end{pmatrix} \end{aligned} \tag{170}$$

We then obtain

$$a_1 = r_1 \frac{1 - r_{12}\dfrac{r_2}{r_1}}{1 - r_{12}^{\;2}} \tag{171}$$

$$a_2 = r_2 \frac{1 - r_{12}\dfrac{r_1}{r_2}}{1 - r_{12}^{\;2}} \tag{172}$$

We assume that $r_1 \geq r_2$ which do not induce loosing generality since the form is symmetrical.

If there is no correlation between two variables, that is, $r_{12} = 0$, we obtain

$$a_1 = r_1 \tag{173}$$

$$a_2 = r_2 \tag{174}$$

The factors are equal to the corresponding correlation factors as are expected.

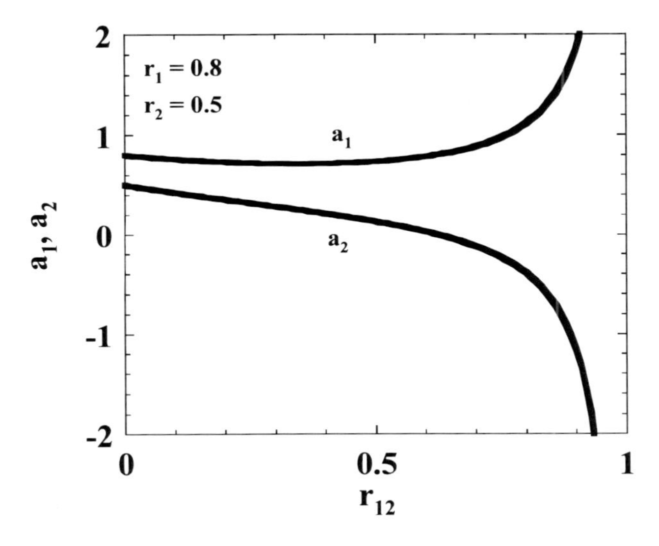

Figure 1. Relationship between multiple factor and the correlation factors.

Figure 1 shows the relationship between the coefficients and correlation factors with respect to the correlation factor between two variables.

The variable a_1 with large r_1 decreases a little with increasing r_{12}, and increases significantly for larger r_{12}.

On the other hand, the variable a_2 with small r_2 decreases little with decreasing r_{12}, and decreases significantly for larger r_{12}.

Differentiating a_1 with respect to r_{12}, we obtain

$$\begin{aligned}\frac{\partial a_1}{\partial r_{12}} &= r_1 \frac{-\frac{r_2}{r_1}\left(1-r_{12}^2\right)-\left(1-r_{12}\frac{r_2}{r_1}\right)\left(-2r_{12}\right)}{\left(1-r_{12}^2\right)^2} \\ &= \frac{-r_2}{\left(1-r_{12}^2\right)^2}\left(r_{12}^2-2\frac{r_1}{r_2}r_{12}+1\right)\end{aligned} \tag{175}$$

Setting 0 for Eq. (175), we obtain

$$r_{12}^2-2\frac{r_1}{r_2}r_{12}+1=0 \tag{176}$$

Solving this equation we obtain

$$r_{12}=\frac{r_1}{r_2}\pm\sqrt{\left(\frac{r_1}{r_2}\right)^2-1} \tag{177}$$

Since r_{12} is smaller than 1, we select the root with negative sign root before the root, and then it is reduced to

$$\begin{aligned}r_{12} &= \frac{r_1}{r_2}-\sqrt{\left(\frac{r_1}{r_2}\right)^2-1} \\ &= \frac{1}{\frac{r_1}{r_2}+\sqrt{\left(\frac{r_1}{r_2}\right)^2-1}}\end{aligned} \tag{178}$$

a_1 has the minimum value for the above r_{12}, and then increases with further increasing.

On the other hand, differentiating a_2 with respect to r_{12}, we obtain

$$r_{12}^2 - 2\frac{r_2}{r_1} r_{12} + 1 = 0 \tag{179}$$

Solving this with respect to r_{12}, we obtain

$$r_{12} = \frac{r_2}{r_1} \pm \sqrt{\left(\frac{r_2}{r_1}\right)^2 - 1} \tag{180}$$

Since $r_2 < r_1$, the term in the root become negative, and hence, we have no root. This means that the factor decreases monotonically as is expected from the figure.

Inspecting the above, we can expect multiple factors not so far from the correlation factor within this minimum r_{12}. In this region, we can roughly image that the both factors are smaller than the independent correlation factors due to the interaction between the explanatory variables. However, they are significantly different from the ones with exceeding the minimum value of r_{12}, and the feature of the multiple factor is complicated. We cannot have a clear image for the value of the factors under this strong interaction between explanatory variables. It hence should be minded that we should use independent variables for the explanatory as possible as we can.

3.7. Partial Correlation Factor

We can extend the partial correlation factor which is discussed in the previous chapter to the multiple variables.

We consider an objective variable y and m kinds of explanatory variables $x_1, x_2, \cdots, x_m$.

Among the explanatory variables, we consider the correlation between y and x_1. In this case, both y and x_1 have relationship between the explanatory variables.

Neglecting a variable x_1, we can evaluate the regression relationship of y as

$$Y_k = c_0 + c_2 x_{k2} + c_3 x_{k3} + \cdots + c_m x_{km} \tag{181}$$

where

$$\begin{pmatrix} c_2 \\ c_3 \\ \cdots \\ c_m \end{pmatrix} = \begin{pmatrix} S_{22}^{(2)} & S_{23}^{(2)} & \cdots & S_{2m}^{(2)} \\ S_{32}^{(2)} & S_{33}^{(2)} & \cdots & S_{3m}^{(2)} \\ \cdots & \cdots & \ddots & \cdots \\ S_{m2}^{(2)} & S_{m3}^{(2)} & \cdots & S_{mm}^{(2)} \end{pmatrix}^{-1} \begin{pmatrix} S_{2y}^{(2)} \\ S_{3y}^{(2)} \\ \cdots \\ S_{my}^{(2)} \end{pmatrix} \tag{182}$$

We can decide $c_2, c_3, \cdots, c_m$ using Eq. (182). After obtaining $c_2, c_3, \cdots, c_m$, we can decide c_0 from the equation below.

$$\bar{y} = c_0 + c_2\bar{x}_2 + c_3\bar{x}_3 + \cdots + c_m\bar{x}_m \tag{183}$$

On the other hand, we regard x_1 as a subject variable, and it is expressed by

$$X_{k1} = d_0 + d_2x_{k2} + d_3x_{k3} + \cdots + d_mx_{km} \tag{184}$$

where

$$\begin{pmatrix} d_2 \\ d_3 \\ \cdots \\ d_p \end{pmatrix} = \begin{pmatrix} S_{22}^{(2)} & S_{23}^{(2)} & \cdots & S_{2m}^{(2)} \\ S_{32}^{(2)} & S_{33}^{(2)} & \cdots & S_{3m}^{(2)} \\ \cdots & \cdots & \ddots & \cdots \\ S_{m2}^{(2)} & S_{m3}^{(2)} & \cdots & S_{mm}^{(2)} \end{pmatrix}^{-1} \begin{pmatrix} S_{21}^{(2)} \\ S_{31}^{(2)} \\ \cdots \\ S_{m1}^{(2)} \end{pmatrix} \tag{185}$$

$d_2, d_3, \cdots, d_m$ can be decided from the equation. d_0 can be evaluate as

$$\bar{x}_1 = d_0 + d_2\bar{x}_2 + d_3\bar{x}_3 + \cdots + d_m\bar{x}_m \tag{186}$$

We then introduce the following two variables given by

$$u_k = y_k - Y_k \tag{187}$$

$$v_{k1} = x_{k1} - X_{k1} \tag{188}$$

These variables can be regarded as ones eliminating the influence of the other explanatory variables. We can then evaluate the followings.

$$\bar{u} = \frac{1}{n}\sum_{k=1}^{n} u_k \tag{189}$$

$$\bar{v}_1 = \frac{1}{n}\sum_{k=1}^{n} v_{k1} \tag{190}$$

$$S_{uu}^{(2)} = \frac{1}{n}\sum_{k=1}^{n}\left(u_k - \bar{u}\right)^2 \tag{191}$$

$$S_{v_1 v_1}^{(2)} = \frac{1}{n}\sum_{k=1}^{n}\left(v_{k1} - \bar{v}_1\right)^2 \tag{192}$$

$$S_{uv_1}^{(2)} = \frac{1}{n}\sum_{k=1}^{n}\left(u_k - \bar{u}\right)\left(v_{k1} - \bar{v}_1\right) \tag{193}$$

Therefore, the corresponding partial correlation factor $r_{(1y),(2,3,\cdots,m)}$ can be evaluated as

$$r_{(1y),(2,3,\cdots,m)} = \frac{S_{uv_1}^{(2)}}{\sqrt{S_{v_1 v_1}^{(2)}}\sqrt{S_{uu}^{(2)}}} \tag{194}$$

We can evaluate the partial correlation factors for other explanatory variables.

3.8. Selection of Explanatory Variable

We used many explanatory variables. However, some of them may not contribute to express the subject variable.

Furthermore, if the interaction between explanatory variables is significant, the determinant of the matrix becomes close to zero, and the related elements in the inverse matrix become unstable or we cannot obtain results with the related numerical error, which is called as multicollinearity.

One difficulty exists that the determinant factor increases with increasing number of explanatory variables. Therefore, we cannot use the determinant factor to evaluate the validity for the regression.

We start with a regression without any explanatory variable, and denote it as model 0. The regression is given by

$$Model0: Y_i = \bar{y} \tag{195}$$

The corresponding variance $S_{e(M0)}^{(2)}$ is given by

$$S_{e(M0)}^{(2)} = \frac{1}{n}\sum_{i=1}^{n}\left(y_i - Y_i\right)^2 = \frac{1}{n}\sum_{i=1}^{n}\left(y_i - \bar{y}\right)^2 = S_{yy}^{(2)} \tag{196}$$

In the next step, we evaluate the validity of x_1, x_2, and x_m, and denote the model as model 1.

The regression using explanatory variable is given by x_l

$$Y_i = a_0 + a_1 x_{il_1} \tag{197}$$

The corresponding variance $S_{e(M1)}^{(2)}$ is given by

$$S_{e(M1)}^{(2)} = \frac{1}{n}\sum_{i=1}^{n}\left(y_i - Y_i\right)^2 = \frac{1}{n}\sum_{i=1}^{n}\left[y_i - \left(a_0 + a_1 x_{il_1}\right)\right]^2 \tag{198}$$

Then the variable

$$F_1 = \frac{\left(nS_{e(M0)}^{(2)} - nS_{e(M1)}^{(2)}\right)\Big/\left(\phi_{e(M0)} - \phi_{e(M1)}\right)}{nS_{e(M1)}^{(2)}\Big/\phi_{e(M1)}} \tag{199}$$

follows F distribution with $F\left(\phi_{e(M0)} - \phi_{e(M1)}, \phi_{e(M1)}\right)$ where $\phi_{e(M0)}$ and $\phi_{e(M1)}$ are the freedom and are given by

$$\phi_{e(M0)} = n - 1 \tag{200}$$

$$\phi_{e(M1)} = n - 2 \tag{201}$$

We can judge the validity of the explanatory variable as

$$\begin{cases} F_1 \geq F\left(\phi_{e(M0)} - \phi_{e(M1)}, \phi_{e(M1)}\right) & valid \\ F_1 < F\left(\phi_{e(M0)} - \phi_{e(M1)}, \phi_{e(M1)}\right) & invalid \end{cases} \tag{202}$$

We evaluate F_1 for x_1 and x_2, that is, $l_1 = 1, 2, \cdots, m$, and evaluate the corresponding F_1.

If both F_1 are invalid, we use the model 0 and the process is end.

If at least one of the F_1 is valid, we select the maximum one, and the model 1 regression is determined as

$$Y_i = a_0 + a_1 x_{il_1} \tag{203}$$

Where l_1 correspond to the selected variable in the above step.

We then evaluate the second variable denote it as model 2. The second variables are residual ones where the first one is selected in the above step.

The corresponding regression is given by

$$Y_i = a_0 + a_1 x_{il_1} + a_2 x_{il_2} \tag{204}$$

The related variance is given by

$$S_{e(M2)}^{(2)} = \frac{1}{n}\sum_{i=1}^{n}\left(y_i - Y_i\right)^2 = \frac{1}{n}\sum_{i=1}^{n}\left[y_i - \left(a_0 + a_1 x_{il_1} + a_2 x_{il_2}\right)\right]^2 \tag{205}$$

Then the variable

$$F_2 = \frac{\left(nS_{e(M1)}^{(2)} - nS_{e(M2)}^{(2)}\right) \Big/ \left(\phi_{e(M1)} - \phi_{e(M2)}\right)}{nS_{e(M2)}^{(2)} \Big/ \phi_{e(M2)}} \tag{206}$$

follows F distribution with $F\left(\phi_{e(M1)} - \phi_{e(M2)}, \phi_{e(M2)}\right)$ where $\phi_{e(M1)}$ and $\phi_{e(M2)}$ are the freedom and are given by

$$\phi_{e(M1)} = n-2 \tag{207}$$

$$\phi_{e(M2)} = n-3 \tag{208}$$

We can judge the validity of the explanatory variable as

$$\begin{cases} F_2 \geq F\left(\phi_{e(M1)} - \phi_{e(M2)}, \phi_{e(M2)}\right) & valid \\ F_2 < F\left(\phi_{e(M1)} - \phi_{e(M2)}, \phi_{e(M2)}\right) & invalid \end{cases} \tag{209}$$

If both F_2 are invalid, we use the model 1 and the process is end.

If at least one of the F_2 is valid, we select the maximum one, and the model 2 regression is determined as

$$Y_i = a_0 + a_1 x_{il_1} + a_2 x_{il_2} \tag{210}$$

where l_2 correspond to the selected variable in the above step.

We repeat the process until all F are invalid, or check all variables.

4. Matrix Expression for Multiple Regression

The multiple regression can be expressed with a matrix form. This gives no new information. However, the expression is simple and one can use the method alternatively.

The objective and p kinds of explanatory variables are related as follows.

$$y_i = a_0 + a_1\left(x_{i1} - \bar{x}_1\right) + a_2\left(x_{i2} - \bar{x}_2\right) + \cdots + a_p\left(x_{ip} - \bar{x}_p\right) + e_i \tag{211}$$

where we have n samples. Therefore, all data are related as

$$\begin{cases} y_1 = a_0 + a_1\left(x_{11} - \bar{x}_1\right) + a_2\left(x_{12} - \bar{x}_2\right) + \cdots + a_p\left(x_{1p} - \bar{x}_p\right) + e_1 \\ y_2 = a_0 + a_1\left(x_{21} - \bar{x}_1\right) + a_2\left(x_{22} - \bar{x}_2\right) + \cdots + a_p\left(x_{2p} - \bar{x}_p\right) + e_2 \\ \cdots \\ y_n = a_0 + a_1\left(x_{n1} - \bar{x}_1\right) + a_2\left(x_{n2} - \bar{x}_2\right) + \cdots + a_p\left(x_{np} - \bar{x}_p\right) + e_1 \end{cases} \tag{212}$$

We introduce matrixes and vectors as

$$Y = \begin{pmatrix} y_1 \\ y_2 \\ \vdots \\ y_n \end{pmatrix} \tag{213}$$

$$X = \begin{pmatrix} 1 & x_{11} - \overline{x}_1 & x_{12} - \overline{x}_2 & \cdots & x_{1p} - \overline{x}_p \\ 1 & x_{21} - \overline{x}_1 & x_{22} - \overline{x}_2 & \cdots & x_{2p} - \overline{x}_p \\ \vdots & \vdots & \vdots & \ddots & \vdots \\ 1 & x_{n1} - \overline{x}_1 & x_{n2} - \overline{x}_2 & \cdots & x_{np} - \overline{x}_p \end{pmatrix} \tag{214}$$

$$\boldsymbol{\beta} = \begin{pmatrix} a_0 \\ a_1 \\ \vdots \\ a_p \end{pmatrix} \tag{215}$$

$$\mathbf{e} = \begin{pmatrix} e_1 \\ e_2 \\ \vdots \\ e_n \end{pmatrix} \tag{216}$$

Using the matrixes and vectors, the corresponding equation is expressed by

$$Y = X\boldsymbol{\beta} + \mathbf{e} \tag{217}$$

The total sum of the square of error Q is then expressed by

$$\begin{aligned} Q &= \mathbf{e}^T \mathbf{e} \\ &= (Y - X\boldsymbol{\beta})^T (Y - X\boldsymbol{\beta}) \\ &= (Y^T - \boldsymbol{\beta}^T X^T)(Y - X\boldsymbol{\beta}) \\ &= Y^T Y - Y^T X\boldsymbol{\beta} - \boldsymbol{\beta}^T X^T Y + \boldsymbol{\beta}^T X^T X\boldsymbol{\beta} \\ &= Y^T Y - 2\boldsymbol{\beta}^T X^T Y + \boldsymbol{\beta}^T X^T X\boldsymbol{\beta} \end{aligned} \tag{218}$$

We obtain the optimum value by minimizing Q. This can be done by vector differentiating (see Appendix 1-10 of volume 3) with respect to $\boldsymbol{\beta}$ it given by

$$\begin{aligned}\frac{\partial Q}{\partial \boldsymbol{\beta}} &= -2\frac{\partial}{\partial \boldsymbol{\beta}}\left(\boldsymbol{\beta}^T X^T Y\right) + \frac{\partial}{\partial \boldsymbol{\beta}}\left(\boldsymbol{\beta}^T X^T X \boldsymbol{\beta}\right) \\ &= -2X^T Y + 2X^T X\boldsymbol{\beta} \\ &= \mathbf{0}\end{aligned} \tag{219}$$

Therefore, we obtain

$$\boldsymbol{\beta} = \left(X^T X\right)^{-1} X^T Y \tag{220}$$

Summary

To summarize the results in this chapter—

We assume that the regression is expressed with m explanatory variables given by

$$Y = a_0 + a_1 X_1 + a_2 X_2 + \cdots + a_m X_m$$

The coefficients are given by

$$\begin{pmatrix} \hat{a}_1 \\ \hat{a}_2 \\ \cdots \\ \hat{a}_p \end{pmatrix} = \begin{pmatrix} S_{11}^{(2)} & S_{12}^{(2)} & \cdots & S_{1m}^{(2)} \\ S_{21}^{(2)} & S_{22}^{(2)} & \cdots & S_{2m}^{(2)} \\ \cdots & \cdots & \ddots & \cdots \\ S_{m1}^{(2)} & S_{m2}^{(2)} & \cdots & S_{mm}^{(2)} \end{pmatrix}^{-1} \begin{pmatrix} S_{1y}^{(2)} \\ S_{2y}^{(2)} \\ \cdots \\ S_{py}^{(2)} \end{pmatrix}$$

The accuracy of the regression can be evaluated with a determinant factor given by

$$R^2 = \frac{S_r^{(2)}}{S_{yy}^{(2)}}$$

where $S_{yy}^{(2)}$ and $S_r^{(2)}$ is the variance of Y given by

$$S_{yy}^{(2)} = \frac{1}{n}\sum_{i=1}^{n}\left(y_i - \bar{y}\right)^2$$

$$S_r^{(2)} = \frac{1}{n}\sum_{i=1}^{n}\left(Y_i - \bar{y}\right)^2$$

The validity of the regression can be evaluated with a parameter F given by

$$F = \frac{s_r^{(2)}}{s_e^{(2)}}$$

where

$$s_r^{(2)} = \frac{nS_r^{(2)}}{m}$$

$$s_e^{(2)} = \frac{nS_e^{(2)}}{n-(m+1)}$$

The variance $S_e^{(2)}$ is given by

$$S_e^{(2)} = \frac{1}{n}\sum_{i=1}^{n}(y_i - Y_i)^2$$

We also evaluate the critical F value of $F_p(m, n-(m+1))$, and judge as

$$\begin{cases} F \le F_p(m, n-(m+1)) \Rightarrow invalid \\ F > F_p(m, n-(m+1)) \Rightarrow valid \end{cases}$$

The residual error for each data can be evaluated as

$$t_k = \frac{z_{ek}}{\sqrt{1-h_{kk}}}$$

where

$$z_{ek} = \frac{e_k}{\sqrt{s_e^{(2)}}}$$

$$h_{kk} = \frac{1}{n} + \frac{D_k^2}{n}$$

and

$$D_k^2 = \begin{pmatrix} x_{k1} - \bar{x}_1 & x_{k2} - \bar{x}_2 & \cdots & x_{km} - \bar{x}_m \end{pmatrix} \begin{pmatrix} S^{11(2)} & S^{12(2)} & \cdots & S^{1m(2)} \\ S^{21(2)} & S^{22(2)} & \cdots & S^{2m(2)} \\ \cdots & \cdots & \ddots & \cdots \\ S^{m1(2)} & S^{m2(2)} & \cdots & S^{mm(2)} \end{pmatrix} \begin{pmatrix} x_{k1} - \bar{x}_1 \\ x_{k2} - \bar{x}_2 \\ \cdots \\ x_{km} - \bar{x}_m \end{pmatrix}$$

This D_k^2 is called as square of Mahalanobis' distance.

The error range of regression line are given by

$$\begin{aligned} &\hat{a}_0 + \hat{a}_1 x_1 + \hat{a}_2 x_2 + \cdots + \hat{a}_m x_m - t_p(\phi_e; P)\sqrt{\left(\frac{1}{n} + \frac{D^2}{n}\right) s_e^{(2)}} \\ &\le Y \le \\ &\hat{a}_0 + \hat{a}_1 x_1 + \hat{a}_2 x_2 + \cdots + \hat{a}_m x_m + t_p(\phi_e; P)\sqrt{\left(\frac{1}{n} + \frac{D^2}{n}\right) s_e^{(2)}} \end{aligned}$$

The error range of objective values are given by

$$\begin{aligned} &\hat{a}_0 + \hat{a}_1 x_1 + \hat{a}_2 x_2 + \cdots + \hat{a}_m x_m - t(\phi_e, P)\sqrt{\left(1 + \frac{1}{n} + \frac{D^2}{n}\right) s_e^{(2)}} \\ &\le y \le \\ &\hat{a}_0 + \hat{a}_1 x_1 + \hat{a}_2 x_2 + \cdots + \hat{a}_m x_m + t(\phi_e, P)\sqrt{\left(1 + \frac{1}{n} + \frac{D^2}{n}\right) s_e^{(2)}} \end{aligned}$$

The partial correlation factor, which is the pure correlation factor between factor 1 and the objective variable y, $r_{(1y),(2,3,\cdots,m)}$ can be evaluated as

$$r_{(1y),(2,3,\cdots,m)} = \frac{S_{uv_1}^{(2)}}{\sqrt{S_{v_1 v_1}^{(2)}}\sqrt{S_{uu}^{(2)}}}$$

where the variables u, v is the explanatory and objective variable eliminating the influence to the other variables given by

$$u_k = y_k - Y_k$$

$$v_{k1} = x_{k1} - X_{k1}$$

Y_k and X_k are regression variables and they are expressed by the other explanatory variables given by

$$Y_k = c_0 + c_2 x_{k2} + c_3 x_{k3} + \cdots + c_m x_{km}$$

$$X_{k1} = d_0 + d_2 x_{k2} + d_3 x_{k3} + \cdots + d_m x_{km}$$

The corresponding coefficients are determined by the standard regression analysis and is given by

$$\begin{pmatrix} c_2 \\ c_3 \\ \cdots \\ c_m \end{pmatrix} = \begin{pmatrix} S_{22}^{(2)} & S_{23}^{(2)} & \cdots & S_{2m}^{(2)} \\ S_{32}^{(2)} & S_{33}^{(2)} & \cdots & S_{3m}^{(2)} \\ \cdots & \cdots & \ddots & \cdots \\ S_{m2}^{(2)} & S_{m3}^{(2)} & \cdots & S_{mm}^{(2)} \end{pmatrix}^{-1} \begin{pmatrix} S_{2y}^{(2)} \\ S_{3y}^{(2)} \\ \cdots \\ S_{my}^{(2)} \end{pmatrix}$$

We can decide c_0 as

$$\bar{y} = c_0 + c_2 \bar{x}_2 + c_3 \bar{x}_3 + \cdots + c_m \bar{x}_m$$

$$\begin{pmatrix} d_2 \\ d_3 \\ \cdots \\ d_p \end{pmatrix} = \begin{pmatrix} S_{22}^{(2)} & S_{23}^{(2)} & \cdots & S_{2m}^{(2)} \\ S_{32}^{(2)} & S_{33}^{(2)} & \cdots & S_{3m}^{(2)} \\ \cdots & \cdots & \ddots & \cdots \\ S_{m2}^{(2)} & S_{m3}^{(2)} & \cdots & S_{mm}^{(2)} \end{pmatrix}^{-1} \begin{pmatrix} S_{21}^{(2)} \\ S_{31}^{(2)} \\ \cdots \\ S_{m1}^{(2)} \end{pmatrix}$$

We can decide d_0 as

$$\bar{x}_1 = d_0 + d_2 \bar{x}_2 + d_3 \bar{x}_3 + \cdots + d_m \bar{x}_m$$

We start with the regression without explanatory variable, and denote it as model 0. The regression is given by

$$Model0 : Y_i = \bar{y}$$

The corresponding variance $S^{(2)}_{e(M0)}$ is given by

$$S^{(2)}_{e(M0)} = \frac{1}{n}\sum_{i=1}^{n}\left(y_i - Y_i\right)^2 = \frac{1}{n}\sum_{i=1}^{n}\left(y_i - \bar{y}\right)^2 = S^{(2)}_{yy}$$

In the next step, we pick up one explanatory variable and denote the model as model 1.

The regression using explanatory variable is given by x_l

$$Y_i = a_0 + a_1 x_{il_1}$$

The corresponding variance $S^{(2)}_{e(M1)}$ is given by

$$S^{(2)}_{e(M1)} = \frac{1}{n}\sum_{i=1}^{n}\left(y_i - Y_i\right)^2 = \frac{1}{n}\sum_{i=1}^{n}\left[y_i - \left(a_0 + a_1 x_{il_1}\right)\right]^2$$

Then the variable

$$F_1 = \frac{\left(nS^{(2)}_{e(M0)} - nS^{(2)}_{e(M1)}\right) \Big/ \left(\phi_{e(M0)} - \phi_{e(M1)}\right)}{nS^{(2)}_{e(M1)} \Big/ \phi_{e(M1)}}$$

follows F distribution with $F\left(\phi_{e(M0)} - \phi_{e(M1)}, \phi_{e(M1)}\right)$ where $\phi_{e(M0)}$ and $\phi_{e(M1)}$ are the freedom and are given by

$$\phi_{e(M0)} = n - 1$$

$$\phi_{e(M1)} = n - 2$$

We can judge the validity of the explanatory variable as

$$\begin{cases} F_1 \geq F\left(\phi_{e(M0)} - \phi_{e(M1)}, \phi_{e(M1)}\right) & valid \\ F_1 < F\left(\phi_{e(M0)} - \phi_{e(M1)}, \phi_{e(M1)}\right) & invalid \end{cases}$$

We select the variable that hold above evaluation and the maximum one.
We repeat this process as follows.
We validate the explanatory variables.

$$F_{i+1} = \frac{\left(nS_{e(Mi)}^{(2)} - nS_{e(Mi+1)}^{(2)}\right) / \left(\phi_{e(Mi)} - \phi_{e(Mi+1)}\right)}{nS_{e(Mi+1)}^{(2)} \Big/ \phi_{e(MI+1)}}$$

follows F distribution with $F\left(\phi_{e(Mi)} - \phi_{e(Mi+1)}, \phi_{e(Mi+1)}\right)$ where $\phi_{e(Mi)}$ and $\phi_{e(Mi+1)}$ are the freedom and are given by

$$\phi_{e(Mi)} = n - i - 1$$

$$\phi_{e(Mi+1)} = n - i - 2$$

We select the variable that has the maximum value holding the equations.
The evaluation is all invalid, the selection of the variable finish.
The multiple regression can be expressed using a matrix form as

$$\boldsymbol{\beta} = \left(X^T X\right)^{-1} X^T Y$$

where

$$y_i = a_0 + a_1\left(x_{i1} - \bar{x}_1\right) + a_2\left(x_{i2} - \bar{x}_2\right) + \cdots + a_p\left(x_{ip} - \bar{x}_p\right) + e_i$$

The matrixes and vectors are defined as

$$Y = \begin{pmatrix} y_1 \\ y_2 \\ \vdots \\ y_n \end{pmatrix}$$

$$X = \begin{pmatrix} 1 & x_{11} - \overline{x}_1 & x_{12} - \overline{x}_2 & \cdots & x_{1p} - \overline{x}_p \\ 1 & x_{21} - \overline{x}_1 & x_{22} - \overline{x}_2 & \cdots & x_{2p} - \overline{x}_p \\ \vdots & \vdots & \vdots & \ddots & \vdots \\ 1 & x_{n1} - \overline{x}_1 & x_{n2} - \overline{x}_2 & \cdots & x_{np} - \overline{x}_p \end{pmatrix}$$

$$\boldsymbol{\beta} = \begin{pmatrix} a_0 \\ a_1 \\ \vdots \\ a_p \end{pmatrix}$$

$$\mathbf{e} = \begin{pmatrix} e_1 \\ e_2 \\ \vdots \\ e_n \end{pmatrix}$$

Chapter 5

STRUCTURAL EQUATION MODELING

ABSTRACT

Latent variables are introduced and the relationship between the variables and obtained variable data are related quantitatively in structural equation modeling (SEM). Therefore, SEM enables us to draw a systematic total relationship structure of the subject. We start with k kinds of observed variables, and propose a path diagram which shows the relationship of the observed variables and their cause or results introducing latent variables. We evaluate the variances and covariances using the observed variable data, and they are also expressed with parameters related to the latent variables. We then determine the parameters so that the deviation between two kinds of expressions for variances and covariances is the minimum.

Keywords: structural equation modeling, observed variable, latent variable, path diagram

1. INTRODUCTION

Structural equation modeling (SEM) shows a cause of the obtained data introducing variables which depends on an analysist's image. Therefore, it is vital to search the cause of the obtained data, and SEM is intensively investigated these days. It is developed in psychology field, where a doctor must search the origin of the action, observation, or medical examination of the patient. This technology is also applied to marketing field and the accommodating fields are extended to more and more.

On the other hand, this technology depends on the analysists' image, and we have not unique results with this technology. We need to understand the meaning of the analytical process although it is quite easy to handle.

In this chapter, we focus on the basis of the technology. The application of it should be referred with the other books.

2. Path Diagram

A path diagram is used in a structural equation modeling. Figure 1 shows the elements of figure used in a path diagram: Correlation, arrow from cause and results, observed variable, latent variable, and error. The data structure is described using these elements.

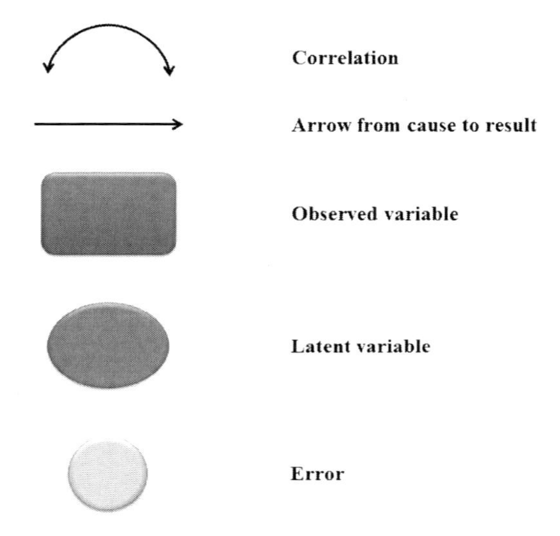

Figure 1. Figure elements used in a path diagram.

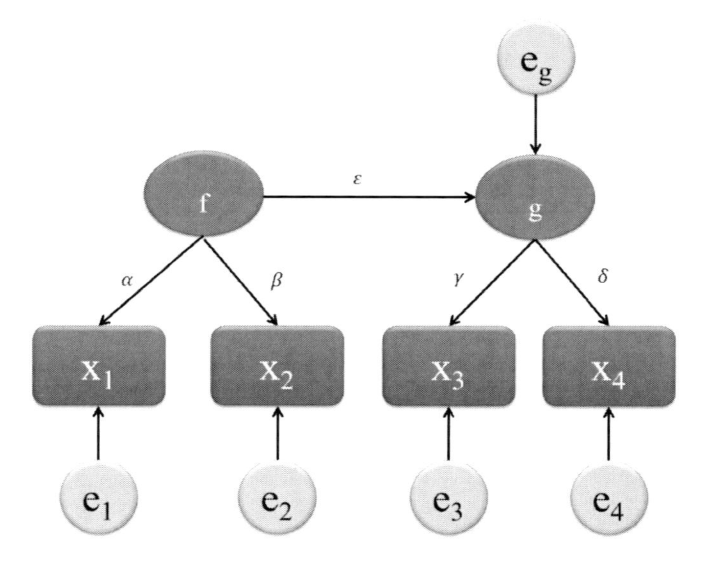

Figure 2. Example of a path diagram.

3. Basic Equation

We form a path diagram using the elements as shown in Figure 2. We investigate the corresponding basic equation.

We have the following equations from the path diagram, which are given by.

$$x_{1i} = \alpha f_i + e_{1i} \tag{1}$$

$$x_{2i} = \beta f_i + e_{2i} \tag{2}$$

$$x_{3i} = \gamma g_i + e_{3i} \tag{3}$$

$$x_{4i} = \delta g_i + e_{4i} \tag{4}$$

$$g_i = \varepsilon f_i + e_{gi} \tag{5}$$

where i is the data ID and we assume n data, that is, $i = 1, 2, \cdots, n$.

The variance for a variable x_{1i} is defined

$$S_{11}^{(2)} = \frac{1}{n}\sum_{i=1}^{n}\left(x_{1i} - \bar{x}_1\right)^2 \tag{6}$$

We can evaluate this variance using the observed data. This variance can also be expressed with the latent variable as

$$\begin{aligned} S_{11}^{(2)} &= \frac{1}{n}\sum_{i=1}^{n}\left(x_{1i} - \bar{x}_1\right)^2 \\ &= \frac{1}{n}\sum_{i=1}^{n}\left(\alpha f_i + e_{1i} - \bar{x}_1\right)^2 \end{aligned} \tag{7}$$

Further, we obtain from Eq. (1) as

$$\begin{aligned} \bar{x}_1 &= \frac{1}{n}\sum_{i=1}^{n} x_{1i} \\ &= \alpha \frac{1}{n}\sum_{i=1}^{n} f_i + \frac{1}{n}\sum_{i=1}^{n} e_{1i} \\ &= \alpha \bar{f} + \bar{e}_1 \end{aligned} \tag{8}$$

where

$$\bar{f} = \frac{1}{n}\sum_{i=1}^{n} f_i \tag{9}$$

$$\bar{e}_1 = \frac{1}{n}\sum_{i=1}^{n} e_{1i} \tag{10}$$

Substituting these into Eq. (7), we obtain

$$\begin{aligned} S_{11}^{(2)} &= \frac{1}{n}\sum_{i=1}^{n}\left(\alpha f_i + e_{1i} - \bar{x}_1\right)^2 \\ &= \frac{1}{n}\sum_{i=1}^{n}\left[\alpha\left(f_i - \bar{f}\right) + \left(e_{1i} - \bar{e}_1\right)\right]^2 \\ &= \alpha^2 \frac{1}{n}\sum_{i=1}^{n}\left(f_i - \bar{f}\right)^2 + 2\alpha\frac{1}{n}\sum_{i=1}^{n}\left(f_i - \bar{f}\right)\left(e_{1i} - \bar{e}_1\right) + \frac{1}{n}\sum_{i=1}^{n}\left(e_{1i} - \bar{e}_1\right)^2 \\ &= \alpha^2 S_{ff}^{(2)} + S_{e_1}^{(2)} \end{aligned} \tag{11}$$

We assume that the correlation between latent variable and error is zero.
We can obtain other variables with similar way as

$$S_{22}^{(2)} = \beta^2 S_{ff}^{(2)} + S_{e_2}^{(2)} \tag{12}$$

$$S_{33}^{(2)} = \gamma^2 S_{gg}^{(2)} + S_{e_3}^{(2)} \tag{13}$$

$$S_{44}^{(2)} = \delta^2 S_{gg}^{(2)} + S_{e_4}^{(2)} \tag{14}$$

$$S_{gg}^{(2)} = \varepsilon^2 S_{ff}^{(2)} + S_{e_g}^{(2)} \tag{15}$$

Latent variable g can be expressed with the other latent variable f. Therefore, we eliminate $S_{gg}^{(2)}$ from $S_{33}^{(2)}$ and $S_{44}^{(2)}$ as

$$
\begin{aligned}
S_{33}^{(2)} &= \gamma^2 S_{gg}^{(2)} + S_{e_3}^{(2)} \\
&= \gamma^2 \left(\varepsilon^2 S_{ff}^{(2)} + S_{e_g}^{(2)} \right) + S_{e_3}^{(2)} \\
&= \gamma^2 \varepsilon^2 S_{ff}^{(2)} + \gamma^2 S_{e_g}^{(2)} + S_{e_3}^{(2)}
\end{aligned}
\tag{16}
$$

$$
\begin{aligned}
S_{44}^{(2)} &= \delta^2 S_{gg}^{(2)} + S_{e_4}^{(2)} \\
&= \delta^2 \left(\varepsilon^2 S_{ff}^{(2)} + S_{e_g}^{(2)} \right) + S_{e_4}^{(2)} \\
&= \delta^2 \varepsilon^2 S_{ff}^{(2)} + \delta^2 S_{e_g}^{(2)} + S_{e_4}^{(2)}
\end{aligned}
\tag{17}
$$

Next, we consider covariances as shown below.

$$
\begin{aligned}
S_{12}^{(2)} &= \frac{1}{n}\sum_{i=1}^{n}\left(x_{1i} - \bar{x}_1\right)\left(x_{2i} - \bar{x}_2\right) \\
&= \frac{1}{n}\sum_{i=1}^{n}\left(\alpha f_i + e_{1i} - \bar{x}_1\right)\left(\beta f_i + e_{2i} - \bar{x}_2\right) \\
&= \frac{1}{n}\sum_{i=1}^{n}\left[\alpha\left(f_i - \bar{f}\right) + \left(e_{1i} - \bar{e}_1\right)\right]\left[\beta\left(f_i - \bar{f}\right) + \left(e_{2i} - \bar{e}\right)_2\right] \\
&= \alpha\beta\frac{1}{n}\sum_{i=1}^{n}\left(f_i - \bar{f}\right)^2 \\
&= \alpha\beta S_{ff}^{(2)}
\end{aligned}
\tag{18}
$$

$$
\begin{aligned}
S_{13}^{(2)} &= \frac{1}{n}\sum_{i=1}^{n}\left(x_{1i} - \bar{x}_1\right)\left(x_{3i} - \bar{x}_3\right) \\
&= \frac{1}{n}\sum_{i=1}^{n}\left(\alpha f_i + e_{1i} - \bar{x}_1\right)\left(\gamma g_i + e_{3i} - \bar{x}_3\right) \\
&= \frac{1}{n}\sum_{i=1}^{n}\left[\alpha\left(f_i - \bar{f}\right) + \left(e_{1i} - \bar{e}_1\right)\right]\left[\gamma\left(g_i - \bar{g}\right) + \left(e_{3i} - \bar{e}_3\right)\right] \\
&= \alpha\gamma\frac{1}{n}\sum_{i=1}^{n}\left(f_i - \bar{f}\right)\left(g_i - \bar{g}\right)
\end{aligned}
\tag{19}
$$

Substituting g_i of Eq. (5) into Eq. (19), we further obtain

$$
\begin{aligned}
S_{13}^{(2)} &= \alpha\gamma\frac{1}{n}\sum_{i=1}^{n}\left(f_i - \bar{f}\right)\left(g_i - \bar{g}\right) \\
&= \alpha\gamma\frac{1}{n}\sum_{i=1}^{n}\left(f_i - \bar{f}\right)\left[\varepsilon\left(f_i - \bar{f}\right) + \left(e_{gi} - \bar{e}_g\right)\right] \\
&= \alpha\gamma\varepsilon S_{ff}^{(2)}
\end{aligned}
\tag{20}
$$

$$\begin{aligned}
S_{14}^{(2)} &= \frac{1}{n}\sum_{i=1}^{n}\left(x_{1i}-\bar{x}_1\right)\left(x_{4i}-\bar{x}_4\right) \\
&= \frac{1}{n}\sum_{i=1}^{n}\left(\alpha f_i + e_{1i}-\bar{x}_1\right)\left(\delta g_i + e_{4i}-\bar{x}_4\right) \\
&= \frac{1}{n}\sum_{i=1}^{n}\left[\alpha\left(f_i-\bar{f}\right)+\left(e_{1i}-\bar{e}_1\right)\right]\left[\delta\left(g_i-\bar{g}\right)+\left(e_{4i}-\bar{e}_4\right)\right] \\
&= \alpha\delta\frac{1}{n}\sum_{i=1}^{n}\left(f_i-\bar{f}\right)\left(g_i-\bar{g}\right) \\
&= \alpha\delta\varepsilon S_{ff}^{(2)}
\end{aligned} \tag{21}$$

$$\begin{aligned}
S_{23}^{(2)} &= \frac{1}{n}\sum_{i=1}^{n}\left(x_{2i}-\bar{x}_2\right)\left(x_{3i}-\bar{x}_3\right) \\
&= \frac{1}{n}\sum_{i=1}^{n}\left(\beta f_i + e_{2i}-\bar{x}_2\right)\left(\gamma g_i + e_{3i}-\bar{x}_3\right) \\
&= \frac{1}{n}\sum_{i=1}^{n}\left[\beta\left(f_i-\bar{f}\right)+\left(e_{2i}-\bar{e}_2\right)\right]\left[\gamma\left(g_i-\bar{g}\right)+\left(e_{3i}-\bar{e}_3\right)\right] \\
&= \beta\gamma\frac{1}{n}\sum_{i=1}^{n}\left(f_i-\bar{f}\right)\left[\varepsilon\left(f_i-\bar{f}\right)+\left(e_{gi}-\bar{e}_g\right)\right] \\
&= \beta\gamma\varepsilon S_{ff}^{(2)}
\end{aligned} \tag{22}$$

$$\begin{aligned}
S_{24}^{(2)} &= \frac{1}{n}\sum_{i=1}^{n}\left(x_{2i}-\bar{x}_2\right)\left(x_{4i}-\bar{x}_4\right) \\
&= \frac{1}{n}\sum_{i=1}^{n}\left(\beta f_i + e_{2i}-\bar{x}_2\right)\left(\delta g_i + e_{4i}-\bar{x}_4\right) \\
&= \frac{1}{n}\sum_{i=1}^{n}\left[\beta\left(f_i-\bar{f}\right)+\left(e_{2i}-\bar{e}_2\right)\right]\left[\delta\left(g_i-\bar{g}\right)+\left(e_{4i}-\bar{e}_4\right)\right] \\
&= \beta\delta\frac{1}{n}\sum_{i=1}^{n}\left(f_i-\bar{f}\right)\left(g_i-\bar{g}\right) \\
&= \beta\delta\frac{1}{n}\sum_{i=1}^{n}\left(f_i-\bar{f}\right)\left[\varepsilon\left(f_i-\bar{f}\right)+\left(e_{gi}-\bar{e}_g\right)\right] \\
&= \beta\delta\varepsilon S_{ff}^{(2)}
\end{aligned} \tag{23}$$

$$
\begin{aligned}
S_{34}^{(2)} &= \frac{1}{n}\sum_{i=1}^{n}\left(x_{3i}-\bar{x}_3\right)\left(x_{4i}-\bar{x}_4\right) \\
&= \frac{1}{n}\sum_{i=1}^{n}\left(\gamma g_i + e_{3i}-\bar{x}_3\right)\left(\delta g_i + e_{4i}-\bar{x}_4\right) \\
&= \frac{1}{n}\sum_{i=1}^{n}\left[\gamma\left(g_i-\bar{g}\right)+\left(e_{3i}-\bar{e}_3\right)\right]\left[\delta\left(g_i-\bar{g}\right)+\left(e_{4i}-\bar{e}_4\right)\right] \\
&= \gamma\delta\frac{1}{n}\sum_{i=1}^{n}\left(g_i-\bar{g}\right)\left(g_i-\bar{g}\right) \\
&= \gamma\delta S_{gg}^{(2)} \\
&= \gamma\delta\left(\varepsilon^2 S_{ff}^{(2)} + S_{e_g}^{(2)}\right) \\
&= \gamma\delta S_{ff}^{(2)} + \gamma\delta S_{e_g}^{(2)}
\end{aligned}
\tag{24}
$$

We can evaluate all $S_{ij}^{(2)}$ using the observed data, and they are also expressed with latent variables and related to parameters. The deviation between the observed one and theoretical one is denoted as Q and is expressed with

$$
\begin{aligned}
Q &= \left[S_{11}^{(2)} - \left(\alpha^2 S_{ff}^{(2)} + S_{e_1}^{(2)}\right)\right]^2 + 2\left[S_{12}^{(2)} - \alpha\beta S_{ff}^{(2)}\right]^2 + 2\left[S_{13}^{(2)} - \alpha\gamma\varepsilon S_{ff}^{(2)}\right]^2 + 2\left[S_{14}^{(2)} - \alpha\delta\varepsilon S_{ff}^{(2)}\right]^2 \\
&+ \left[S_{22}^{(2)} - \left(\beta^2 S_{ff}^{(2)} + S_{e_2}^{(2)}\right)\right]^2 + 2\left[S_{23}^{(2)} - \beta\gamma\varepsilon S_{ff}^{(2)}\right]^2 + 2\left[S_{24}^{(2)} - \beta\delta\varepsilon S_{ff}^{(2)}\right]^2 \\
&+ \left[S_{33}^{(2)} - \left(\gamma^2\varepsilon^2 S_{ff}^{(2)} + \gamma^2 S_{e_g}^{(2)} + S_{e_3}^{(2)}\right)\right]^2 + 2\left[S_{34}^{(2)} - \left(\gamma\delta S_{ff}^{(2)} + \gamma\delta S_{e_g}^{(2)}\right)\right]^2 \\
&+ \left[S_{44}^{(2)} - \left(\delta^2\varepsilon^2 S_{ff}^{(2)} + \delta^2 S_{e_g}^{(2)} + S_{e_4}^{(2)}\right)\right]^2
\end{aligned}
\tag{25}
$$

We impose that the parameters should be determined so that Q is the minimum. The parameters we should decide are as follows.

$$
\alpha, \beta, \gamma, \delta, \varepsilon, S_{e_1}^{(2)}, S_{e_2}^{(2)}, S_{e_3}^{(2)}, S_{e_4}^{(2)}, S_{e_g}^{(2)}, S_{ff}^{(2)} \tag{26}
$$

We express the parameters as symbolically as ξ.

Partial differentiate Q with respect to ξ, we impose that it is zero.

$$
\frac{\partial Q}{\partial \xi} = 0 \tag{27}
$$

Therefore, we obtain 11 equations for 11 parameters.

The solution for the parameters can be evaluated simply as below.

We assume that the kinds of observed variable as k. In this case, we have x_1, x_2, x_3, x_4 ,and k is 4. We have variances with the kinds of

$$\frac{1}{2}k(k+1) \tag{28}$$

In this case we have variables of number of

$$\frac{1}{2}k(k+1) = \frac{1}{2} \times 4 \times 5 = 10 \tag{29}$$

We express the number of parameters as p. The following should be held to decide parameters uniquely.

$$p \leq \frac{1}{2}k(k+1) \tag{30}$$

In this case, $p = 11$, and hence the condition is invalid, and we cannot decide parameters in the above case.

When the imposed condition is not valid, we have two ways.

One way is to decide the parameter values so that the imposed condition becomes valid, and decide the other parameters.

The other one is to change the model.

SUMMARY

To summarize the results in this chapter—

We evaluate the variances and covariances of k kinds of observed data. That is, we obtain $S_{ij}^{(2)}$ based on the observed data.

We also evaluate the $S_{ij}^{(2)}$ including parameters assumed in the path diagram. We denote them as a_{ij}.

We then evaluate a parameter associated with the deviation as

$$Q = \sum_{i,j} \left(S_{ij}^{(2)} - a_{ij} \right)^2$$

Denoting parameters included in a_{ij} as ξ, we impose that

$$\frac{\partial Q}{\partial \xi} = 0$$

We have number of $n[\xi]$ equations which is the number of the parameters in the assumed path diagram. We set the number of observed variables as k. The followings condition must be held to decide parameters.

$$n[\xi] \le \frac{1}{2} k(k+1)$$

When the imposed condition is not valid, we have two ways. One way is to decide the parameter values so that the imposed condition becomes valid, and decide the other parameters. The other one is to change the model.

Chapter 6

FUNDAMENTALS OF A BAYES' THEOREM

ABSTRACT

We assume that an event is related to some causes. When we obtain some event, the Bayes' theorem enables us to specify the probability of the cause of the event. The related cause must be a certain one, and hence the related probability must be 1. However, we do not know which is the real cause, and predict it based on the obtained data. We can make the value of the probability close to 1 or 0 using Bayesian updating based on the obtained data. Many evens are related to each other in many cases. The events are expressed by nodes, and the relationship is expressed by a network connected by lines with arrows, which is called as a Bayesian network. All nodes have their conditional probabilities. A relationship between any two nodes in the network can be evaluated quantitatively using the conditional probabilities.

Keywords: Bayes' theorem, conditional probability, Bayesian updating, priori probability, posterior probability, Bayesian network, parent node, child node

1. INTRODUCTION

A Bayes' theorem is based on the conditional probability, which is well established in standard statistics. Therefore, the Bayes' theorem is not so different from the standard statistics in that standpoint of view. In the standard statistics, we basically obtain data associated with causes and predict results based on the data. The Bayes' theorem discusses the subjects in the opposite direction with incomplete data set. We obtain the results and predict the related cause in the Bayes' theorem. Further, even when the prediction is not so accurate at the beginning, it is improved by updating related data. We can get vast data and accumulate them easily these days. Therefore, the Bayesian theory utilizing updating data plays an important role in these days. We further treat the case where many events are related to each other, which can be analyzed with a Bayesian network.

2. A Bayes' Theorem

Let us consider an example of a medical check as shown in Figure 1.

When we check a sick people, we obtain a result of positive reaction with a ratio of 98%. The event of the positive reaction is denoted as E , and the corresponding probability is denoted as

$$P(E) = 0.98 \tag{1}$$

Note that we consider only sick people. The high percentage is good itself, but it is not enough for the medical check. It is important that it does not react to non-sick people.

Therefore, we assume the probability where the non-sick people react with the medical check as

$$P(E) = 0.05 \tag{2}$$

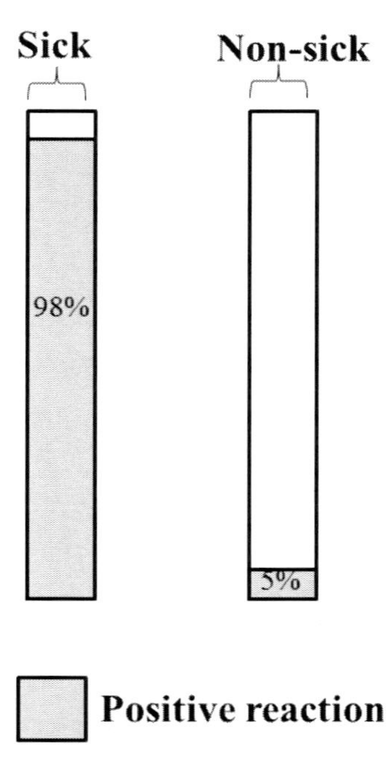

Figure 1. Reaction ratio of sick and non-sick people.

The sick people are denoted as A and the non-sick people as B. The probability of 98% of Eq. (1) is the one among the sick people, and hence the notation should be modified to express the situation clearly. We then denote Eq. (1) as

$$P(E|A) = 0.98 \tag{3}$$

The right side of A in Eq. (3) expresses the population for the probability where event E occurs.

The probability for the non-sick people is then denoted as $P(E|B)$ and is expressed as

$$P(E|B) = 0.05 \tag{4}$$

This expresses that the event E occurs for the non-sick people B.

The medical check is required to be high for $P(E|A)$ and to be low for $P(E|B)$. Therefore, this example is an expected one.

We further need a ratio of the sick people to the total one, which is denoted as $P(A)$, and assume it to be 5% here. Therefore, we have

$$P(A) = 0.05 \tag{5}$$

There are only sick or non-sick people, and hence the ratio of the non-sick people is given by

$$P(B) = 1 - P(A) = 0.95 \tag{6}$$

Strictly speaking, $P(A)$ means the probability that even A occurs among the whole events. The whole events are A and $\overline{A}$ in this case. We denote the whole case as Ω. Therefore, $P(A)$ should be expressed with

$$P(A) \rightarrow P(A|\Omega) \tag{7}$$

We frequently ignore this Ω. This is implicit assumption and simplifies the expression. We then have all information associated with this subject.

The ratio that people obtain positive reaction in the medical check is given by

$$\begin{aligned} P(E) &= P(E|A)P(A) + P(E|B)P(B) \\ &= 0.98 \times 0.05 + 0.05 \times 0.95 \\ &= 0.0965 \end{aligned} \tag{8}$$

Note that this total ratio is as low as about 10%. However, this ratio is not so interesting. This simply means that the sick people ratio is low.

We study the probability that the Bayes' theorem treats.

We select one person from a set, and perform the medical check, and obtain the positive reaction. What is the probability that the person is really sick? The probability is denoted as $P(A|E)$. The corresponding situation is shown in Figure 2. The probability is the ratio of the positively reacted people in A to the total reacted people, and we do not care about the non-reacted people. We discuss the situation under where we obtain the positive reaction.

The ratio of the positively reacted people in A is given by

$$P(E|A)P(A) \tag{9}$$

The ratio of the positively reacted people in B is given by

$$P(E|B)P(B) \tag{10}$$

Therefore, the target probability is evaluated as

$$\begin{aligned} P(A|E) &= \frac{P(E|A)P(A)}{P(E|A)P(A) + P(E|B)P(B)} \\ &= \frac{0.98 \times 0.05}{0.98 \times 0.05 + 0.05 \times 0.95} \\ &= 0.51 \end{aligned} \tag{11}$$

This is called as Bayes' probability, which expresses the probability of the cause for obtained result. This may be rather lower value than we expect since $P(E|A) = 0.98$. The appreciation of the result is shown below.

The positively reacted people constitution is schematically shown in Figure 2. The evaluated ratio is the reacted sick people to the total reacted people. Although the reaction ratio for the non-sick people is low, the number of the non-sick people is much larger than the sick people. Therefore, the number of the reacted people among non-sick people is comparable with the number of the reacted people among the sick people. Consequently, we have a rather low value ratio.

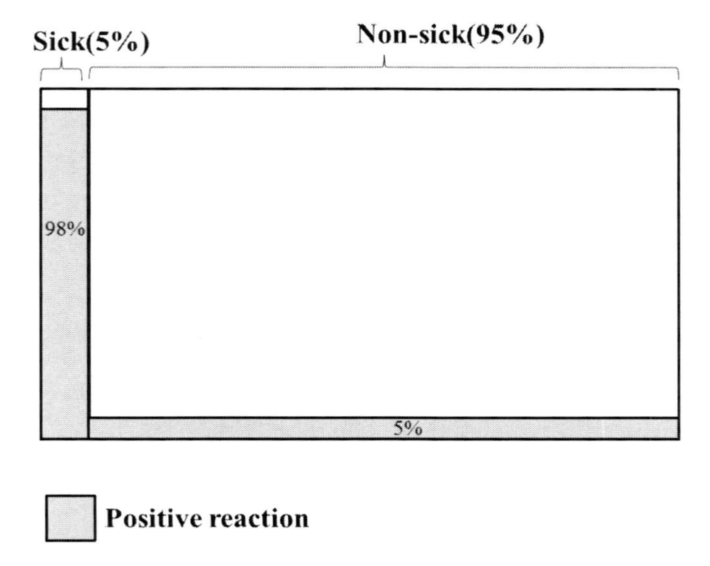

Figure 2. Reaction ratio of sick and non-sick people.

We can easily extend this theory to a general case where the causes associated with the obtained event are many.

m kinds of subjects are denoted as $A_i \ (i = 1, 2, \cdots, m)$. When we have an event E, the probability that the cause of the event A_i is expressed by

$$\begin{aligned} P(A_i|E) &= \frac{P(E|A_i)P(A_i)}{P(E)} \\ &= \frac{P(E|A_i)P(A_i)}{P(E|A_1)P(A_1) + P(E|A_2)P(A_2) + \cdots + P(E|A_m)P(A_m)} \\ &= \frac{P(E|A_i)P(A_i)}{\sum_{j=1}^{m} P(E|A_j)P(A_j)} \end{aligned} \tag{12}$$

This is called as a Bayes' theorem.

3. Similar Famous Examples

There are other two famous examples of the Bayes' theorem, which we introduce here briefly.

3.1. Three Prisoners

There are three prisoners, A, B, and C. One of them is decided to be pardoned although the selection is at random and we do not know who he is beforehand.

After the decision, the jailer knows who is pardoned. The prisoner A asks the jailer to notice the unpardoned one among the prisoners B and C. The jailer answers to the prisoner A that the prisoner B is unpardoned.

The prisoner A thought that the probability that he is pardoned was $1/3$ before he heard the jailer's answer. The prisoner A then heard that the prisoner B is unpardoned and hence the prisoners A or C is pardoned. Therefore, the prisoner A thought that the probability that he (the prisoner A) is pardoned is increased from $1/3$ to $1/2$.

In this problem, we assume that the jailer is honest and tells a prisoner name B or C with the same probability of $1/2$ when the prisoner A is pardoned.

We study whether the above discussion is valid or not.

The events are denoted as follows.

A: The prisoner A is pardoned.
B: The prisoner B is pardoned.
C: The prisoner C is pardoned.

S_A: The jailer talks that the prisoner A is not pardoned.
S_B: The jailer talks that the prisoner B is not pardoned.
S_C: The jailer talks that the prisoner C is not pardoned.

The probability that each member is pardoned is the same and is given by

$$P(A) = P(B) = P(C) = \frac{1}{3} \tag{13}$$

We assume that the jailer is honest and does not tell a lie as mentioned before, and hence

$$P(S_B | B) = 0 \tag{14}$$

$$P(S_B|C) = 1 \tag{15}$$

Let us consider the case where the prisoner A is pardoned. We then have

$$P(S_B|A) = \frac{1}{2} \tag{16}$$

$$P(S_C|A) = \frac{1}{2} \tag{17}$$

The probability $P(A|S_B)$ is given by

$$\begin{aligned} P(A|S_B) &= \frac{P(S_B|A)P(A)}{P(S_B|A)P(A) + P(S_B|B)P(B) + P(S_B|C)P(C)} \\ &= \frac{\frac{1}{2} \times \frac{1}{3}}{\frac{1}{2} \times \frac{1}{3} + 0 \times \frac{1}{3} + 1 \times \frac{1}{3}} \\ &= \frac{1}{3} \end{aligned} \tag{18}$$

which is not $1/2$. Therefore, the probability that the prisoner A is pardoned is invariant.

3.2. A Monty Hall Problem

We treated the Monty Hall problem in Chapter 2 of volume 1, and study again the problem in the standpoint of the Bayes' theorem. We explain the problem again.

There are three rooms where a car is in one room and goats are in the other rooms. The doors are shut at the beginning. We call the room and the door where a car is in as winning ones.

The guest selects one room first and does not check whether it is a winning one or not, that is, the door is not opened. Monty Hall checks the other rooms. There must be one non-winning room at least, that is, there is a room where a goat is in. Monty Hall deletes the non-winning room, that is, he opens the door where a goat is in. The problem is whether the guest should change the selected room or keep his first selected room.

The door that the guest selects is denoted as A and the door that Monty Hall opens is denoted as C, and the rest door is denoted as B.

The probability that the door A is the winning one is denoted as $P(A)$, the probability that the door B is the winning one is denoted as $P(B)$, and the probability that the door C is the winning one is denoted as $P(C)$.

The event that Monty Hall opens the door is denoted as D.

The important probability is $P(D|A)$ and $P(D|B)$.

$P(D|A)$ is the probability that Monty Hall opens the door C under the condition that the door A is the winning one. In this case, non-winning door may be B or C, and hence the probability that Monty Hall opens door C is $1/2$. Therefore, we obtain

$$P(D|A) = \frac{1}{2} \tag{19}$$

$P(D|B)$ is the probability that Monty Hall open the door C under the condition that the door C is the winning one. In this case, non-winning door must be C, and hence the probability that Monty Hall opens door C is 1. Therefore, we obtain

$$P(D|B) = 1 \tag{20}$$

Finally we can evaluate the probability where the guest does not change the room as

$$\begin{aligned} P(A|D) &= \frac{P(D|A)P(A)}{P(D|A)P(A) + P(D|B)P(B)} \\ &= \frac{\frac{1}{2} \times \frac{1}{3}}{\frac{1}{2} \times \frac{1}{3} + 1 \times \frac{1}{3}} \\ &= \frac{1}{3} \end{aligned} \tag{21}$$

If the guest changes the selected room from A to B, the corresponding probability is given by

$$P(B|D) = \frac{P(D|B)P(B)}{P(D|A)P(A) + P(D|B)P(B)}$$
$$= \frac{1 \times \frac{1}{3}}{\frac{1}{2} \times \frac{1}{3} + 1 \times \frac{1}{3}}$$
$$= \frac{2}{3} \tag{22}$$

Therefore, the guest should change the room from A to B.

What is the difference between the three prisoner problem and the Monty Hall problem.

One can change the decision in Monty Hall problem after he gets information, while one cannot change it in the three prisoner problem. If the prisoner A can be changed to prisoner B, the situation becomes identical to the Monty Hall problem.

4. Bayesian Updating

The probabilities shown up to here may be against our intuition. However, one of our targets is specify the cause related to the event. For example, the resulted probability for medical check was about 0.5 even when we obtain positive reaction. The value of the probability may be interesting. However, what can we do with the probability of 0.5? We need to obtain extreme probability values close to 1 or 0 to do something further.

Bayesian updating is a procedure to make the probability values to the extreme ones.

4.1. Medical Check

Let us consider the situation of the sick people example in more detail.

We obtain positive reaction E with the medical check, and the probability for the person to be sick is given by

$$P(A|E) = \frac{P(E|A)P(A)}{P(E|A)P(A) + P(E|B)P(B)}$$
$$= \frac{0.98 \times 0.05}{0.98 \times 0.05 + 0.05 \times 0.95}$$
$$= 0.51 \tag{23}$$

We can also evaluate the probability for the person to be non-sick after we obtain event E as

$$\begin{aligned} P(B|E) &= \frac{P(E|B)P(B)}{P(E|A)P(A)+P(E|B)P(B)} \\ &= \frac{0.05\times 0.95}{0.98\times 0.05+0.05\times 0.95} \\ &= 0.49 \end{aligned} \tag{24}$$

We do not know the detail of the checked person before the medical check. Therefore, we use a general probability values for $P(A)$ and $P(B)$. However, we now know that his medical check result is positive. Therefore, we can regard $P(A)$ and $P(B)$ for the person as $P(A|E)$, and $P(B|E)$, respectively. We can then regard

$$P(A|E) \to P(A) \tag{25}$$

$$P(B|E) \to P(B) \tag{26}$$

The probabilities of $P(A)=0.05$ and $P(B)=0.95$ before the medical check are called as priori probabilities, and the probabilities $P(A|E)\to P(A)=0.51$ and $P(B|E)\to P(B)=0.49$ are called as posterior probabilities, which is shown in Figure 3.

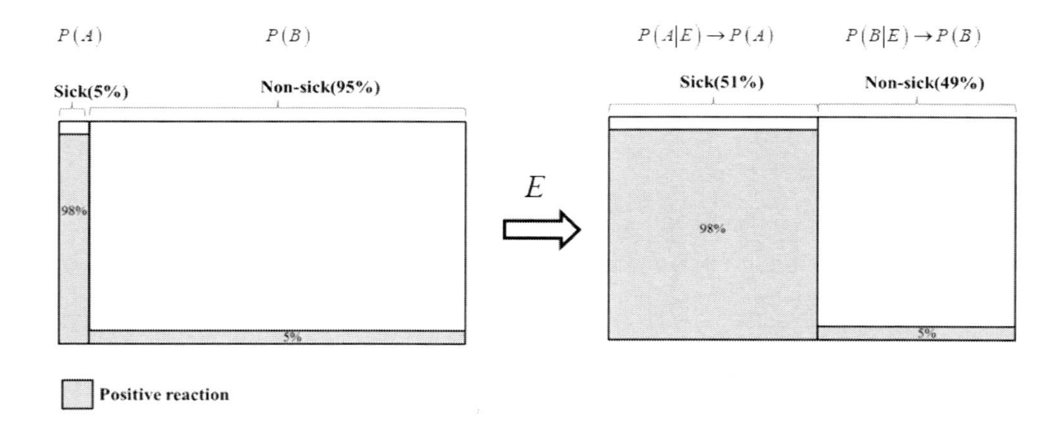

Figure 3. Priori and posterior probability where we have a positive reaction in the medical check.

4.2. Double Medical Checks

We cannot do clearly with posterior probability of Eqs. (25) and (26) since they are not so extreme values.

We try the same medical check again, and assume that we have again the positive reaction. It is not the same as the first one. We use posterior probabilities in this case.

Let us consider the situation of the sick people example in more detail.

We obtain positive reaction E with the medical check, and the probability for the person to be sick is given by

$$\begin{aligned} P(A|E) &= \frac{P(E|A)P(A)}{P(E|A)P(A)+P(E|B)P(B)} \\ &= \frac{0.98 \times 0.51}{0.98 \times 0.51 + 0.05 \times 0.49} \\ &= 0.95 \end{aligned} \tag{27}$$

We can also evaluate the probability for the person to be non-sick after we obtain event E as

$$\begin{aligned} P(B|E) &= \frac{P(E|B)P(B)}{P(E|A)P(A)+P(E|B)P(B)} \\ &= \frac{0.05 \times 0.49}{0.98 \times 0.51 + 0.05 \times 0.49} \\ &= 0.05 \end{aligned} \tag{28}$$

Therefore, the person should perform further process as shown in Figure 4.

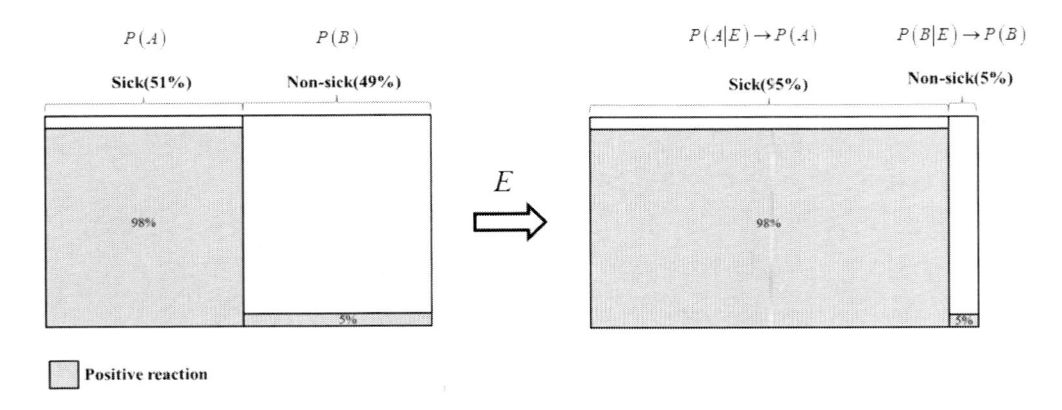

Figure 4. Priori and posterior probability after we have positive reaction in the second medical check.

4.3. Two Kinds of Medical Checks

We consider two kinds of medical checks here. The first one is alternatively rough estimation, and the second accurate one, which is shown in Figure 5.

The positive reaction probabilities for the test 1 is denoted as $P_1(E|A)=0.8$ and $P_1(E|B)=0.1$, and the positive reaction probabilities for the test 2 as $P_2(E|A)=0.98$ and $P_1(E|B)=0.03$. We assume the prior probabilities of $P(A)=0.05$ and $P(B)=0.95$.

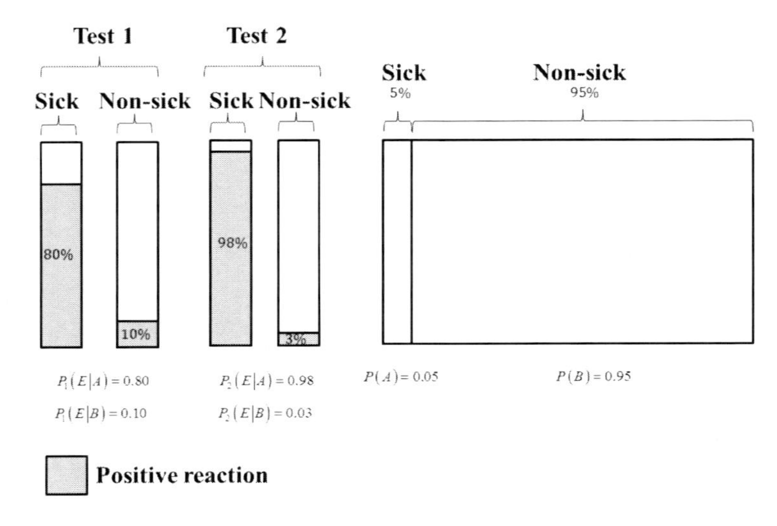

Figure 5. Two kinds of medical check. The test 1 is relatively rough check compared with the test 2 since the related probabilities are not so extreme.

We perform the test 1 medical check first.

If we obtain negative reaction, we can evaluate the posterior probabilities as

$$P_1(A|\bar{E}) = \frac{P_1(\bar{E}|A)P(A)}{P_1(\bar{E}|A)P(A)+P_1(\bar{E}|B)P(B)}$$
$$= \frac{0.2\times 0.05}{0.2\times 0.05+0.90\times 0.95}$$
$$= 0.01 \tag{29}$$

$$P_1(B|\bar{E}) = \frac{P_1(\bar{E}|B)P(B)}{P_1(\bar{E}|A)P(A)+P_1(\bar{E}|B)P(B)}$$
$$= \frac{0.90\times 0.95}{0.2\times 0.05+0.90\times 0.95}$$
$$= 0.99 \tag{30}$$

Therefore, we can convincingly judge that the person is not sick, and finish the medical check. The corresponding probabilities update feature is shown in Figure 6.

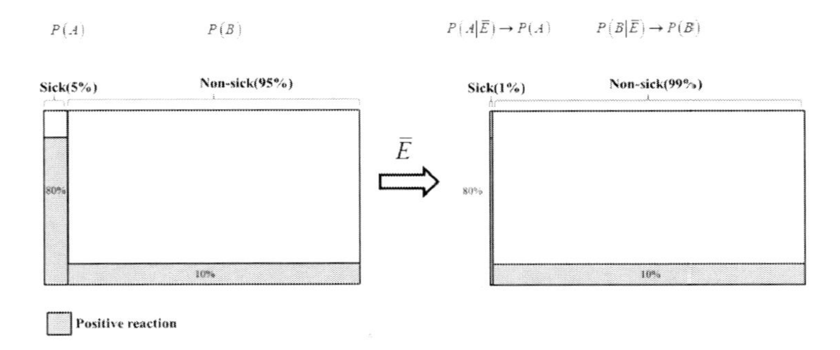

Figure 6. Priori and posterior probability after we have negative reaction in the test1 medical check.

We then consider a case where we obtain positive reaction in the test 1 medical check. The corresponding posterior probabilities are given by

$$\begin{aligned} P_1(A|E) &= \frac{P_1(E|A)P(A)}{P_1(E|A)P(A)+P_1(E|B)P(B)} \\ &= \frac{0.80\times 0.05}{0.80\times 0.05+0.10\times 0.95} \\ &= 0.30 \end{aligned} \tag{31}$$

$$\begin{aligned} P_1(B|E) &= \frac{P_1(E|B)P(B)}{P_1(E|A)P(A)+P_1(E|B)P(B)} \\ &= \frac{0.10\times 0.95}{0.80\times 0.05+0.10\times 0.95} \\ &= 0.70 \end{aligned} \tag{32}$$

The corresponding Bayesian updating process is shown in Figure 7. We cannot judge clearly with these posterior probabilities, and perform the test 2 medical check.

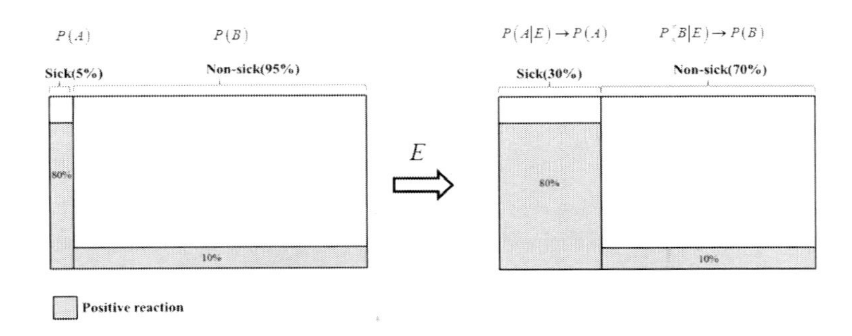

Figure 7. Priori and posterior probability after we have positive reaction in the test 1 medical check.

We use Eqs. (31) and (32) as the priori probabilities for the test 2.

If we obtain positive reaction in this medical check, the posterior probabilities are evaluated as follows.

$$
\begin{aligned}
P_2(A|E) &= \frac{P_2(E|A)P(A)}{P_2(E|A)P(A)+P_2(E|B)P(B)} \\
&= \frac{0.98\times 0.30}{0.98\times 0.30+0.03\times 0.70} \\
&= 0.93
\end{aligned}
\tag{33}
$$

$$
\begin{aligned}
P_2(B|E) &= \frac{P_2(E|B)P(B)}{P_2(E|A)P(A)+P_2(E|B)P(B)} \\
&= \frac{0.03\times 0.70}{0.98\times 0.30+0.03\times 0.70} \\
&= 0.07
\end{aligned}
\tag{34}
$$

Therefore, the person is probably sick.

If we obtain negative reaction in this medical check, the posterior probabilities are evaluated as follows.

$$
\begin{aligned}
P_2(A|\bar{E}) &= \frac{P_2(\bar{E}|A)P(A)}{P_2(\bar{E}|A)P(A)+P_2(\bar{E}|B)P(B)} \\
&= \frac{0.02\times 0.30}{0.02\times 0.30+0.97\times 0.70} \\
&= 0.01
\end{aligned}
\tag{35}
$$

$$
\begin{aligned}
P_2(B|\bar{E}) &= \frac{P_2(\bar{E}|B)P(B)}{P_2(\bar{E}|A)P(A)+P_2(\bar{E}|B)P(B)} \\
&= \frac{0.97\times 0.70}{0.02\times 0.30+0.97\times 0.70} \\
&= 0.99
\end{aligned}
\tag{36}
$$

Therefore, the person is probably non-sick.

4.4. A Punctual Person Problem

We can treat some ambiguous problem using the procedure shown in the previous section. The theory is exactly the same as the one in the previous section, but we should define the ambiguous problem by making some assumptions.

Let us consider the case for a punctual person, where we want to decide whether a person is punctual or not punctual. The ambiguous point in this problem is that a punctual person is not clearly defined in general. We feel that a punctual person is apt to be on time with high probability and not punctual people are not apt to keep a time. The probability is not 1 or 0, but some values between 0 and 1 for both cases.

We need to define punctual quantitatively, although it is rather difficult.

We form two groups: One group consists of persons who are thought to be punctual, and the other group consists of persons who are thought to be not punctual. We then gather data of on time or off time and finally get the data such as shown in Table 1. This is a definition of punctual or not punctual. That is,

Punctual: $\text{On time probability} = \frac{4}{5}; \text{Off time probability} = \frac{1}{5}$

Not punctual: $\text{On time probability} = \frac{1}{4}; \text{Off time probability} = \frac{3}{4}$

Table 1. Definition of punctual or not punctual

Character	Off time(D_1)	On time(D_2)
Not Punctual (H_1)	$\frac{3}{4}$	$\frac{1}{4}$
Punctual (H_2)	$\frac{1}{5}$	$\frac{4}{5}$

We investigate a person S whether he is punctual or not.

The unpunctual case is denoted as H_1, and punctual case as H_2. The off-time is denoted as D_1 and on time as D_2, which is also shown in Table 1.

S' s score in recent five meetings are D_2, D_2, D_1, D_1, D_1.

Step 0

We set the initial value for a member S. We have an impression that he is not punctual, and hence set the initial one as

$$P(H_1) = 0.6 \tag{37}$$

$$P(H_2) = 0.4 \tag{38}$$

Step 1

Next we re-evaluate S considering the obtained data.

In the first meeting, he was on time, and hence, we have

$$P(H_1|D_2) = \frac{P(D_2|H_1)P(H_1)}{P(D_2|H_1)P(H_1) + P(D_2|H_2)P(H_2)}$$
$$= \frac{\frac{1}{4}\times 0.6}{\frac{1}{4}\times 0.6 + \frac{4}{5}\times 0.4}$$
$$= 0.319 \tag{39}$$

$$P(H_2|D_2) = \frac{P(D_2|H_2)P(H_2)}{P(D_2|H_1)P(H_1) + P(D_2|H_2)P(H_2)}$$
$$= \frac{\frac{4}{5}\times 0.4}{\frac{1}{4}\times 0.6 + \frac{4}{5}\times 0.4}$$
$$= 0.681 \tag{40}$$

Therefore, the updating $P(H_1)$ and $P(H_2)$ are

$$P(H_1) = 0.319 \tag{41}$$

$$P(H_2) = 0.681 \tag{42}$$

Step 2

He was also on time in the second meeting, and we have

$$P(H_1|D_2) = \frac{P(D_2|H_1)P(H_1)}{P(D_2|H_1)P(H_1) + P(D_2|H_2)P(H_2)}$$
$$= \frac{\frac{1}{4}\times 0.310}{\frac{1}{4}\times 0.310 + \frac{4}{5}\times 0.681}$$
$$= 0.125 \tag{43}$$

$$P(H_2|D_2) = \frac{P(D_2|H_2)P(H_2)}{P(D_2|H_1)P(H_1)+P(D_2|H_2)P(H_2)}$$
$$= \frac{\frac{4}{5}\times 0.681}{\frac{1}{4}\times 0.310+\frac{4}{5}\times 0.681}$$
$$= 0.875 \tag{44}$$

Therefore, the updating $P(H_1)$ and $P(H_2)$ are

$$P(H_1) = 0.125 \tag{45}$$

$$P(H_2) = 0.875 \tag{46}$$

Step 3

He was off time in the third meeting, and we have

$$P(H_1|D_1) = \frac{P(D_1|H_1)P(H_1)}{P(D_1|H_1)P(H_1)+P(D_1|H_2)P(H_2)}$$
$$= \frac{\frac{3}{4}\times 0.125}{\frac{3}{4}\times 0.125+\frac{1}{5}\times 0.875}$$
$$= 0.349 \tag{47}$$

$$P(H_2|D_1) = \frac{P(D_1|H_2)P(H_2)}{P(D_1|H_1)P(H_1)+P(D_1|H_2)P(H_2)}$$
$$= \frac{\frac{1}{5}\times 0.875}{\frac{3}{4}\times 0.125+\frac{1}{5}\times 0.875}$$
$$= 0.651 \tag{48}$$

Therefore, the updating $P(H_1)$ and $P(H_2)$ are

$$P(H_1) = 0.349 \tag{49}$$

$$P(H_2) = 0.651 \tag{50}$$

Step 4

He was off time in the fourth meeting, and we have

$$\begin{aligned} P(H_1|D_1) &= \frac{P(D_1|H_1)P(H_1)}{P(D_1|H_1)P(H_1) + P(D_1|H_2)P(H_2)} \\ &= \frac{\frac{3}{4} \times 0.349}{\frac{3}{4} \times 0.349 + \frac{1}{5} \times 0.651} \\ &= 0.668 \end{aligned} \tag{51}$$

$$\begin{aligned} P(H_2|D_1) &= \frac{P(D_1|H_2)P(H_2)}{P(D_1|H_1)P(H_1) + P(D_1|H_2)P(H_2)} \\ &= \frac{\frac{1}{5} \times 0.651}{\frac{3}{4} \times 0.349 + \frac{1}{5} \times 0.651} \\ &= 0.332 \end{aligned} \tag{52}$$

Therefore, the updating $P(H_1)$ and $P(H_2)$ are

$$P(H_1) = 0.668 \tag{53}$$

$$P(H_2) = 0.332 \tag{54}$$

Step 5

He was off time in the fifth meeting, and we have

$$P(H_1|D_1) = \frac{P(D_1|H_1)P(H_1)}{P(D_1|H_1)P(H_1)+P(D_1|H_2)P(H_2)}$$
$$= \frac{\frac{3}{4}\times 0.668}{\frac{3}{4}\times 0.668+\frac{1}{5}\times 0.332}$$
$$= 0.883 \tag{55}$$

$$P(H_2|D_1) = \frac{P(D_1|H_2)P(H_2)}{P(D_1|H_1)P(H_1)+P(D_1|H_2)P(H_2)}$$
$$= \frac{\frac{1}{5}\times 0.332}{\frac{3}{4}\times 0.668+\frac{1}{5}\times 0.332}$$
$$= 0.117 \tag{56}$$

Therefore, the updating $P(H_1)$ and $P(H_2)$ are

$$P(H_1) = 0.883 \tag{57}$$

$$P(H_2) = 0.117 \tag{58}$$

Therefore, we are convincing that he is not punctual based on the data.

It should be noted that the results depend on the definition of the punctual people. However, we can treat such ambiguous issue quantitatively using the procedure.

4.5. A Spam Mail Problem

We receive many mails every day. In the huge mails, we have sometimes spam mails. We want to divide the spam mail definitely by evaluating words in the mail. This is one famous example that the Bayesian updating procedure accommodates.

We make a dictionary where key words and related probabilities are listed as word 1, 2, 3, • • •, where three of which are shown in Table 2.

Table 2. Dictionary for evaluating spam mails

Word	Spam	Non-spam
W01	r_{s1}	r_{ns1}
W02	r_{s2}	r_{ns2}
W03	r_{s3}	r_{ns3}

We denote the event of a spam mail as S, and a non-spam mail as $\bar{S}$. We assume that we find words 1, 2, and 3 in a mail, and want to evaluate whether the mail is spam or not.

We need to define the initial probability for $P(S)$ and $P(\bar{S})$. We assume that we know that $P(S)=r_{s0}$. Therefore, we can assume that $P(\bar{S})=1-r_{s0}$.

We find the word 1, and the corresponding probability is given by

$$\begin{aligned} P(S|W01) &= \frac{P(W01|S)P(S)}{P(W01|S)P(S)+P(W01|\bar{S})P(\bar{S})} \\ &= \frac{r_{s1}r_{s0}}{r_{s1}r_{s0}+r_{ns1}(1-r_{s0})} \to P(S) \end{aligned} \tag{59}$$

$$\begin{aligned} P(\bar{S}|W01) &= \frac{P(W01|\bar{S})P(\bar{S})}{P(W01|S)P(S)+P(W01|\bar{S})P(\bar{S})} \\ &= \frac{r_{ns1}(1-r_{s0})}{r_{s1}r_{s0}+r_{ns1}(1-r_{s0})} \to P(\bar{S}) \end{aligned} \tag{60}$$

We find the word 2, and the corresponding probability is given by

$$\begin{aligned} P(S|W02) &= \frac{P(W02|S)P(S)}{P(W02|S)P(S)+P(W02|\bar{S})P(\bar{S})} \\ &= \frac{r_{s2}\dfrac{r_{s1}r_{s0}}{r_{s1}r_{s0}+r_{ns1}(1-r_{s0})}}{r_{s2}\dfrac{r_{s1}r_{s0}}{r_{s1}r_{s0}+r_{ns1}(1-r_{s0})}+r_{ns2}\dfrac{r_{ns1}(1-r_{s0})}{r_{s1}r_{s0}+r_{ns1}(1-r_{s0})}} \\ &= \frac{r_{s2}r_{s1}r_{s0}}{r_{s2}r_{s1}r_{s0}+r_{ns2}r_{ns1}(1-r_{s0})} \to P(S) \end{aligned} \tag{61}$$

$$P(\bar{S}|W02)=\frac{P(W02|\bar{S})P(\bar{S})}{P(W02|S)P(S)+P(W02|\bar{S})P(\bar{S})}$$
$$=\frac{r_{ns2}\dfrac{r_{ns1}(1-r_{s0})}{r_{s1}r_{s0}+r_{ns1}(1-r_{s0})}}{r_{s2}\dfrac{r_{s1}r_{s0}}{r_{s1}r_{s0}+r_{ns1}(1-r_{s0})}+r_{ns2}\dfrac{r_{ns1}(1-r_{s0})}{r_{s1}r_{s0}+r_{ns1}(1-r_{s0})}}$$
$$=\frac{r_{ns2}r_{ns1}(1-r_{s0})}{r_{s2}r_{s1}r_{s0}+r_{ns2}r_{ns1}(1-r_{s0})}\rightarrow P(\bar{S}) \tag{62}$$

We finally find the word 3, and corresponding probabilities are given by

$$P(S|W03)=\frac{P(W03|S)P(S)}{P(W03|S)P(S)+P(W03|\bar{S})P(\bar{S})}$$
$$=\frac{r_{s3}\dfrac{r_{s2}r_{s1}r_{s0}}{r_{s2}r_{s1}r_{s0}+r_{ns2}r_{ns1}(1-r_{s0})}}{r_{s3}\dfrac{r_{s2}r_{s1}r_{s0}}{r_{s2}r_{s1}r_{s0}+r_{ns2}r_{ns1}(1-r_{s0})}+r_{ns3}\dfrac{r_{ns2}r_{ns1}(1-r_{s0})}{r_{s2}r_{s1}r_{s0}+r_{ns2}r_{ns1}(1-r_{s0})}}$$
$$=\frac{r_{s3}r_{s2}r_{s1}r_{s0}}{r_{s3}r_{s2}r_{s1}r_{s0}+r_{ns3}r_{ns2}r_{ns1}(1-r_{s0})}\rightarrow P(S) \tag{63}$$

$$P(\bar{S}|W03)=\frac{P(W03|\bar{S})P(\bar{S})}{P(W03|S)P(S)+P(W03|\bar{S})P(\bar{S})}$$
$$=\frac{r_{ns3}\dfrac{r_{ns2}r_{ns1}(1-r_{s0})}{r_{s2}r_{s1}r_{s0}+r_{ns2}r_{ns1}(1-r_{s0})}}{r_{s3}\dfrac{r_{s2}r_{s1}r_{s0}}{r_{s2}r_{s1}r_{s0}+r_{ns2}r_{ns1}(1-r_{s0})}+r_{ns3}\dfrac{r_{ns2}r_{ns1}(1-r_{s0})}{r_{s2}r_{s1}r_{s0}+r_{ns2}r_{ns1}(1-r_{s0})}}$$
$$=\frac{r_{ns3}r_{ns2}r_{ns1}(1-r_{s0})}{r_{s3}r_{s2}r_{s1}r_{s0}+r_{ns3}r_{ns2}r_{ns1}(1-r_{s0})}\rightarrow P(\bar{S}) \tag{64}$$

We can easily extend this to the general case where we find m kinds of keywords in a mail.

The probabilities that the mail is spam or non-spam are given by

$$P(S) \leftarrow \frac{r_{s0}\prod_{i=1}^{m} r_{si}}{r_{s0}\prod_{i=1}^{m} r_{si} + (1-r_{s0})\prod_{i=1}^{m} r_{nsi}} \tag{65}$$

$$P(\bar{S}) \leftarrow \frac{(1-r_{s0})\prod_{i=1}^{m} r_{nsi}}{r_{s0}\prod_{i=1}^{m} r_{si} + (1-r_{s0})\prod_{i=1}^{m} r_{nsi}} \tag{66}$$

It should be noted that we neglect the correlation between each word in the above analysis.

4.6. A Vase Problem

We consider a vase problem, which is frequently treated and is convenient to extend the formula.

We know the two kinds of vases where black and gray balls are in, and we know the each ratio of black ball number to the total one. We select one vase where we do not know which one it is. We take a ball from the vase and return it. Getting the data, we want to guess which vase we use.

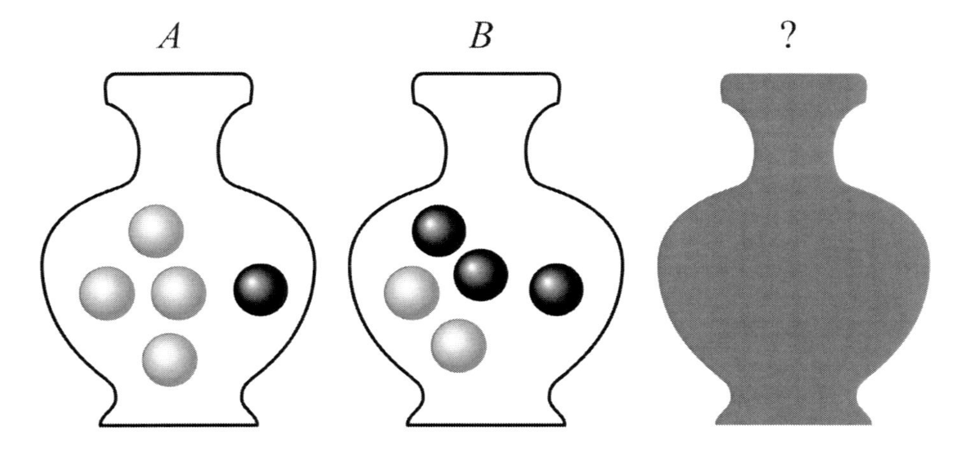

Figure 8. A vase problem with two kinds of balls.

Let us consider two vases A and B as shown in Figure 8. The vase A contains balls with a ratio of gray to black is 4:1, and B with a ratio of 2:3. We express the event of obtaining a black ball as E and a gray balls as $\bar{E}$, and hence the probabilities are expressed as

$$P(E|A) = \frac{1}{5} \tag{67}$$

$$P(E|B) = \frac{3}{5} \tag{68}$$

These probabilities characterize the vases.

The probabilities that we obtain a black ball are given by

$$P(A|E) = \frac{P(E|A)P(A)}{P(E|A)P(A) + P(E|B)P(B)} \tag{69}$$

$$P(B|E) = \frac{P(E|B)P(B)}{P(E|A)P(A) + P(E|B)P(B)} \tag{70}$$

The probabilities that we obtain a gray ball are given by

$$P(A|\bar{E}) = \frac{P(\bar{E}|A)P(A)}{P(\bar{E}|A)P(A) + P(\bar{E}|B)P(B)} \tag{71}$$

$$P(B|\bar{E}) = \frac{P(\bar{E}|B)P(B)}{P(\bar{E}|A)P(A) + P(\bar{E}|B)P(B)} \tag{72}$$

We select one type of a vase and cannot distinguish which vase it is by its appearance. We do not know which vase it is. Therefore, one of the probability $P(A)$ or $P(B)$ is 1 and the other is 0. We pick up one ball and return it to the vase, and perform five times and obtain the results of $\bar{E}EE\bar{E}E$. Therefore, the number of black balls is larger and hence we can guess that the vase B is more plausible. We want to evaluate the probability more quantitatively. The procedures are as follows.

Step 0

If we have some information associated with the number of vase A and B, and hence its ratio, we can use them as the initial probability for $P(A)$ and $P(B)$. If we do not have any information associated with the ratio, we assume the probabilities as

$$P(A) = P(B) = \frac{1}{2} \tag{73}$$

Step 1

We obtain a gray ball $(\bar{E})$ in the first trial, and hence the corresponding probabilities are given by

$$\begin{aligned} P(A|\bar{E}) &= \frac{P(\bar{E}|A)P(A)}{P(\bar{E}|A)P(A) + P(\bar{E}|B)P(B)} \\ &= \frac{\frac{4}{5} \times \frac{1}{2}}{\frac{4}{5} \times \frac{1}{2} + \frac{2}{5} \times \frac{1}{2}} \\ &= 0.667 \end{aligned} \tag{74}$$

$$\begin{aligned} P(B|\bar{E}) &= \frac{P(\bar{E}|B)P(B)}{P(\bar{E}|A)P(A) + P(\bar{E}|B)P(B)} \\ &= \frac{\frac{2}{5} \times \frac{1}{2}}{\frac{4}{5} \times \frac{1}{2} + \frac{2}{5} \times \frac{1}{2}} \\ &= 0.333 \end{aligned} \tag{75}$$

We regard these as $P(A)$ and $P(B)$, that is,

$$\begin{cases} P(A) = P(A|\bar{E}) = 0.667 \\ P(B) = P(B|\bar{E}) = 0.333 \end{cases} \tag{76}$$

Step 2

We obtain a black ball (E) in the second trial, and hence the corresponding probabilities are given by

$$\begin{aligned} P(A|E) &= \frac{P(E|A)P(A)}{P(E|A)P(A)+P(E|B)P(B)} \\ &= \frac{\frac{1}{5}\times 0.667}{\frac{1}{5}\times 0.667+\frac{3}{5}\times 0.333} \\ &= 0.4 \end{aligned} \tag{77}$$

$$\begin{aligned} P(B|E) &= \frac{P(E|B)P(B)}{P(E|A)P(A)+P(E|B)P(B)} \\ &= \frac{\frac{2}{5}\times 0.333}{\frac{4}{5}\times 0.667+\frac{2}{5}\times 0.333} \\ &= 0.6 \end{aligned} \tag{78}$$

We regard these as $P(A)$ and $P(B)$, that is,

$$\begin{cases} P(A) = P(A|E) = 0.4 \\ P(B) = P(B|E) = 0.6 \end{cases} \tag{79}$$

Step 3

We obtain a black ball (E) in the third trial, and hence the corresponding probabilities are given by

$$\begin{aligned} P(A|E) &= \frac{P(E|A)P(A)}{P(E|A)P(A)+P(E|B)P(B)} \\ &= \frac{\frac{1}{5}\times 0.4}{\frac{1}{5}\times 0.4+\frac{3}{5}\times 0.6} \\ &= 0.182 \end{aligned} \tag{80}$$

$$P(B|E) = \frac{P(E|B)P(B)}{P(E|A)P(A) + P(E|B)P(B)}$$
$$= \frac{\frac{3}{5} \times 0.6}{\frac{1}{5} \times 0.4 + \frac{3}{5} \times 0.6}$$
$$= 0.818 \tag{81}$$

We regard these as $P(A)$ and $P(B)$, that is,

$$\begin{cases} P(A) = P(A|E) = 0.182 \\ P(B) = P(B|E) = 0.818 \end{cases} \tag{82}$$

Step 4

We obtain a gray ball $(\bar{E})$ in the third trial, and hence the corresponding probabilities are given by

$$P(A|\bar{E}) = \frac{P(\bar{E}|A)P(A)}{P(\bar{E}|A)P(A) + P(\bar{E}|B)P(B)}$$
$$= \frac{\frac{4}{5} \times 0.182}{\frac{4}{5} \times 0.182 + \frac{2}{5} \times 0.818}$$
$$= 0.308 \tag{83}$$

$$P(B|\bar{E}) = \frac{P(\bar{E}|B)P(B)}{P(\bar{E}|A)P(A) + P(\bar{E}|B)P(B)}$$
$$= \frac{\frac{2}{5} \times 0.818}{\frac{4}{5} \times 0.182 + \frac{2}{5} \times 0.818}$$
$$= 0.692 \tag{84}$$

We regard these as $P(A)$ and $P(B)$, that is,

$$\begin{cases} P(A) = P(A|\bar{E}) = 0.308 \\ P(B) = P(B|\bar{E}) = 0.692 \end{cases} \tag{85}$$

Step 5

We obtain a black ball (E) in the third trial, and hence the corresponding probabilities are given by

$$\begin{aligned} P(A|E) &= \frac{P(E|A)P(A)}{P(E|A)P(A)+P(E|B)P(B)} \\ &= \frac{\frac{1}{5}\times 0.308}{\frac{1}{5}\times 0.308+\frac{3}{5}\times 0.692} \\ &= 0.129 \end{aligned} \tag{86}$$

$$\begin{aligned} P(B|E) &= \frac{P(E|B)P(B)}{P(E|A)P(A)+P(E|B)P(B)} \\ &= \frac{3}{5}\times 0.692 \\ &= 0.871 \end{aligned} \tag{87}$$

We regard these as $P(A)$ and $P(B)$, that is,

$$\begin{cases} P(A) = P(A|E) = 0.182 \\ P(B) = P(B|E) = 0.818 \end{cases} \tag{88}$$

The final probability of $P(A)$ is much smaller than $P(B)$, and hence we guess the vase is type B. When we update the probability, the accuracy of the probability improves, which is shown later.

We treat two types of vase A and B here. However, we can easily extend it to many kinds of A_i where i is 1, 2, ・・・, m. The events are expressed by E_j. We need to know the conditional probability of

$$P(E_j|A_i) \tag{89}$$

In this analysis, E_j is not limited to E or $\bar{E}$ as is the case for the example. We can use any number kinds of the event, but need to know the corresponding conditional probabilities of Eq. (89).

We can set initial probabilities as

$$P(A_i) = \frac{1}{m} \tag{90}$$

If we obtain an event of E_1, the corresponding probability under the event E_1 are given by

$$P(A_i|E_1) = \frac{P(E_1|A_i)P(A_i)}{\sum_{i=1}^{m} P(E_1|A_i)P(A_i)} \to P(A_i) \tag{91}$$

We can update the probabilities of $P(A_i)$ using this equation.

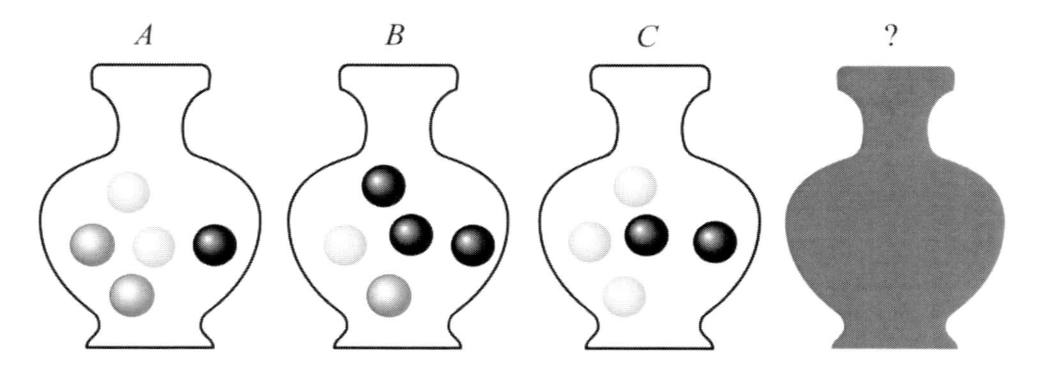

Figure 9. A vase problem with three kinds of balls.

4.7. Extension of Vase Problem

We can easily extend the vase problem, where we can increase the number of the vase, and kinds of balls. We increase the vase number from 2 to 3 denoted as A_1, A_2, and A_3, and kind of balls form black and gray to black, gray, and white ones denoted as E_1, E_2, and E_3, which is shown in Figure 9.

We characterize the vases as the ratio of each ball, which is given by

$$P(E_1|A_1)=\frac{1}{5};P(E_2|A_1)=\frac{2}{5};P(E_3|A_1)=\frac{2}{5} \tag{92}$$

$$P(E_1|A_2)=\frac{3}{5};P(E_2|A_2)=\frac{1}{5};P(E_3|A_3)=\frac{1}{5} \tag{93}$$

$$P(E_1|A_3)=\frac{2}{5};P(E_2|A_3)=\frac{0}{5};P(E_3|A_3)=\frac{3}{5} \tag{94}$$

If we have some information associated with the initial condition for $P(A_i)$, we should use them. If we do not have any information associated with the ratio, we assume the identical probabilities given by

$$P(A_1)=P(A_2)=P(A_3)=\frac{1}{3} \tag{95}$$

If we obtain a certain ball which is expressed as E_i, then we can evaluate posterior probabilities as

$$P(A_1|E_i)=\frac{P(E_i|A_1)}{P(E_i|A_1)P(A_1)+P(E_i|A_2)P(A_2)+P(E_i|A_3)P(A_3)} \tag{96}$$

$$P(A_2|E_i)=\frac{P(E_i|A_2)}{P(E_i|A_1)P(A_1)+P(E_i|A_2)P(A_2)+P(E_i|A_3)P(A_3)} \tag{97}$$

$$P(A_3|E_i)=\frac{P(E_i|A_3)}{P(E_i|A_1)P(A_1)+P(E_i|A_2)P(A_2)+P(E_i|A_3)P(A_3)} \tag{98}$$

We can easily extend the number of the vase to m and kinds number of balls to l. We can then evaluate the posterior probabilities as

$$P(A_k|E_i)=\frac{P(E_i|A_k)}{\sum_{j=1}^{m}P(E_i|A_j)P(A_j)} \tag{99}$$

where $i=1,2,\cdots,l$. Note that we need to know the initial priori probabilities $P(A_j)$.

4.8. Comments on Prior Probability

In the Bayesian updating, we need to know the prior probabilities.

Let us consider it using the two kinds of medical check example. We use the same probability associated with the medical check, but two different sets of prior probabilities.

We assume that we obtain positive reactions for the test 1 and the test 2. The related posterior probabilities are given by

$$P(A|EE)=\frac{P_2(E|A)P_1(E|A)P_0(A)}{P_2(E|A)P_1(E|A)P_0(A)+P_2(E|B)P_1(E|B)P_0(B)} \tag{100}$$

$$P(B|EE)=\frac{P_2(E|B)P_1(E|B)P_0(B)}{P_2(E|A)P_1(E|A)P_0(A)+P_2(E|B)P_1(E|B)P_0(B)} \tag{101}$$

We use two different prior probability sets of $P_0(A)=0.05$ and $P_0(B)=0.95$, and $P_0(A)=0.5$ and $P_0(B)=0.5$. The related posterior profanities for sick are 0.93 and 1.00, respectively. The values are different, but there is no difference in the standpoint of the judge for next doing subjects. Therefore, the initial prior probabilities are not so important if we repeat the check many times.

If we know the prior probabilities, we use it, and use certain ones if we do not know them in detail, and the identical one divided by the number of the cause m if we do not about it anything.

4.9. Mathematical Study for Bayesian Updating

We observed that we obtain proper probability when we repeat Bayesian updating. We study the corresponding mathematics here.

We consider the vase problem.

A vase A contains black and gray balls and the ratio of black ball to the total is p_1. We have many other vases of which ratios are different from p_1. We do not know the vase from which we take balls although it is the vase A. When we perform N trials, we can expect p_1N black balls and $(1-p_1)N$ gray balls. What we should do is to prove the probability has the maximum value for the vase A.

The probability that we take a black ball is denoted as p in general. The probability that after the N times trial f is given by

$$f \propto p^{Np_1}(1-p)^{N(1-p_1)} \tag{102}$$

We differentiate this equation with respect to p and set it to 0, and obtain

$$\begin{aligned}\frac{df}{dp} &\propto Np_1 p^{Np_1-1}(1-p)^{N(1-p_1)} - N(1-p_1)p^{Np_1}(1-p)^{N(1-p_1)-1} \\ &= Np^{Np_1-1}(1-p)^{N(1-p_1)-1}\left[p_1(1-p)-p(1-p_1)\right] \\ &= Np^{Np_1-1}(1-p)^{N(1-p_1)-1}(p_1-p) \\ &= 0\end{aligned} \tag{103}$$

Therefore, we obtain the maximum value for

$$p = p_1 \tag{104}$$

Figure 10 shows the dependence of on probability on p for various trial numbers N and p_1: $p_1 = 0.2$ for (a) and $p_1 = 0.5$ for (b). The probability has the maximum value for p_1, and it is limited to p_1 with increasing N. The features are the same for $p_1 = 0.2$ and 0.5.Therefore, we can expect one significant probability value compared with the others and can state clearly which vase it is with increasing updating numbers.

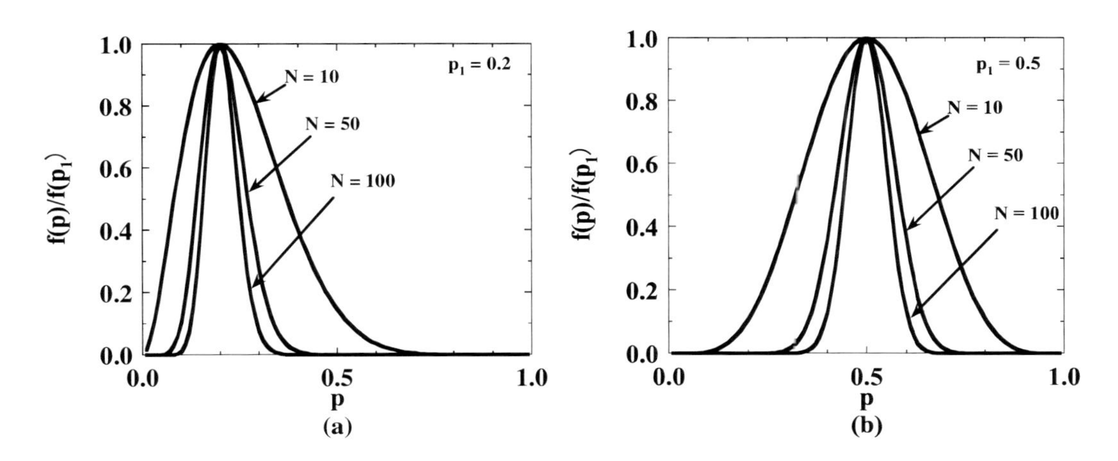

Figure 10. Dependence of probability on the probability for taking a black ball (a) $p_1 = 0.2$, (b) $p_1 = 0.5$.

5. A BAYESIAN NETWORK

Many events are related to each other and some events are causes of the other events in a real world. Further, the cause events are also sometimes the results of other events. Consequently, the related events can be expressed by a network. The event in the network is denoted as a node, and a conditional probability is assigned to each node, which is called as the Bayesian network. Using the Bayesian network, we can predict the quantitative probability between any two events in the network.

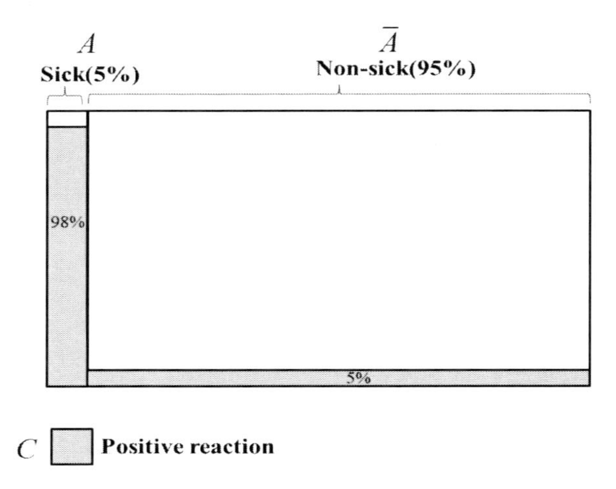

Figure 11. Schematic figure for medical check.

5.1. A Child Node and a Parent Node

Let us consider first the previous example of a medical check as shown in Figure 11.

We denote the sick or non-sick as an event of A_i. If the person is sick, we denote $A_i = A = 1$, and if the person is non-sick, we denote $A_i = \overline{A} = 0$. We denote the positive or negative reaction event as C_i. If the person obtains positive reaction, we denote $C_i = C = 1$, and if the person obtains negative reaction, we denote $C_i = \overline{C} = 0$.

The event A_i can be regarded as the cause of the event C_i and C_i is the result of the event A_i. The event A_i is also called as the parent event for the event C_i, and the event C_i is also called as the child event for event A_i. The events A_i and C_i are called as nodes in the network.

We also need the related probabilities of the nodes, which are given by

$$P(A) = 0.05 \tag{105}$$

$$\begin{aligned} P(\overline{A}) &= 1 - P(A) \\ &= 0.95 \end{aligned} \tag{106}$$

$$P(C|A) = 0.98 \tag{107}$$

$$P(C|\overline{A}) = 0.05 \tag{108}$$

We regard this as a Bayesian network. We regard A_i as a cause of event C_i. Therefore, the corresponding network is expressed as shown in Figure 12. We note that the subscript i in A_i and i in C_i are independent each other. We use this expression for simplicity, and this is applied to the other figures.

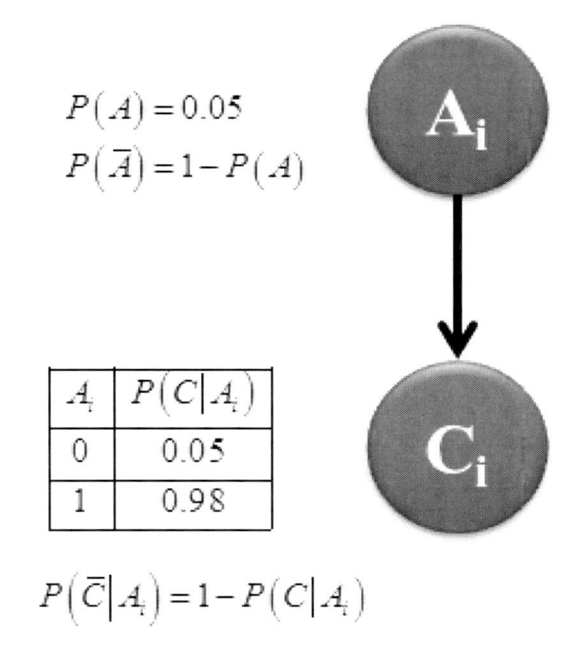

Figure 12. A Bayesian network for the medical check.

$P(C_i)$ are not described in the figure. However, we can evaluate them as follows.

$$\begin{aligned} P(C) &= P(C|\bar{A})P(\bar{A}) + P(C|A)P(A) \\ &= P(C|\bar{A})[1-P(A)] + P(C|A)P(A) \\ &= 0.05\times[1-0.05] + 0.98\times0.05 \\ &= 0.0965 \end{aligned} \tag{109}$$

$$\begin{aligned} P(\bar{C}) &= 1-P(C) \\ &= 1-0.0965 \\ &= 0.9305 \end{aligned} \tag{110}$$

We want to obtain the conditional probability $P(A|C)$ and $P(\bar{A}|C)$, which is given by

$$\begin{aligned} P(A|C) &= \frac{P(C|A)P(A)}{P(C)} \\ &= \frac{0.98\times0.05}{0.0965} \\ &= 0.508 \end{aligned} \tag{111}$$

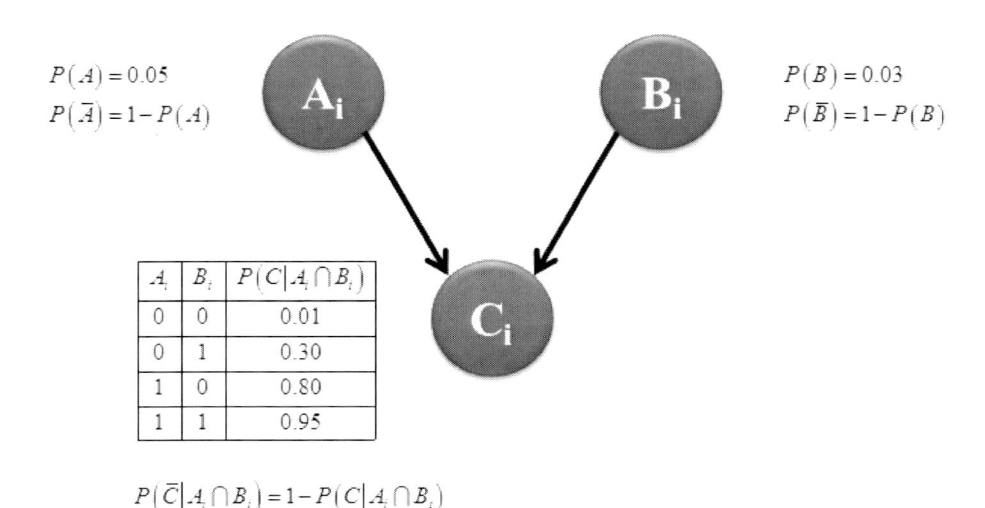

A_i	B_i	$P(C\|A_i \cap B_i)$
0	0	0.01
0	1	0.30
1	0	0.80
1	1	0.95

Figure 13. A Bayesian network with one hierarchy and one child node and two parents nodes.

We extend the Bayesian network of one hierarchy and one child node and two parents nodes as shown in Figure 13.

The nodes are denoted as A_i, B_i, and C_i. A_i and B_i are causes of C_i. C_i is called as a child node for nodes A_i and B_i, and nodes A_i and B_i are called as parents nodes for the node A_i. The events B_i, and C_i are regarded as the causes of the result A_i.

When the event related to A_i occurs, we express it as $A_i = A = 1$. When it does not occur, we express it as $A_i = \overline{A} = 0$. Summarizing above, we can express the status of each node as

$$A_i = \begin{cases} A = 1 \\ \overline{A} = 0 \end{cases} \tag{112}$$

$$B_i = \begin{cases} B = 1 \\ \overline{B} = 0 \end{cases} \tag{113}$$

$$C_i = \begin{cases} C = 1 \\ \overline{C} = 0 \end{cases} \tag{114}$$

We assign probabilities to each node as follows.

The data associated with the node A_i is given by

$$P(A) = 0.05 \tag{115}$$

and

$$\begin{aligned} P(\overline{A}) &= 1 - P(A) \\ &= 0.95 \end{aligned} \tag{116}$$

The data associated with the node B_i is given by

$$P(B) = 0.03 \tag{117}$$

and

$$P(\bar{B}) = 1 - P(B) = 0.97 \tag{118}$$

The data associated with the node C should be $P(C|A_i \cap B_i)$ which is related to the events of A_i and B_i. $P(\bar{C}|A_i \cap B_i)$ is then given by

$$P(\bar{C}|A_i \cap B_i) = 1 - P(C|A_i \cap B_i) \tag{119}$$

The above is shown in in Figure 13
Using the above fundamental data, we then evaluate the probabilities below.

$$\begin{aligned} P(C) &= P(C|\bar{A} \cap \bar{B})P(\bar{A})P(\bar{B}) + P(C|\bar{A} \cap B)P(\bar{A})P(B) \\ &\quad + P(C|A \cap \bar{B})P(A)P(\bar{B}) + P(C|A \cap B)P(A)P(B) \\ &= 0.01 \times (1-0.05) \times (1-0.03) + 0.30 \times (1-0.05) \times 0.03 \\ &\quad + 0.80 \times 0.05 \times (1-0.03) + 0.95 \times 0.05 \times 0.03 \\ &= 0.058 \end{aligned} \tag{120}$$

We implicitly assume that the events A_i and B_j are independent and use

$$P(A_i \cap B_j) = P(A_i)P(B_j) \tag{121}$$

We further obtain

$$P(\bar{C}) = 1 - P(C) = 0.942 \tag{122}$$

We also evaluate conditional probabilities.
We first evaluate $P(A|C)$, which is given by

$$P(A|C) = \frac{P(C|A)P(A)}{P(C)} \tag{123}$$

We know $P(C)$ and $P(A)$, but do not know $P(C|A)$ which can be expanded as

$$\begin{aligned} P(C|A) &= P(C|A \cap \bar{B})P(\bar{B}) + P(C|A \cap B)P(B) \\ &= 0.80 \times (1 - 0.03) + 0.95 \times 0.03 \\ &= 0.805 \end{aligned} \tag{124}$$

Therefore, we obtain

$$\begin{aligned} P(A|C) &= \frac{P(C|A)P(A)}{P(C)} \\ &= \frac{0.805 \times 0.05}{0.058} \\ &= 0.694 \end{aligned} \tag{125}$$

We can obtain the other conditional probabilities with a similar way as

$$\begin{aligned} P(B|C) &= \frac{P(C|B)P(B)}{P(C)} \\ &= \frac{\left[P(C|\bar{A} \cap B)P(\bar{A}) + P(C|A \cap B)P(A)\right]P(B)}{P(C)} \\ &= \frac{[0.30 \times 0.95 + 0.95 \times 0.05] \times 0.03}{0.058} \\ &= 0.665 \end{aligned} \tag{126}$$

We then automatically obtain

$$\begin{aligned} P(\bar{A}|C) &= 1 - P(A|C) \\ &= 0.306 \end{aligned} \tag{127}$$

$$\begin{aligned} P(\bar{B}|C) &= 1 - P(B|C) \\ &= 0.335 \end{aligned} \tag{128}$$

We can further obtain

$$P\left(A\middle|\bar{C}\right)=\frac{P\left(\bar{C}\middle|A\right)P(A)}{P\left(\bar{C}\right)}$$
$$=\frac{\left[1-P\left(C\middle|A\right)\right]P(A)}{P\left(\bar{C}\right)}$$
$$=\frac{[1-0.805]\times 0.05}{0.942}$$
$$=0.010 \tag{129}$$

$$P\left(B\middle|\bar{C}\right)=\frac{P\left(\bar{C}\middle|B\right)P(B)}{P\left(\bar{C}\right)}$$
$$=\frac{\left[1-P\left(C\middle|B\right)\right]P(B)}{P\left(\bar{C}\right)}$$
$$=\frac{\left[1-\left[P\left(C\middle|B\cap\bar{A}\right)P\left(\bar{A}\right)+P\left(C\middle|B\cap A\right)P(A)\right]\right]P(B)}{P\left(\bar{C}\right)}$$
$$=\frac{\left[1-\left(0.30\times 0.95+0.95\times 0.05\right)\right]\times 0.03}{0.942}$$
$$=0.021 \tag{130}$$

We can then automatically obtain

$$P\left(\bar{A}\middle|\bar{C}\right)=1-P\left(A\middle|\bar{C}\right)$$
$$=0.09 \tag{131}$$

$$P\left(\bar{B}\middle|\bar{C}\right)=1-P\left(B\middle|\bar{C}\right)$$
$$=0.979 \tag{132}$$

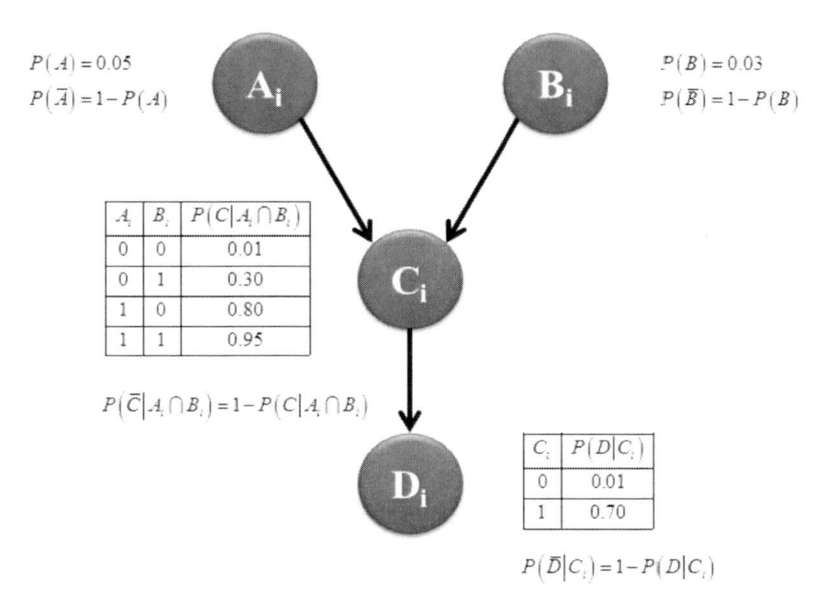

Figure 14. A Bayesian network which is added node D_i to the one in Figure 13.

5.2. A Bayesian Network with Multiple Hierarchy

We considered one hierarchy in the previous section, where there is one child node and its two parents nodes, that is, there is only one child node.

We then extend the network by adding a node D_i as shown Figure 14. Node C_i is a child node for nodes A_i and B_i, but it is also a child node for node D_i.

Since we add the node D_i, we need to know the related probability; we assume which is given by

$$P(D|\bar{C})=0.01 \tag{133}$$

$$P(D|C)=0.70 \tag{134}$$

We can easily evaluate $P(D)$ as

$$\begin{aligned} P(D) &= P(D|\bar{C})P(\bar{C})+P(D|C)P(C) \\ &= 0.01\times 0.942+0.70\times 0.058 \\ &= 0.050 \end{aligned} \tag{135}$$

We also need a conditional probability such as $P(A|D)$, which can be evaluated as

$$P(A|D) = \frac{P(D|A)P(A)}{P(D)} \tag{136}$$

In this equation, the unknown term is only $P(D|A)$. The route is $A_i \to C_i \to D_i$, and we fix $A_i = A$ and $D_i = D$. The problem is how we treat the C_i, which can be C or $\bar{C}$. Therefore, $P(D|A)$ can be expanded as

$$\begin{aligned} P(D|A) &= P(D|C)P(C|A)P(C) + P(D|\bar{C})P(\bar{C}|A)P(\bar{C}) \\ &= P(D|C)P(C|A)P(C) + P(D|\bar{C})\left[1 - P(C|A)\right]P(\bar{C}) \end{aligned} \tag{137}$$

where $P(C|A)$ also depend on B_i, we do not know its explicit expression, but can obtain as follows.

$$\begin{aligned} P(C|A) &= P(C|A \cap \bar{B})P(\bar{B}) + P(C|A \cap B)P(B) \\ &= 0.80 \times (1 - 0.03) + 0.95 \times 0.03 \\ &= 0.8045 \end{aligned} \tag{138}$$

Therefore, we obtain $P(D|A)$ as

$$\begin{aligned} P(D|A) &= P(D|C)P(C|A)P(C) + P(D|\bar{C})\left[1 - P(C|A)\right]P(\bar{C}) \\ &= 0.70 \times 0.8045 \times 0.058 + 0.01 \times (1 - 0.8045) \times (1 - 0.058) \\ &= 0.0345 \end{aligned} \tag{139}$$

This can be generally expressed as

$$P(D|A) = \sum_i P(D|C_i)P(C_i|A)P(C_i) \tag{140}$$

We can easily extend this procedure to any network system.

We consider the network shown in Figure 15, where we add a node F_i. The conditional probability $P(A|F)$ can be evaluated as below.

$$P(A|F) = \frac{P(F|A)P(A)}{P(F)} \tag{141}$$

where

$$P(A|F) = \sum_{i,j} P(F|D_i)P(D_i|C_j)P(C_j|A)P(D_i)P(C_j) \tag{142}$$

One problem for the long Bayesian network is that the number of term increases significantly as shown in Eq. (142).

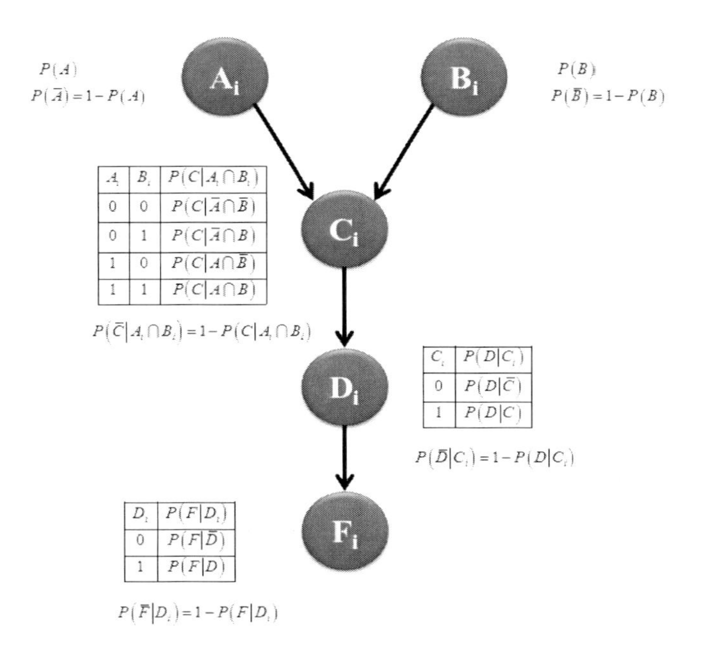

Figure 15. A Bayesian network, which is added node F_i to the one in Figure 14.

We treat one more example as shown in Figure 16.

The target conditional probability is $P(A|G)$. The corresponding routes are $A_i \to C_i \to D_i \to G_i \to F_i \to G_i$ and $A_i \to C_i \to E_i \to G_i \to F_i \to G_i$. We can then evaluate it as

$$P(A|G) = \sum_{i,j,k} P(G|F_i)P(F_i|D_j)P(D_j|C_k)P(C_k|A)P(F_i)P(D_j)P(C_k)$$
$$+\sum_{i,j,k} P(G|F_i)P(F_i|E_j)P(E_j|C_k)P(C_k|A)P(F_i)P(E_j)P(C_k) \tag{143}$$

where F depends on both D and E, and hence the related term is expanded as

$$P(F_i|D_j) = P(F_i|D_j \cap \bar{E})P(\bar{E}) + P(F_i|D_j \cap E)P(E) \tag{144}$$

$$P(F_i|E_j) = P(F_i|E_j \cap \bar{D})P(\bar{D}) + P(F_i|E_j \cap D)P(D) \tag{145}$$

C depends on both A and B, and hence the related term is expanded as

$$P(C_k|A) = P(C_k|A \cap \bar{B})P(\bar{B}) + P(C_k|A \cap B)P(B) \tag{146}$$

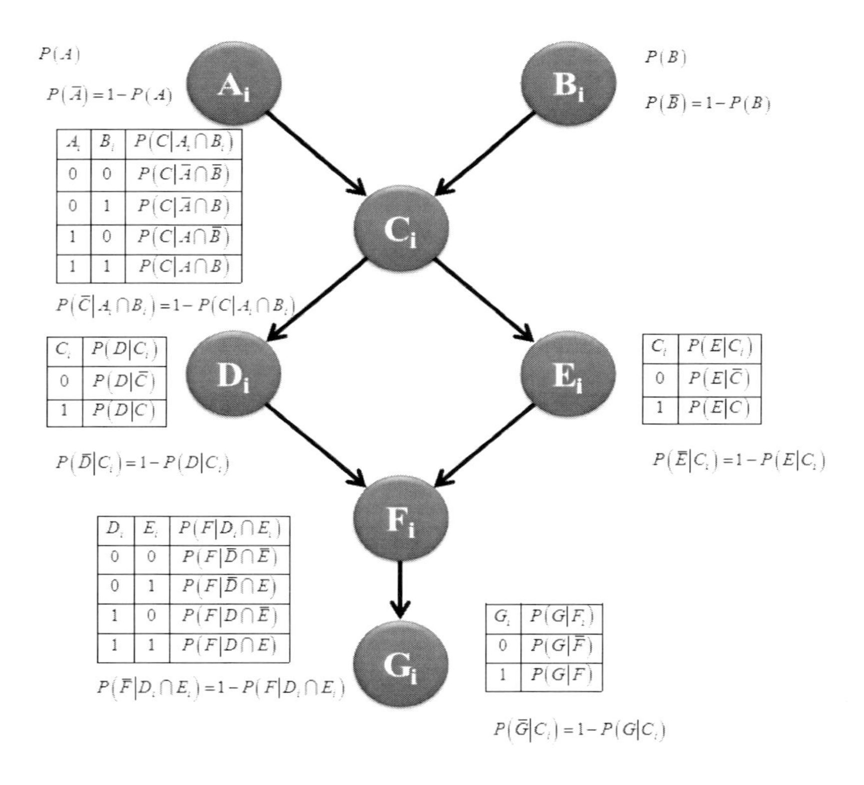

Figure 16. A Bayesian network, which has various routs.

SUMMARY

To summarize the results in this chapter:

We consider m kinds of events $A_i \ (i = 1, 2, \cdots, m)$. When we have an event E, the probability for the related cause of A_i is expressed by

$$P(A_i|E) = \frac{P(A_i)P(E|A_i)}{P(A_1)P(E|A_1) + P(A_2)P(E|A_2) + \cdots + P(A_m)P(E|A_m)}$$

We select one cause or member k. The corresponding probability $P(k)$ is 1 and the other probability is 0. However, we do not know what is the cause or who is the one, that is, we do not know which i is k. We update the $P(A_i)$ with obtaining the results of E. We have some kinds of E which are denoted as E_a and E_b for example and we know the conditional probability $P(E_a|A_i)$ and $P(E_b|A_i)$. The resultant events $E_a, E_b, \cdots$ reflect k.

We assume that we obtain events E_a, E_a, E_b. Based on the results, we update the $P(A_i)$ as

$$P(A_i|E_a) = \frac{P(A_i)P(E_a|A_i)}{P(A_1)P(E_a|A_1) + P(A_2)P(E_a|A_2) + \cdots + P(A_m)P(E_a|A_m)} \to P(A_i)$$

$$P(A_i|E_a) = \frac{P(A_i)P(E_a|A_i)}{P(A_1)P(E_a|A_1) + P(A_2)P(E_a|A_2) + \cdots + P(A_m)P(E_a|A_m)} \to P(A_i)$$

$$P(A_i|E_b) = \frac{P(A_i)P(E_b|A_i)}{P(A_1)P(E_b|A_1) + P(A_2)P(E_b|A_2) + \cdots + P(A_m)P(E_b|A_m)} \to P(A_i)$$

We can expect that the predominant $P(A_i)$ is the required k.

A Bayesian network expresses the relationship between causes and results graphically. The probability for the event is related to the events in the previous step where the pervious event is one or many. We can evaluate probabilities related to any two event in the network.

Chapter 7

A Bayes' Theorem for Predicting Population Parameters

Abstract

We predict a probability distribution associated with macro parameters of a set, where we use priori/likelihood/posterior distributions. We show first the likelihood distributions and multiply it to a priori distribution, and the resultant posterior distribution gives the information associated with the macro parameters of the population set. We show analytical approach first, where conjugate distributions are recommended to use. We then show numerical approach of Marcov Chain Monte Carlo (MCMC) method. In the MCMC method, there are two prominent ways of Gibbs sampling and Metropolis-Hastings algorithm. This process is extended to the procedures with various parameters in likelihood function which is constrained and controlled by the priori distribution function, which is called as a hierarchical Bayesian theory.

Keywords: Bayes' theorem, conditional probability, priori probability, likelihood function, posterior probability; binomial distribution, beta distribution, beta function, multinomial distribution, dirichlet distribution, normal function, inverse gamma distribution, gamma function, Markov chain Monte Carlo, Gibbs sampling, Metropolis-Hastings algorithm, hierarchical Bayesian theory

1. Introduction

Figure 1 shows the schematic targets of the Bayesian theory up to this chapter and the one of this chapter. In the Bayesian updating, we should know the probability of a certain cause. For example, we need to specify the ratio of black balls in a vase. However, we want to specify the ratio itself without assuming the ratio. In general, we want to predict macro ammeters of a population set. We show analytical and numerical models to decide the parameters in this chapter.

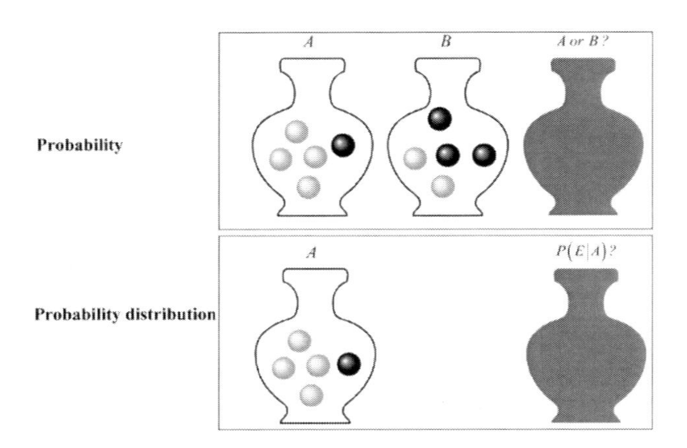

Figure 1. The target of Bayesian updating is to identify the vase kinds. The target of the theory in this chapter is to predict the black ball ratio in the vase.

2. A Maximum Likelihood Method

A method to predict macro parameters of a population set consists of likelihood and priori distributions. We first discuss the likelihood function.

The maximum likelihood method is utilized to predict parameters of a population set in general. We treat the macro parameter of the population set as a variable, and evaluate the probability based on obtained data. We impose that the parameter value gives the maximum probability. We start with this maximum likelihood method.

2.1. Example for a Maximum Likelihood Method

We show two examples for a maximum likelihood method here.

2.2. Number of Rabbits on a Hill

We want to know a total number of rabbits on a hill. We assume that the total number of the rabbits is θ which is unknown and should be evaluated as follows.

We get $a = 15$ rabbits, mark, and release them.

After a while, we get $n = 30$ rabbits, and found $x = 5$ marked ones among them.

The corresponding probability that we get x marked rabbit can be evaluated as

$$f(x) = \frac{{}_{15}C_x \times {}_{\theta-15}C_{30-x}}{{}_{\theta}C_{30}} \tag{1}$$

This is the probability distribution for a variable x. However, we obtain $x=5$, and the unknown variable is the total number of rabbit θ. Therefore, we regard that the function of Eq. (1) expresses the probability with respect to θ, and hence it is denoted as

$$L(\theta) = \frac{{}_{15}C_5 \times {}_{\theta-15}C_{30-5}}{{}_{\theta}C_{30}} \tag{2}$$

Figure 2 shows the dependence of the probability on θ. It has a peak for 89.5. We regard that the situation occurs with the maximum probability, and hence we guess that there are about 89.5 rabbits on the hill. The corresponding distribution is shown in the figure. It should be noted that the form of Eq. (2) is the one for x, and hence it is normalized with respect to x not to θ. Therefore, the integration of Figure 2 deviates from 1. Therefore, the rigorous expression for $L(\theta)$ is given by

$$L(\theta) \propto \frac{{}_{15}C_5 \times {}_{\theta-15}C_{30-5}}{{}_{\theta}C_{30}} \tag{3}$$

If we do not care about the normalization and focus on the peak position, Eq. (2) is OK.

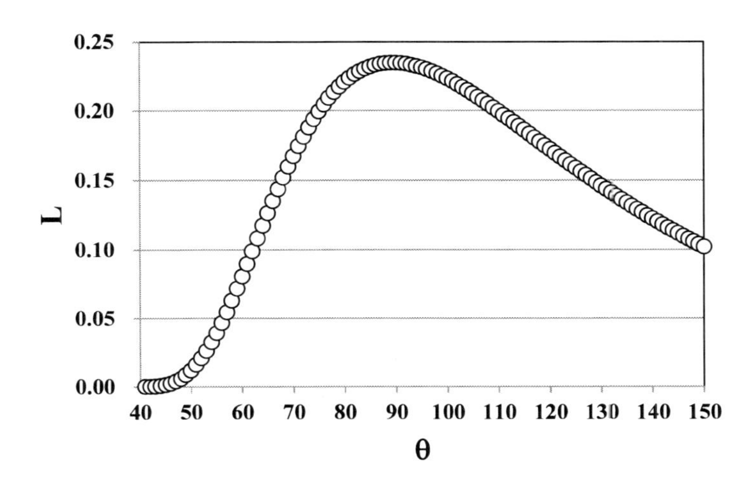

Figure 2. Dependence of L associated with a rabbit number on θ.

2.3. Decompression Sickness

Decompression sickness is a fatal problem for divers. We want to know the probability where a diver feels a decompression problem as a function of water depth.

The data we can get are water depths and related feelings associated with the decompression. If the diver feels a problem, the data is YES and NO for vice versa.

We can then obtain data such as Table 1.

Table 1. Data of water depth and the divers feelings associated with decompression

Member ID	Depth (m)	Result
1	34	YES
2	34	NO
3	34	NO
4	34	NO
5	35	NO
6	35	NO
7	36	YES
8	36	NO
9	37	YES
10	37	YES
11	38	YES
12	38	YES

Next, we should define a probability where a diver feels a decompression problem. The probability should be small with small water depth, increase with the increasing depth, and then saturate. We therefore assume the form of the probability as

$$f(x,\theta)=\frac{x}{x+\theta} \tag{4}$$

where x is the water depth. θ corresponds to the one where 50 % divers feel decompression problems.

On the other hand, the probability where a diver does not feel a decompression problem is given by

$$1-f(x,\theta) \tag{5}$$

That is, the data YES is related to $f(x)$, and the data NO is related to $1-f(x)$.

The probability where we obtain the data on Table 1 is denoted as $L(\theta)$ and is given by

$$L(\theta)=f(34,\theta)\times\left[1-f(34,\theta)\right]^3\times\left[1-f(35,\theta)\right]^2 \times f(36,\theta)\times\left[1-f(36,\theta)\right]\times\left[f(37,\theta)\right]^2\times\left[f(38,\theta)\right]^2 \quad (6)$$

Figure 3 shows the dependence of L on θ. L has a peek for $\theta=35.5$. Therefore, the probability is given by

$$f(x)=\frac{x}{x+35.5} \quad (7)$$

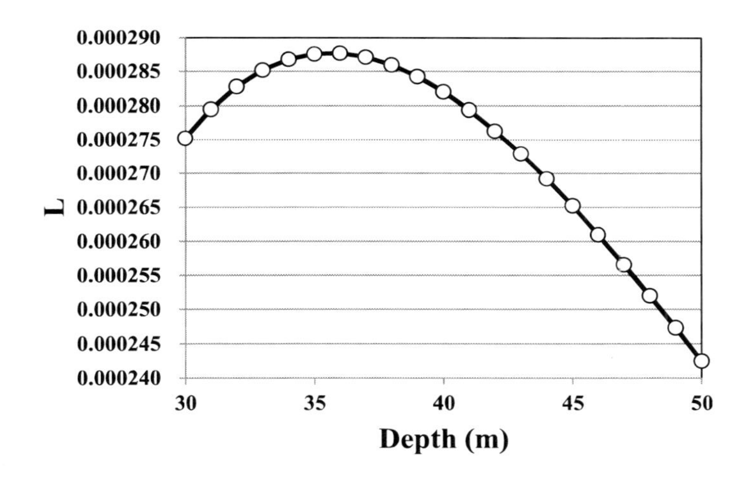

Figure 3. Dependence of L associated with decompression sickness on depth.

3. Prediction of Macro Parameters of a Set with Conjugate Probability Functions

3.1. Theoretical Framework

We studied Bayesian updating in the previous chapter, where we evaluate the probability for the subject, and eventually decide which is the cause for the result.

We treat a different situation in this chapter. We want to know macro parameters of a set based on extracted data and our experience. The prominent macro parameters for the set are average, variance, or ratio. We show how we can evaluate them in this chapter.

The procedure is similar to the Bayesian updating discussed in the previous chapter. However, we should use probability distributions instead of probabilities.

We re-consider the Bayes' theorem which is given by

$$P(H|D) = \frac{P(D|H)P(H)}{P(D)} \tag{8}$$

where H is the cause (Hypothesis), D is the result (Data) caused by H.

We regard that H is decided by the macroscopic parameter of the set such as average. Therefore, H can be regarded as the macroscopic parameter for the set, and we denote it as θ. We modify the Bayes' theory as

$$P(H|D) = \frac{P(D|H)P(H)}{P(D)} \rightarrow \pi(\theta|D) = \frac{f(D|\theta)\pi(\theta)}{P(D)} \tag{9}$$

We call $\pi(\theta)$ as the priori probability distribution which expresses the probability that we obtain θ. $f(D|\theta)$ is the likelihood probability distribution where we have the data D under the macroscopic value of θ. $\pi(\theta|D)$ is the posterior probability distribution which is the product of $f(D|\theta)$ and $\pi(\theta)$. $P(D)$ is independent of θ and we neglect it and express Eq. (9) as

$$\pi(\theta|D) = Kf(D|\theta)\pi(\theta) \tag{10}$$

where we can decide K so that the integration of the probability with respect to θ is one.

In the above theory, there are some ambiguous points. We do not know accurately how we define $\pi(\theta)$, which depends on the person's image or experience. We use a simple one where we can easily obtain $\pi(\theta|D)$ with used $\pi(\theta)$ and $f(D|\theta)$, and impose that the form of the all distributions of $\pi(\theta)$, $f(D|\theta)$, and $\pi(\theta|D)$ are the same. That is, the product of two same form functions is the same as the two functions. There are limited functions which have such characteristics. We show such combinations as shown in Table 2. We basically know the likelihood function, and require that $\pi(\theta)$ and $\pi(\theta|D)$ are the same type distribution functions, which is called as conjugate functions. If we regard the likelihood function as one with a variable of θ, the corresponding function is different, which is shown in the bracket in the table. In those cases, all the type of the distributions of $\pi(\theta)$, $f(D|\theta)$, and $\pi(\theta|D)$ are the same.

Table 2. The combinations of priori, likelihood, and posterior functions

$\pi(\theta)$	$f(D\|\theta)$	$\pi(\theta\|D)$
Beta function	Binominal function (Beta function)	Beta function
Dirichlet distribution	Multinomial distribution (Dirichlet distribution)	Dirichlet distribution
Normal function	Normal function (Normal function)	Normal function
Inverse Gamma function	Normal function (Inverse Gamma function)	Inverse Gamma function
Gamma function	Poisson function (Gamma function)	Gamma function

3.2. Beta/Binominal (Beta)/Beta Distribution

We study an example for the Bernoulli trails and evaluate the population ratio. We obtain the two kinds of events 1 or 0. θ corresponds to the probability that we obtain the event 1 for each event.

We assume the probability θ where we obtain x times target event of 1 among n trials. The corresponding probability with respect to x is a binominal function and is given by

$$f(x) = {}_nC_x\theta^x(1-\theta)^{n-1} \tag{11}$$

We then convert this probability distribution function to the one for probability to obtain θ under the data $D = x$, which is the likelihood function denoted as $f(D|\theta)$. Since the factor ${}_nC_x$ does not include the variable parameter θ for the likelihood function, it should be changed so that the total probability with respect to θ is one. The factor is denoted as K. We then obtain the likelihood function as

$$f(D|\theta) = K\theta^x (1-\theta)^{n-1} \tag{12}$$

We can decide K by imposing that the total integration of the probability is one, which is given by

$$\begin{aligned}\int_0^\infty f(D|\theta)d\theta &= K\int_0^\infty \theta^x (1-\theta)^{n-1} d\theta \\ &= 1\end{aligned} \tag{13}$$

We then obtain the corresponding likelihood function, which is the Beta distribution B_e and is given by

$$\begin{aligned}f(D|\theta) &= B_e(\alpha_L, \beta_L) \\ &= \frac{\theta^{\alpha_L - 1}(1-\theta)^{\beta_L - 1}}{B(\alpha_L, \beta_L)}\end{aligned} \tag{14}$$

where

$$\alpha_L = x + 1 \tag{15}$$

$$\beta_L = n - x + 1 \tag{16}$$

$B(\alpha, \beta)$ is a Beta function given by

$$B(\alpha, \beta) = \int_0^1 \theta^{\alpha-1}(1-\theta)^{\beta-1} d\theta \tag{17}$$

The Beta distribution has parameters of α and β.

We use a Beta distribution function $B_e(\alpha_0, \beta_0)$ as the priori probability distribution, which is given by

$$\pi(\theta) = B_e(\alpha_0, \beta_0) = \frac{1}{B(\alpha_0, \beta_0)}\theta^{\alpha_0 - 1}(1-\theta)^{\beta_0 - 1} \tag{18}$$

There is not a clear reason to decide the parameters of α_0 and β_0. Further, it is not so easy to decide the parameters. It is convenient to decide the corresponding average and standard deviation and convert them to α_0 and β_0.

The relationship between the average and variance and parameters given by

$$\mu_0 = \frac{\alpha_0}{\alpha_0 + \beta_0} \tag{19}$$

$$\sigma_0^2 = \frac{\alpha_0 \beta_0}{(\alpha_0 + \beta_0 + 1)(\alpha_0 + \beta_0)^2} \tag{20}$$

We then obtain

$$\alpha_0 = \frac{\mu_0^2}{\sigma_0^2}(1 - \mu_0) - \mu_0 \tag{21}$$

$$\beta_0 = \frac{1 - \mu_0}{\mu_0}\left[\frac{\mu_0^2}{\sigma_0^2}(1 - \mu_0) - \mu_0\right] \tag{22}$$

We obtain a posterior distribution as

$$\begin{aligned} \pi(\theta|D) &\propto f(D|\theta)\pi(\theta) \\ &\propto \theta^{\alpha_L - 1}(1-\theta)^{\beta_L - 1}\theta^{\alpha_0 - 1}(1-\theta)^{\beta_0 - 1} \\ &= \theta^{(\alpha_0 + \alpha_L - 1) - 1}(1-\theta)^{(\beta_0 + \beta_L - 1) - 1} \end{aligned} \tag{23}$$

Finally, we obtain

$$\pi(\theta|D) = \frac{\theta^{\alpha^* - 1}(1-\theta)^{\beta^* - 1}}{B(\alpha^*, \beta^*)} \tag{24}$$

where

$$\alpha^* = \alpha_0 + \alpha_L - 1 \tag{25}$$

$$\beta^* = \beta_0 + \beta_L - 1 \tag{26}$$

The average and variance of θ denoted as μ_θ and σ_θ^2 are given by

$$\mu_\theta = \frac{\alpha^*}{\alpha^* + \beta^*} \tag{27}$$

$$\sigma_\theta^2 = \frac{\alpha^* \beta^*}{\left(\alpha^* + \beta^* + 1\right)\left(\alpha^* + \beta^*\right)^2} \tag{28}$$

3.3. Dirichlet/Dirichlet (Multinomial)/Dirichlet Distribution

The binomial distribution is extended to a multinomial distribution where the event is more than two. In general, we need to consider multiple events of more than two kinds. Let us consider m kinds of events and a corresponding probability for $P\left(X_1 = x_1, X_2 = x_2, \cdots, X_m = x_m\right)$ is given by

$$f\left(x_1, x_2, \cdots, x_m\right) = \frac{n!}{x_1! x_2! \cdots x_m!} p^{x_1} p^{x_2} \cdots p^{x_m} \tag{29}$$

where

$$p_1 + p_2 + \cdots + p_m = 1 \tag{30}$$

$$x_1 + x_2 + \cdots + x_m = n \tag{31}$$

We then convert this probability distribution to the one for the probability to obtain θ_i under the data $D = \begin{pmatrix} x_1 & x_2 & \cdots & x_m \end{pmatrix}$, which is the likelihood function denoted as $f\left(D \middle| \theta_i\right)$.

We then obtain the corresponding likelihood function, which is the Dirichlet distribution given by

$$f\left(\theta_1, \theta_2, \cdots, \theta_m \middle| D\right) = \frac{\Gamma(\alpha)}{\Gamma(\alpha_1)\Gamma(\alpha_2)\cdots\Gamma(\alpha_m)} \theta_1^{\alpha_1 - 1} \theta_2^{\alpha_2 - 1} \cdots \theta_m^{\alpha_m - 1} \tag{32}$$

where

$$\alpha_i = x_i + 1 \tag{33}$$

We use a Dirichlet distribution as the priori probability distribution, which is given by

$$\pi(\theta_1, \theta_2, \cdots, \theta_m) = \frac{\Gamma(\alpha_0)}{\Gamma(\alpha_{10})\Gamma(\alpha_{20})\cdots\Gamma(\alpha_{m0})} \theta_1^{\alpha_{10}-1}\theta_2^{\alpha_{20}-1}\cdots\theta_m^{\alpha_{m0}-1} \tag{34}$$

We obtain a posterior distribution as

$$\begin{aligned}\pi(\theta_1, \theta_2, \cdots, \theta_m | D) &\propto \theta_1^{\alpha_1-1}\theta_2^{\alpha_2-1}\cdots\theta_m^{\alpha_m-1} \times \theta_1^{\alpha_{10}-1}\theta_2^{\alpha_{20}-1}\cdots\theta_m^{\alpha_{m0}-1} \\ &= \theta_1^{(\alpha_1+\alpha_{10}-1)-1}\theta_2^{(\alpha_2+\alpha_{20}-1)-1}\cdots\theta_m^{(\alpha_m+\alpha_{m0}-1)-1}\end{aligned} \tag{35}$$

Finally, we obtain

$$\pi(\theta_1, \theta_2, \cdots, \theta_m | D) = \frac{\Gamma(\alpha^*)}{\Gamma(\alpha_1^*)\Gamma(\alpha_2^*)\cdots\Gamma(\alpha_m^*)} \theta_1^{\alpha_1^*-1}\theta_2^{\alpha_2^*-1}\cdots\theta_m^{\alpha_m^*-1} \tag{36}$$

where

$$\alpha_i^* = \alpha_i + \alpha_{i0} - 1 \tag{37}$$

3.4. Gamma/Poisson (Gamma)/Gamma Distribution

We also study an example for the Bernoulli trials and do not evaluate the population ratio, but the average number that the target event occurs where the probability that one event occurs is quite small. In this situation, we can regard that the likelihood function follows the Poisson distribution.

We assume that we obtain an event x times target events among n trials, and the probability that we obtain a target event is as small as p. The corresponding probability with respect to x $f(x)$ is a Poisson distribution given by

$$f(x) = \frac{\theta^x}{x!} e^{-\theta} \tag{38}$$

where θ is the average target event number and is given by

$$\theta = np \tag{39}$$

We then convert this probability distribution function to the one for the probability to obtain the average target event number θ under the data of $D = x$, which is the likelihood function denoted as $f(D|\theta)$. Since the factor $x!$ does not include the variable parameter θ for the likelihood function, it should be changed so that the total probability with respect to θ is one. That is, we denote the factor as K. We then obtain the likelihood function as

$$f(D|\theta) = K\theta^x e^{-\theta} \tag{40}$$

We can decide K by imposing that the total integration of the probability is one, which is given by

$$\begin{aligned}\int_0^\infty f(D|\theta)d\theta &= K\int_0^\infty \theta^x e^{-\theta} d\theta \\ &= 1\end{aligned} \tag{41}$$

We then obtain the corresponding likelihood function. It is a Gamma distribution and is given by

$$f(D|\theta) = \frac{\theta^{n_L - 1} e^{-\lambda_L \theta}}{\Gamma(n_L)} \tag{42}$$

where

$$n_L = x + 1 \tag{43}$$

$$\lambda_L = 1 \tag{44}$$

$\Gamma(n)$ is a Gamma function given by

$$\Gamma(n) = \int_0^\infty \theta^{n-1} e^{-\theta} d\theta \tag{45}$$

We usually use a Gamma distribution $\pi(\theta)$ as the priori probability distribution, which is given by

$$\pi(\theta) = \frac{\lambda_0^{n_0} \theta^{n-1} e^{-\lambda_0 \theta}}{\Gamma(n_0)} \tag{46}$$

It is not so easy to decide the priori probability parameters of α_0 and β_0. It is convenient to decide the corresponding average and standard deviation and convert them to α_0 and β_0.

The relationship between the average and variance and parameters given by

$$\mu_0 = \frac{n_0}{\lambda_0} \tag{47}$$

$$\sigma_0^2 = \frac{n_0}{\lambda_0^2} \tag{48}$$

We then obtain

$$\lambda_0 = \frac{\mu_0}{\sigma_0^2} \tag{49}$$

$$n_0 = \frac{\mu_0^2}{\sigma_0^2} \tag{50}$$

We obtain the posterior probability distribution as

$$\begin{aligned} \pi(\theta|D) &\propto f(D|\theta)\pi(\theta) \\ &\propto \theta^{n_L - 1} e^{-\lambda_L \theta} \theta^{n_0 - 1} e^{-\lambda_0 \theta} \\ &= \theta^{(n_0 + n_L - 1) - 1} e^{-(\lambda_0 + \lambda_L)\theta} \end{aligned} \tag{51}$$

Finally, we obtain the posterior distribution as

$$\pi(\theta|D) = \frac{\theta^{n^*-1}e^{-\lambda^*\theta}}{\Gamma(n^*)} \tag{52}$$

where

$$n^* = n_0 + n_L - 1 \tag{53}$$

$$\lambda^* = \lambda_0 + \lambda_L \tag{54}$$

The average and variance are given by

$$\mu = \frac{n^*}{\lambda^*} \tag{55}$$

$$\sigma^2 = \frac{n^*}{\lambda^{*2}} \tag{56}$$

3.5. Normal /Normal/Normal Distribution

If we obtain a data set of $x_1, x_2, \cdots, x_n$, we want to predict the average of the set. We only know that the data follow the normal distributions. The corresponding probability that we obtain $D = x_1, x_2, \cdots, x_n$ is given by

$$f(D) = \frac{1}{\sqrt{2\pi\sigma^2}}\exp\left[-\frac{(x_1-\mu)^2}{2\sigma^2}\right]\frac{1}{\sqrt{2\pi\sigma^2}}\exp\left[-\frac{(x_2-\mu)^2}{2\sigma^2}\right]\cdots\frac{1}{\sqrt{2\pi\sigma^2}}\exp\left[-\frac{(x_n-\mu)^2}{2\sigma^2}\right] \tag{57}$$

We assume that we know the variance σ^2, but do not know the average. We therefore replace μ as θ. Then the probability is expressed as

$$f(D) = \frac{1}{\sqrt{2\pi\sigma^2}}\exp\left[-\frac{(x_1-\theta)^2}{2\sigma^2}\right]\frac{1}{\sqrt{2\pi\sigma^2}}\exp\left[-\frac{(x_2-\theta)^2}{2\sigma^2}\right]\cdots\frac{1}{\sqrt{2\pi\sigma^2}}\exp\left[-\frac{(x_n-\theta)^2}{2\sigma^2}\right] \tag{58}$$

We then convert this probability distribution function to the one for probability to obtain the average θ under the data D, which is the likelihood function denoted as $f(D|\theta)$. Since the factor $1/\sqrt{2\pi\sigma^2}$ does not include the variable parameter θ for the likelihood function, it should be changed so that the total probability with respect to θ is one. That is, we denote the factor as K and obtain the likelihood function as

$$\begin{aligned}
f(D|\theta) &= K\exp\left[-\frac{(x_1-\theta)^2}{2\sigma^2}\right]\exp\left[-\frac{(x_2-\theta)^2}{2\sigma^2}\right]\cdots\exp\left[-\frac{(x_n-\theta)^2}{2\sigma^2}\right] \\
&= K\exp\left[-\frac{1}{2\sigma^2}\sum_{i=1}^{n}(x_i-\theta)^2\right] \\
&= K\exp\left[-\frac{1}{2\sigma^2}\left(\sum_{i=1}^{n}x_i^2-2\theta\sum_{i=1}^{n}x_i+n\theta^2\right)\right] \\
&= K\exp\left[-\frac{n}{2\sigma^2}\left(-2\theta\bar{x}+\theta^2-2\theta\bar{x}\right)\right]\exp\left[-\frac{n}{2\sigma^2}\frac{1}{n}\sum_{i=1}^{n}x_i^2\right] \\
&= K\exp\left[-\frac{n}{2\sigma^2}\left((\theta-\bar{x})^2-\bar{x}^2\right)\right]\exp\left[-\frac{n}{2\sigma^2}\frac{1}{n}\sum_{i=1}^{n}x_i^2\right] \\
&= K\exp\left[-\frac{n}{2\sigma^2}\frac{1}{n}\sum_{i=1}^{n}x_i^2\right]\exp\left[\frac{n\bar{x}^2}{2\sigma^2}\right]\exp\left[-\frac{(\theta-\bar{x})^2}{2\frac{\sigma^2}{n}}\right] \\
&= K'\exp\left[-\frac{(\theta-\bar{x})^2}{2\frac{\sigma^2}{n}}\right]
\end{aligned} \tag{59}$$

where

$$K' = K\exp\left[-\frac{n}{2\sigma^2}\frac{1}{n}\sum_{i=1}^{n}x_i^2\right]\exp\left[\frac{n\bar{x}^2}{2\sigma^2}\right] \tag{60}$$

We can decide the K' by imposing that the total integration of the probability is one, which is given by

$$\int_0^\infty f(D|\theta)d\theta = 1 \tag{61}$$

Therefore, we obtain a corresponding likelihood function as

$$f(D|\theta) = \frac{1}{\sqrt{2\pi\sigma_L^2}} \exp\left[-\frac{(\theta - \mu_L)^2}{2\sigma_L^2} \right] \tag{62}$$

where

$$\sigma_L^2 = \frac{\sigma^2}{n} \tag{63}$$

$$\mu_L = \bar{x} = \frac{1}{n} \sum_{i=1}^{n} x_i \tag{64}$$

We then assume the priori distribution of

$$\pi(\theta) = \frac{1}{\sqrt{2\pi}\sigma_0} \exp\left[-\frac{(\theta - \mu_0)^2}{2\sigma_0^2} \right] \tag{65}$$

The corresponding posterior distribution is given by

$$\begin{aligned}
\pi(\theta|D) &\propto f(D|\theta)\pi(\theta) \\
&\propto \exp\left[-\frac{(\theta - \mu_0)^2}{2\sigma_0^2} \right] \exp\left[-\frac{(\theta - \mu_L)^2}{2\sigma_L^2} \right] \\
&\propto \exp\left[-\frac{\left(\theta - \dfrac{\dfrac{\mu_0}{\sigma_0^2} + \dfrac{\mu_L}{\sigma_L^2}}{\dfrac{1}{\sigma_0^2} + \dfrac{1}{\sigma_L^2}} \right)^2}{2\left(\dfrac{1}{\dfrac{1}{\sigma_0^2} + \dfrac{1}{\sigma_L^2}} \right)} \right]
\end{aligned} \tag{66}$$

Therefore, we obtain the posterior distribution as

$$\pi(\theta|D) = \frac{1}{\sqrt{2\pi}\sigma^*} \exp\left[-\frac{(\theta - \mu^*)^2}{2\sigma^{*2}} \right] \tag{67}$$

where

$$\frac{1}{\sigma^{*2}} = \frac{1}{\sigma_0^2} + \frac{1}{\sigma_L^2} \tag{68}$$

$$\mu^* = \left(\frac{\mu_0}{\sigma_0^2} + \frac{\mu_L}{\sigma_L^2} \right) \sigma^{*2} \tag{69}$$

3.6. Inverse Gamma/Normal (Inverse Gamma)/Inverse Gamma Distribution

In the previous section, we studied how to predict the average of a set, and we assumed that we know the variance of the set. We study a case where we do not know the variance of a set and know the average μ.

If we obtain a data set of $x_1, x_2, \cdots, x_n$, we want to predict the variance of the set. We only know that the data follow the normal distributions. The corresponding probability that we obtain $D = x_1, x_2, \cdots, x_n$ is given by

$$f(D) = \frac{1}{\sqrt{2\pi\sigma^2}} \exp\left[-\frac{(x_1 - \mu)^2}{2\sigma^2} \right] \frac{1}{\sqrt{2\pi\sigma^2}} \exp\left[-\frac{(x_2 - \mu)^2}{2\sigma^2} \right] \cdots \frac{1}{\sqrt{2\pi\sigma^2}} \exp\left[-\frac{(x_n - \mu)^2}{2\sigma^2} \right] \tag{70}$$

We regard that we know the average μ, but do not know the variance σ^2. We therefore replace σ^2 as θ. Then the probability is expressed as

$$\begin{aligned}
f(D) &= \frac{1}{\sqrt{2\pi\theta}} \exp\left[-\frac{(x_1 - \mu)^2}{2\theta} \right] \frac{1}{\sqrt{2\pi\theta}} \exp\left[-\frac{(x_2 - \mu)^2}{2\theta} \right] \cdots \frac{1}{\sqrt{2\pi\theta}} \exp\left[-\frac{(x_n - \mu)^2}{2\theta} \right] \\
&= \left(\frac{1}{\sqrt{2\pi}} \right)^n \theta^{-\frac{n}{2}} \exp\left[-\frac{\sum_{i=1}^{n} (x_i - \mu)^2}{2\theta} \right] \\
&= \left(\frac{1}{\sqrt{2\pi}} \right)^n \theta^{-\frac{n}{2}} \exp\left[-\frac{\sum_{i=1}^{n} \left(x_i^2 - 2\mu x_i + \mu^2 \right)}{2\theta} \right] \\
&= \left(\frac{1}{\sqrt{2\pi}} \right)^n \theta^{-\left(\frac{n}{2}-1\right)-1} \exp\left[-\frac{n\left(\mu^2 - 2\bar{x}\mu \right) + \sum_{i=1}^{n} x_i^2}{2\theta} \right] \\
&= \left(\frac{1}{\sqrt{2\pi}} \right)^n \theta^{-\left(\frac{n}{2}-1\right)-1} \exp\left[-\frac{n\left[(\mu - \bar{x})^2 - \bar{x}^2 \right] + \sum_{i=1}^{n} x_i^2}{2} \frac{1}{\theta} \right]
\end{aligned} \tag{71}$$

where

$$\bar{x} = \frac{1}{n}\sum_{i=1}^{n} x_i \tag{72}$$

Therefore, we obtain a corresponding likelihood function as

$$f\left(D|\theta\right) = \frac{\theta^{-n_L-1}\exp\left(-\frac{\lambda_L}{\theta}\right)}{\Gamma\left(n_L\right)} \tag{73}$$

where

$$n_L = \frac{n}{2} - 1 \tag{74}$$

$$\lambda_L = \frac{1}{2}\left\{n\left[\left(\mu - \bar{x}\right)^2 - \bar{x}^2\right] + \sum_{i=1}^{n} x_i^2\right\} \tag{75}$$

This is the inverse Gamma distribution.

We assume a priori distribution for variance σ^2 given by

$$\pi\left(\theta\right) = \frac{\lambda_0^{n_0}\theta^{-n_0-1}Exp\left(-\frac{\lambda_0}{\theta}\right)}{\Gamma\left(n_0\right)} \tag{76}$$

It is not so easy to decide the priori probability parameters of n_0 and λ_0. It is convenient to decide the corresponding average and standard deviation and convert them to n_0 and λ_0.

The relationships between the average and variance and parameters are given by

$$\mu_0 = \frac{\lambda_0}{n_0 - 1} \tag{77}$$

$$\sigma_0^2 = \frac{\lambda_0{}^2}{(n_0 - 1)^2 (n_0 - 2)} \tag{78}$$

We then obtain

$$n_0 = 2 + \frac{\mu_0^2}{\sigma_0^2} \tag{79}$$

$$\lambda_0 = \mu_0 \left(1 + \frac{\mu_0^2}{\sigma_0^2} \right) \tag{80}$$

The corresponding posterior distribution is given by

$$\begin{aligned} \pi(\theta|D) &\propto f(D|\theta)\pi(\theta) \\ &\propto \frac{\lambda_L{}^{n_L} \theta^{-n_L - 1} \exp\left(-\frac{\lambda_L}{\theta} \right)}{\Gamma(n_L)} \frac{\lambda_0{}^{n_0} \theta^{-n_0 - 1} Exp\left(-\frac{\lambda_0}{\theta} \right)}{\Gamma(n_0)} \\ &\propto \theta^{-(n_0 + n_L + 1) - 1} \exp\left(-\frac{\lambda_0 + \lambda_L}{\theta} \right) \end{aligned} \tag{81}$$

Therefore, we obtain the posterior distribution as

$$\pi(\theta|D) = \frac{\theta^{-n^* - 1} \exp\left(-\frac{\lambda^*}{\theta} \right)}{\Gamma(n^*)} \tag{82}$$

where

$$n^* = n_0 + n_L + 1 \tag{83}$$

$$\lambda^* = \lambda_0 + \lambda_L \tag{84}$$

The average and variance are given by

$$\mu = \frac{\lambda^*}{n^* - 1} \tag{85}$$

$$\sigma^{*2} = \frac{\lambda^*}{\left(n^* - 1\right)^2 \left(n^* - 2\right)} \tag{86}$$

4. PROBABILITY DISTRIBUTION WITH TWO PARAMETERS

In the previous section, we show how to decide one parameter in a posterior probability distribution. In the analysis, we select a convenient distribution to obtain a rather simple posterior probability distribution with one parameter, which is called as conjugate distributions.

However, we want to treat distributions with many parameters in general. We need to obtain the procedure to obtain many parameters. We show the treatment for two parameters in this section. We will extend the treatment for many parameters of more than two in the later section.

We treat two parameters of an average θ_1 and a variance θ_2. The posterior distribution for the average is given by

$$\pi\left(\theta_1 \middle| D\right) = \frac{1}{\sqrt{2\pi}\sigma^*} \exp\left[-\frac{\left(\theta_1 - \mu^*\right)^2}{2\sigma^{*2}}\right] \tag{87}$$

where

$$\frac{1}{\sigma^{*2}} = \frac{1}{\sigma_0^2} + \frac{1}{\sigma_L^2} \tag{88}$$

$$\mu^* = \left(\frac{\mu_0}{\sigma_0^2} + \frac{\mu_L}{\sigma_L^2}\right)\sigma^{*2} \tag{89}$$

$$\sigma_L^2 = \frac{\theta_2}{n} \tag{90}$$

$$\mu_L = \bar{x} = \frac{1}{n}\sum_{i=1}^{n} x_i \tag{91}$$

The posterior distribution for the variance is given by

$$\pi\left(\theta_2 \middle| D\right) = \frac{\theta^{-n^*-1} \exp\left(-\frac{\lambda^*}{\theta_2}\right)}{\Gamma\left(n^*\right)} \tag{92}$$

where

$$n^* = n_0 + n_L + 1 \tag{93}$$

$$\lambda^* = \lambda_0 + \lambda_L \tag{94}$$

$$n_L = \frac{n}{2} - 1 \tag{95}$$

$$\lambda_L = \frac{1}{2}\left\{ n\left[\left(\theta_1 - \bar{x}\right)^2 - \bar{x}^2\right] + \sum_{i=1}^{n} x_i^{\,2} \right\} \tag{96}$$

The corresponding schematic distribution is shown in Figure 4.

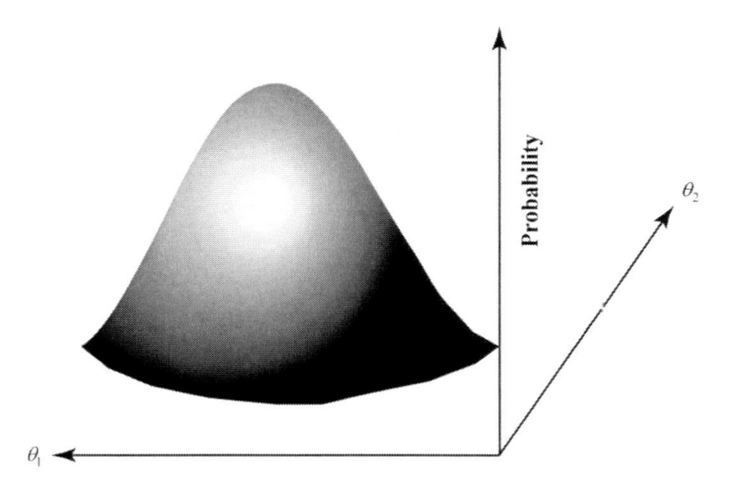

Figure 4. Probability distribution with two parameters of θ_1 and θ_2.

We need to decide the parameters of θ_1 and θ_2 where they are constant numbers in one equation although they are variables in the other equations.

We show various methods to decide them in the following sections.

To perform concrete calculation, we use a data set as shown in Table 3. The data number is 30, and the sample average $\bar{x}$ and unbiased variance s^2 are given by

$$\begin{aligned}\bar{x} &= \sum_{i=1}^{30} x_i \\ &= 5.11\end{aligned} \tag{97}$$

$$\begin{aligned}s^2 &= \frac{1}{30-1}\sum_{i=1}^{30}\left(x_i - \bar{x}\right)^2 \\ &= 4.77\end{aligned} \tag{98}$$

where x_i is data value in the table.

Table 3. Data for evaluating two macro parameters of a set

Data ID	Data	Data ID	Data	Data ID	Data
1	6.0	11	3.3	21	3.4
2	10.0	12	7.6	22	3.8
3	7.6	13	5.8	23	3.3
4	3.5	14	6.7	24	5.7
5	1.4	15	2.8	25	6.3
6	2.5	16	4.8	26	8.4
7	5.6	17	6.3	27	4.6
8	3.0	18	5.3	28	2.8
9	2.2	19	5.4	29	7.9
10	5.0	20	3.3	30	8.9

Average	5.11
Variance	4.77

4.1. MCMC Method (A Gibbs Sampling Method)

A Gibbs sampling method gives us the parameter values with numerical iterative method by generating a random number related to the posterior probability distribution.

We write again the posterior distribution for average given by

$$\pi\left(\theta_1 \middle| D, \theta_2^{(0)}\right) = \frac{1}{\sqrt{2\pi}\sigma^*} \exp\left[-\frac{\left(\theta_1 - \mu^*\right)^2}{2\sigma^{*2}}\right] \tag{99}$$

where

$$\frac{1}{\sigma^{*2}} = \frac{1}{\sigma_0^2} + \frac{1}{\sigma_L^2} \tag{100}$$

$$\mu^* = \left(\frac{\mu_0}{\sigma_0^2} + \frac{\mu_L}{\sigma_L^2}\right)\sigma^{*2} \tag{101}$$

$$\sigma_L^2 = \frac{\theta_2}{n} \tag{102}$$

$$\mu_L = \bar{x} = \frac{1}{n}\sum_{i=1}^{n} x_i \tag{103}$$

We should set the parameters as below.

$$\begin{aligned} \mu_0 &= \bar{x} \\ &= 5.11 \end{aligned} \tag{104}$$

$$\begin{aligned} \sigma_0{}^2 &= \frac{s^2}{n} \\ &= \frac{4.77}{30} \\ &= 0.16 \end{aligned} \tag{105}$$

$$\begin{aligned} \mu_L &= \bar{x} \\ &= 5.11 \end{aligned} \tag{106}$$

We should also set θ_2 for evaluating θ_1. We set its initial value as

$$\theta_2^{(0)} = s^2$$
$$= 4.77 \tag{107}$$

Using the distribution function, we generate a random number for θ_1 which is denoted as $\theta_1^{(0)}$ and is expressed by

$$\theta_1^{(1)} = rand\left[Inv\_\pi\left(\theta_1 \middle| D, \theta_2^{(0)}\right)\right]$$
$$= 5.00 \tag{108}$$

Eq.(108) expresses the random number based on the probability function of $\pi\left(\theta_1 \middle| D, \theta_2^{(0)}\right)$. The procedure to obtain the random number based on a function is described in Chapter 14 of volume 3.

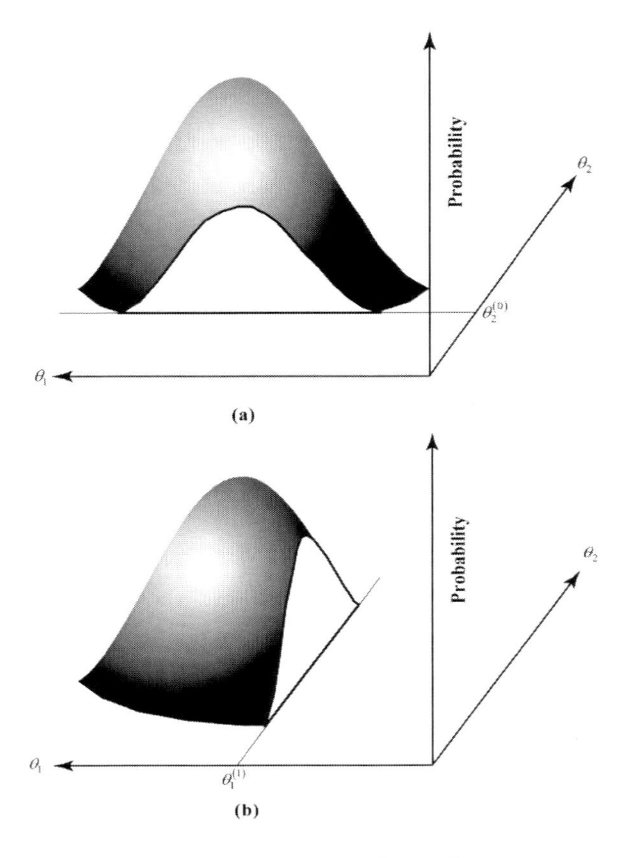

Figure 5. Probability distribution with two parameters of θ_1 and θ_2. (a) θ_2 is set and θ_1 is generated. (b) θ_1 is set and θ_2 is generated.

We can generate the random number associated with a certain probability distribution $P(\theta)$ using a uniform random number as follows. We integrate $P(\theta)$, and form a function $F(\theta)$ as

$$F(\theta)=\int_0^{\theta} P(\theta)d\theta \tag{109}$$

We generate the uniform random number, and obtain a corresponding θ, which is shown in Figure 6.

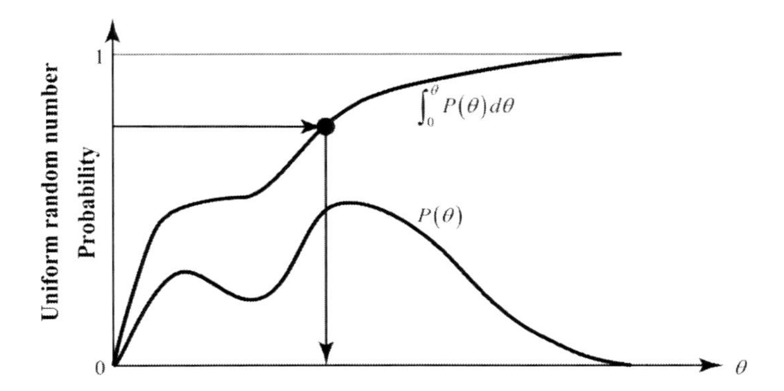

Figure 6. Random number generation related to the probability distribution of $P(\theta)$ using inverse function method.

The posterior probability distribution for the variance is then given by

$$\pi\left(\theta_2 \middle| D, \theta_1^{(1)}\right)=\frac{\theta_2^{-n^*-1}\exp\left(-\frac{\lambda^*}{\theta_2}\right)}{\Gamma\left(n^*\right)} \tag{110}$$

where

$$n^* = n_0 + n_L + 1 \tag{111}$$

$$\lambda^* = \lambda_0 + \lambda_L \tag{112}$$

$$n_L = \frac{n}{2} - 1 \tag{113}$$

$$\lambda_L = \frac{1}{2}\left\{ n\left[\left(\theta_1 - \bar{x}\right)^2 - \bar{x}^2 \right] + \sum_{i=1}^{n} x_i^2 \right\} \tag{114}$$

We should set n_0 and λ_0. We assume that an average of the variance μ_0 as

$$\mu_0 = s^2 \tag{115}$$

and a variance of the variance as

$$\sigma_0^2 = s^4 \tag{116}$$

Therefore, we obtain

$$\begin{aligned} n_0 &= 2 + \frac{\mu_0^2}{\sigma_0^2} \\ &= 3 \end{aligned} \tag{117}$$

$$\begin{aligned} \lambda_0 &= \mu_0 \left(1 + \frac{\mu_0^2}{\sigma_0^2} \right) \\ &= 4.77 \times 2 \\ &= 9.54 \end{aligned} \tag{118}$$

We can further obtain

$$\begin{aligned} n_L &= \frac{n}{2} - 1 \\ &= \frac{30}{2} - 1 \\ &= 14 \end{aligned} \tag{119}$$

$$\begin{aligned} n^* &= n_0 + n_L + 1 \\ &= 3 + 14 + 1 \\ &= 18 \end{aligned} \tag{120}$$

$$\sum_{i=1}^{n} x_i^{\ 2} = 920.6 \tag{121}$$

$$\begin{aligned}\lambda_L &= \frac{1}{2}\left\{ n\left[(\theta_1 - \bar{x})^2 - \bar{x}^2 \right] + \sum_{i=1}^{n} x_i^{\ 2} \right\} \\ &\approx \frac{1}{2}\left\{ n\left[\left(\theta_1^{(1)} - \bar{x}\right)^2 - \bar{x}^2 \right] + \sum_{i=1}^{n} x_i^{\ 2} \right\} \\ &= \frac{1}{2}\left\{ 30\left[(5.07 - 5.11)^2 - 5.11^2 \right] + 920.6 \right\} \\ &= 457.769\end{aligned} \tag{122}$$

Using the distribution function, we generate a random number for θ_2 based on the distribution function and denote it as $\theta_2^{(1)}$ which is given by

$$\begin{aligned}\theta_2^{(1)} &= rand\left[Inv\_\pi\left(\theta_2 \middle| D, \theta_1^{(1)}\right) \right] \\ &= 4.19\end{aligned} \tag{123}$$

We then move to the updated $\pi\left(\theta_1 \middle| D, \theta_2^{(1)}\right)$ where $\theta_2^{(1)}$ is used. We repeat the above process many times.

In the initial stage of the cycle, the results are influenced by the initial values. Therefore; we ignore the first 1000 data for example. The other data express the distribution for μ and σ^2.

We then perform the Gibbs sampling 20000 cycle times, and delete the first 1000 cycle times data, and obtain the results as shown in Figure 7. The corresponding average and variance values are given by

$$\begin{cases} \mu_{\theta_1} = 5.11 \\ \mu_{\theta_2} = 4.61 \end{cases} \tag{124}$$

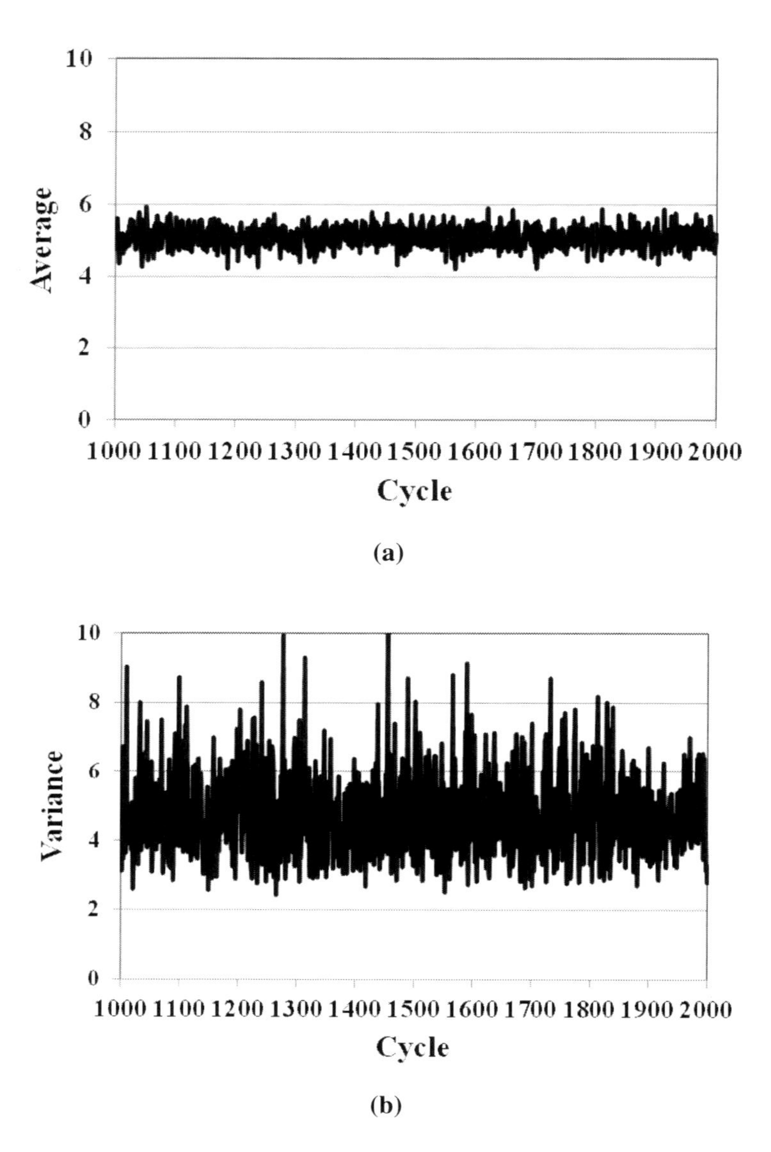

Figure 7. Gibbs sampling. (a) Average, (b) Variance.

4.2. MCMC Method (A MH Method)

In the previous section, we obtained the parameters of posterior probability distributions by generating random numbers related to posterior distributions. However, the posterior distribution changes every time, and we should evaluate functions to generate the random number related to the posterior probability distribution. A Metropolis-Hastings (MH) algorithm method gives us almost the same results without using the generation of random numbers related to the posterior probability distribution.

We utilize detailed balance conditions as follows.

We define a probability distribution $P(x_i)$ where we have a status of x_i. We also define a transition function $W(x_i \rightarrow x_{i+1})$, that expresses the transition probability where the status changes form x_i to x_{i+1}. The detailed balance condition requires the followings.

$$P(x_i)W(x_i \rightarrow x_{i+1}) = P(x_{i+1})W(x_{i+1} \rightarrow x_i) \tag{125}$$

If the probability $P(x_i)$ is larger than $P(x_{i+1})$, the transition from the status x_i to the status x_{i+1} is small and vice versa. In the MH algorithm, the above process is expressed simply. The transition occurs significantly when $P(x_i)$ is smaller than $P(x_{i+1})$, and the transition occurs little when $P(x_i)$ is larger than $P(x_{i+1})$, which is shown below in more detail.

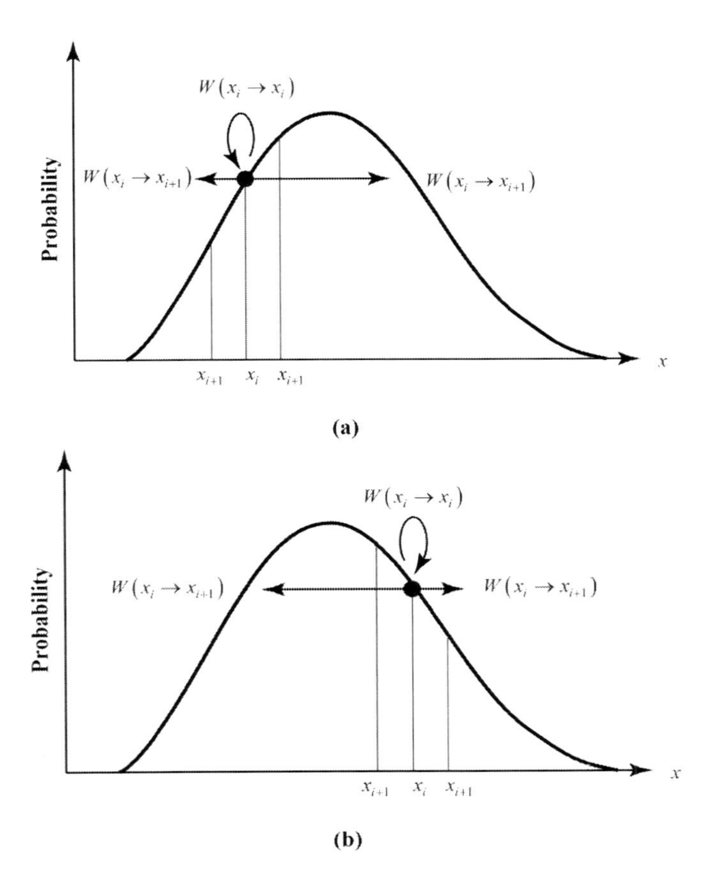

Figure 8. Transition probability W. The length qualitatively expresses the strength of the transition. (a) The probability gradient is positive. (b) The probability gradient is negative.

Figure 8 shows the schematic transition probability. Two cases are shown, one is the probability gradient is positive (a), and negative (b). We select a next step using random number, and hence x_{i+1} have a chance of both sides of the location of x_i. We decide whether the transition occurs or not.

We define the ratio r given by

$$r = \frac{P(x_{i+1})}{P(x_i)} \tag{126}$$

When r is larger than 1, we always change the status from x_i to x_{i+1}.

When r is smaller than 1, we generate a random number q which is uniform one with the range of $[0,1]$. We then compare r with q.

When r is larger than q, we change the status from x_i to x_{i+1}.

When r is smaller than q, we do not change the status.

The above process ensures that we stay long in the status where the corresponding probability is high and vice versa.

We utilize the above process in the MH algorithm shown below.

We use the same posterior probability distribution given by

$$\pi\left(\theta_1^{(0)} \middle| D, \theta_2^{(0)}\right) = \frac{1}{\sqrt{2\pi}\sigma^*} \exp\left[-\frac{\left(\theta_1^{(0)} - \mu^*\right)^2}{2\sigma^{*2}}\right] \tag{127}$$

where we set the initial values for $\theta_1^{(0)}$ and $\theta_2^{(0)}$ as

$$\begin{aligned} \theta_1^{(0)} &= \bar{x} \\ &= 5.11 \end{aligned} \tag{128}$$

$$\begin{aligned} \theta_2^{(0)} &= s^2 \\ &= 4.77 \end{aligned} \tag{129}$$

The related parameters are given by

$$\frac{1}{\sigma^{*2}} = \frac{1}{\sigma_0^2} + \frac{1}{\sigma_L^2} \tag{130}$$

$$\mu^* = \left(\frac{\mu_0}{\sigma_0^2} + \frac{\mu_L}{\sigma_L^2} \right) \sigma^{*2} \tag{131}$$

$$\begin{aligned} \mu_0 &= \bar{x} \\ &= 5.11 \end{aligned} \tag{132}$$

$$\begin{aligned} \sigma_0{}^2 &= \frac{s^2}{n} \\ &= \frac{4.77}{30} \\ &= 0.16 \end{aligned} \tag{133}$$

$$\begin{aligned} \mu_L &= \bar{x} \\ &= 5.11 \end{aligned} \tag{134}$$

$$\begin{aligned} \sigma_L^2 &= \frac{\theta_2^{(0)}}{n} \\ &= \frac{\theta_2^{(0)}}{30} \end{aligned} \tag{135}$$

Using the parameters above, we can evaluate the posterior probability associated with the average of Eq. (127).

The posterior distribution associated with the variance is given by

$$\pi\left(\theta_2^{(0)} \middle| D, \theta_1^{(0)}\right) = \frac{\left(\theta_2^{(0)}\right)^{-n^*-1} \exp\left(-\frac{\lambda^*}{\theta_2^{(0)}}\right)}{\Gamma\left(n^*\right)} \tag{136}$$

where

$$n^* = n_0 + n_L + 1 \tag{137}$$

$$\lambda^* = \lambda_0 + \lambda_L \tag{138}$$

$$n_L = \frac{n}{2} - 1 \tag{139}$$

$$\lambda_L = \frac{1}{2}\left\{ n\left[\left(\theta_1^{(0)} - \bar{x} \right)^2 - \bar{x}^2 \right] + \sum_{i=1}^{n} x_i^{\ 2} \right\} \tag{140}$$

We should set n_0 and λ_0. We assume that an average of the variance μ_0 as

$$\mu_0 = s^2 \tag{141}$$

and a variance of the variance as

$$\sigma_0^2 = s^4 \tag{142}$$

Therefore, we obtain

$$\begin{aligned} n_0 &= 2 + \frac{\mu_0^2}{\sigma_0^2} \\ &= 3 \end{aligned} \tag{143}$$

$$\begin{aligned} \lambda_0 &= \mu_0 \left(1 + \frac{\mu_0^2}{\sigma_0^2} \right) \\ &= 4.77 \times 2 \\ &= 9.54 \end{aligned} \tag{144}$$

We can further obtain

$$\begin{aligned} n_L &= \frac{n}{2} - 1 \\ &= \frac{30}{2} - 1 \\ &= 14 \end{aligned} \tag{145}$$

$$n^* = n_0 + n_L + 1 \\ = 3 + 14 + 1 \\ = 18 \tag{146}$$

$$\sum_{i=1}^{n} x_i^2 = 920.6 \tag{147}$$

$$\lambda_L = \frac{1}{2}\left\{ n\left[\left(\theta_1^{(0)} - \bar{x} \right)^2 - \bar{x}^2 \right] + \sum_{i=1}^{n} x_i^2 \right\} \\ = \frac{1}{2}\left\{ 30\left[\left(\theta_1^{(0)} - 5.11 \right)^2 - 5.11^2 \right] + 920.6 \right\} \tag{148}$$

Using the parameters above, we can evaluate the distribution function associated with the variance.

We then generate increments of the parameters using random numbers related to normal distributions of $N\left(0, \varepsilon_{\theta_1}\right)$, and $N\left(0, \varepsilon_{\sigma}\right)$. ε_{θ_1} and ε_{σ} are variances for the average and variance, respectively. We set the steps as

$$\varepsilon_{\theta_1} = \frac{\theta_1^{(0)}}{\eta_{\theta_1}} \tag{149}$$

$$\varepsilon_{\theta_2} = \frac{\theta_2^{(0)}}{\eta_{\theta_2}} \tag{150}$$

η_{θ_1} and η_{σ} are constant numbers and we use 10 for this case. We then generate the tentative next values as

$$\theta_{1t} = \theta_1^{(0)} + rand\left[Inv\_N\left(0, \varepsilon_{\theta_1}\right) \right] \tag{151}$$

$$\theta_{2t} = \theta_2^{(0)} + rand\left[Inv\_N\left(0, \varepsilon_{\theta_2}\right) \right] \tag{152}$$

We then evaluate the two ratios given by

$$p_{\theta_1} = \frac{\pi\left(\theta_{1t} \middle| D, \theta_2^{(0)}\right)}{\pi\left(\theta_1^{(0)} \middle| D, \theta_2^{(0)}\right)} \tag{153}$$

$$p_{\theta_2} = \frac{\pi\left(\theta_{2t} \middle| D, \theta_1^{(0)}\right)}{\pi\left(\theta_2^{(0)} \middle| D, \theta_1^{(0)}\right)} \tag{154}$$

If p_{θ_1} and p_{θ_2} are larger than 1, we set

$$\theta_1^{(1)} = \theta_{1t} \tag{155}$$

$$\theta_2^{(1)} = \theta_{2t} \tag{156}$$

when p_{θ_1} or p_{θ_2} is smaller than 1, we generate a uniform random number denoted as q, that is,

$$q = rand(\ ) \tag{157}$$

$rand(\ )$ means that it is a uniform random number between 0 and 1. We then compare q with p_{θ_1} and p_{θ_2}, and decide next step parameters as follows.

$$\begin{cases} \theta_1^{(1)} = \theta_{1t} & for\ p_{\theta_1} \geq q \\ \theta_1^{(1)} = \theta_1^{(0)} & for\ p_{\theta_1} < q \end{cases} \tag{158}$$

$$\begin{cases} \left.\begin{array}{l} \sigma^{(1)} = \sigma_t \\ \theta_2^{(1)} = \theta_{2t} \end{array}\right. & for\ p_{\theta_2} \geq q \\ \left.\begin{array}{l} \sigma^{(1)} = \sigma^{(0)} \\ \theta_2^{(1)} = \theta_2^{(0)} \end{array}\right. & for\ p_{\theta_2} < q \end{cases} \tag{159}$$

We repeat the process many times. Figure 9 shows the schematic parameter generation in the MH algorithm method.

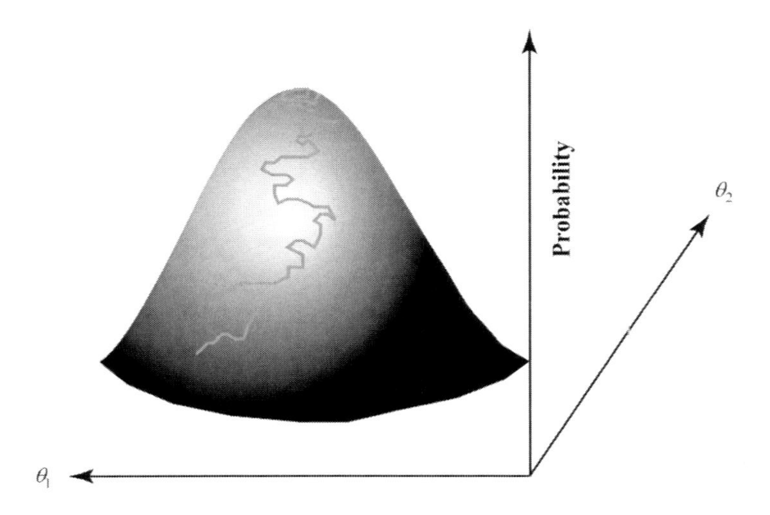

Figure 9. Parameter generation for a MH algorithm method.

In the initial stage of the cycle, the results are influenced by the initial values for μ and σ^2. Therefore, we ignore the first 1000 data for example. The other data express the distribution for μ and σ^2.

The corresponding average and variance values are given by

$$\begin{cases} \mu_{\theta_1} = 5.19 \\ \mu_{\theta_2} = 4.76 \end{cases} \tag{160}$$

Figure 11 shows the algorithm flow for the MH method for one parameter.

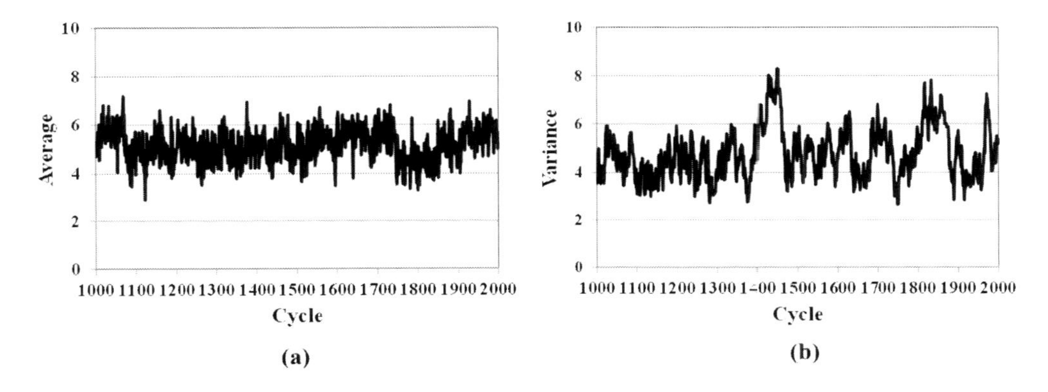

Figure 10. A Metropolis-Hastings (MH) algorithm method. (a) Average, (b) Variance.

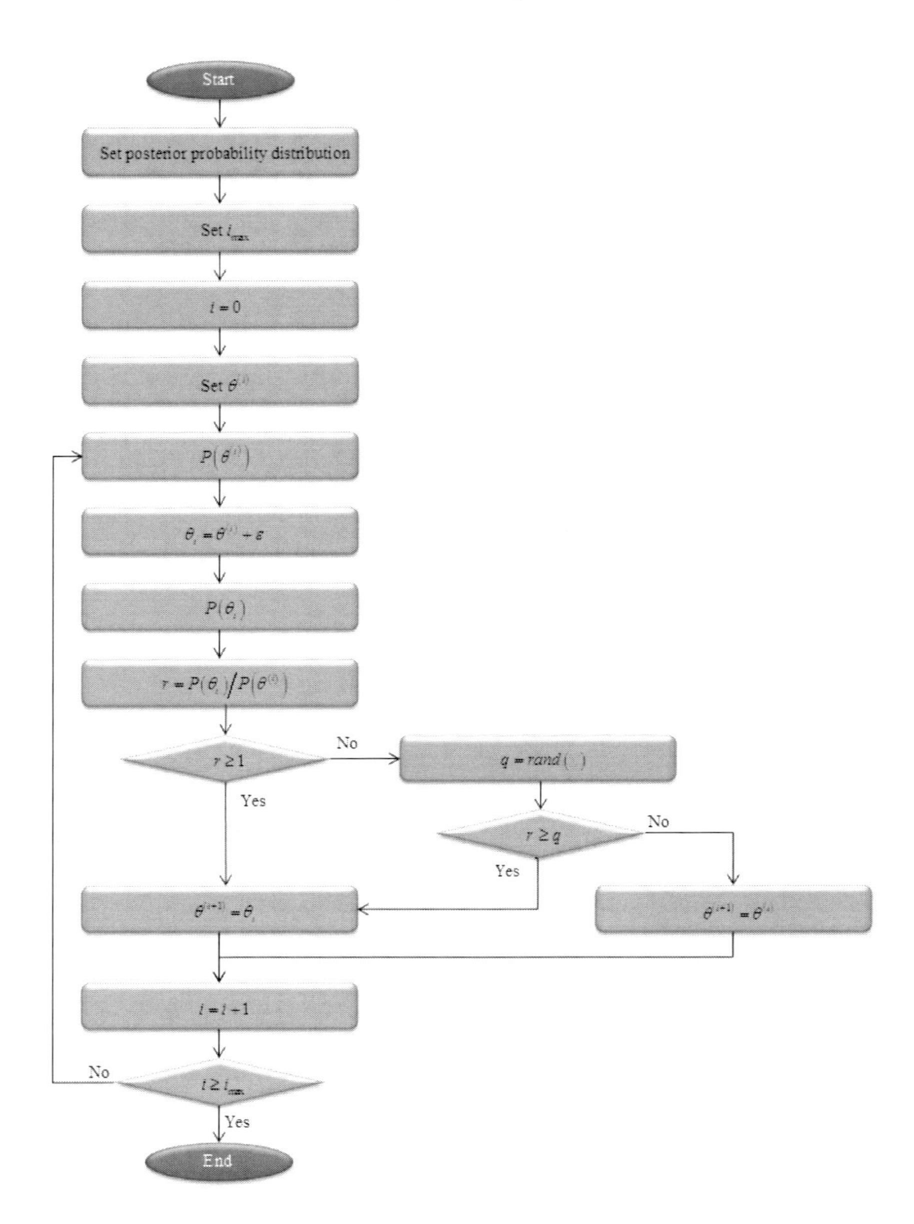

Figure 11. Metropolis-Hastings algorithm flow.

4.3. Posterior Probability Distribution with Many Parameters

We can easily extend the above procedures described in the previous section to many parameters of more than two. The l parameters are denoted as $\theta_1, \theta_2, \cdots, \theta_l$. We need to know the posterior probability distribution for each parameter where the other parameters are included as constant numbers. The posterior distributions are then expressed as

$$\begin{aligned}
&\pi\left(\theta_1 \middle| D, \theta_2, \theta_3, \cdots, \theta_l\right) \\
&\pi\left(\theta_2 \middle| D, \theta_1, \theta_3, \cdots, \theta_l\right) \\
&\cdots \\
&\pi\left(\theta_i \middle| D, \theta_1, \theta_2, \cdots, \theta_{i-1}, \theta_{i+1}, \cdots, \theta_l\right) \\
&\cdots \\
&\pi\left(\theta_l \middle| D, \theta_1, \theta_2, \cdots, \theta_{l-1}\right)
\end{aligned} \tag{161}$$

In the Gibbs sampling method, we use the same posterior probability distributions, and also set initial value for $\theta_2, \theta_3, \cdots, \theta_l$ as $\theta_2^{(0)}, \theta_3^{(0)}, \cdots, \theta_l^{(0)}$. We then generate the random number for the parameter values as

$$\begin{cases}
\theta_1^{(1)} = rand\left[Inv_\pi\left(\theta_1 \middle| D, \theta_2^{(0)}, \theta_3^{(0)}, \cdots, \theta_l^{(0)}\right)\right] \\
\theta_2^{(1)} = rand\left[Inv_\pi\left(\theta_2 \middle| D, \theta_1^{(1)}, \theta_3^{(0)}, \cdots, \theta_l^{(0)}\right)\right] \\
\cdots \\
\theta_i^{(1)} = rand\left[Inv_\pi\left(\theta_i \middle| D, \theta_1^{(1)}, \theta_2^{(1)}, \cdots, \theta_{i-1}^{(1)}, \theta_{i+1}^{(0)}, \cdots, \theta_l^{(0)}\right)\right] \\
\cdots \\
\theta_l^{(1)} = rand\left[Inv_\pi\left(\theta_i \middle| D, \theta_1^{(1)}, \theta_2^{(1)}, \cdots, \theta_{l-1}^{(1)}\right)\right]
\end{cases} \tag{162}$$

where $rand\left[Inv\_\pi(\theta)\right]$ means the generating random number for θ based on the function $\pi(\theta)$.

We then repeat the process.

In the MH MC method, we use the same posterior probability distributions, and also set initial value for $\theta_1, \theta_2, \theta_3, \cdots, \theta_l$ as $\theta_1^{(0)}, \theta_2^{(0)}, \theta_3^{(0)}, \cdots, \theta_l^{(0)}$. We should also set the step for each parameters as $\varepsilon_{\theta_1}, \varepsilon_{\theta_2}, \cdots, \varepsilon_{\theta_l}$.

We generate tentative parameters as

$$\theta_{1t} = \theta_1^{(0)} + rand\left[Inv\_N\left(0, \varepsilon_{\theta_1}\right)\right] \tag{163}$$

$$\theta_{2t} = \theta_2^{(0)} + rand\left[Inv\_N\left(0, \varepsilon_{\theta_2}\right)\right] \tag{164}$$

$$\theta_{lt} = \theta_l^{(0)} + rand\left[Inv\_N\left(0, \varepsilon_{\theta_l}\right)\right] \tag{165}$$

We then evaluate the ratios given by

$$p_{\theta_1} = \frac{\pi\left(\theta_{1t} \middle| D, \theta_2^{(0)}, \theta_3^{(0)}, \cdots, \theta_l^{(0)}\right)}{\pi\left(\theta_1^{(0)} \middle| D, \theta_2^{(0)}, \theta_3^{(0)}, \cdots, \theta_l^{(0)}\right)} \tag{166}$$

$$p_{\theta_2} = \frac{\pi\left(\theta_{2t} \middle| D, \theta_1^{(0)}, \theta_3^{(0)}, \cdots, \theta_l^{(0)}\right)}{\pi\left(\theta_2^{(0)} \middle| D, \theta_1^{(0)}, \theta_3^{(0)}, \cdots, \theta_l^{(0)}\right)} \tag{167}$$

$$p_{\theta_l} = \frac{\pi\left(\theta_{lt} \middle| D, \theta_1^{(0)}, \theta_2^{(0)}, \cdots, \theta_{l-1}^{(0)}\right)}{\pi\left(\theta_l^{(0)} \middle| D, \theta_1^{(0)}, \theta_2^{(0)}, \cdots, \theta_{l-1}^{(0)}\right)} \tag{168}$$

We further generate a random number $q^{(1)}$ as

$$q^{(1)} = rand(\ ) \tag{169}$$

which is a uniform random number between 0 and 1.

We decide the updated parameters as below.

We consider i-th parameter in general. If the i-th parameter is larger than 1, we update the value as

$$\theta_i^{(1)} = \theta_{it} \tag{170}$$

When p_{θ_i} is smaller than 1, we then compare it with $q^{(1)}$, and set θ_{it} when it is larger than $q^{(1)}$, and set $\theta_i^{(0)}$ when it is smaller than $q^{(1)}$. That is, it is summarized as

$$\theta_i^{(1)} = \begin{cases} \theta_{it} & for\ p_{\theta_i} \geq 1\ or\ p_{\theta_i} \geq q^{(1)} \\ \theta_i^{(0)} & for\ p_{\theta_i} < q^{(1)} \end{cases} \tag{171}$$

We repeat the above process and obtain $\theta_1, \theta_2, \theta_3, \cdots, \theta_l$.

In the MH MC method, we need not posterior probability distributions of each parameter, but it is sufficient that we have one posterior probability distribution that includes whole parameters of $\theta_1, \theta_2, \theta_3, \cdots, \theta_l$ as

$$g = \left(D; \theta_1, \theta_2, \theta_3, \cdots, \theta_l\right) \tag{172}$$

We can then regard the posterior probability distribution for each parameter as

$$\begin{aligned}
&\pi\left(\theta_1 \middle| D, \theta_2, \theta_3, \cdots, \theta_l\right) = K_1 g\left(D; \theta_1, \theta_2, \theta_3, \cdots, \theta_l\right) \\
&\pi\left(\theta_2 \middle| D, \theta_1, \theta_3, \cdots, \theta_l\right) = K_2 g\left(D; \theta_1, \theta_2, \theta_3, \cdots, \theta_l\right) \\
&\cdots \\
&\pi\left(\theta_i \middle| D, \theta_1, \theta_2, \cdots, \theta_{i-1}, \theta_{i+1}, \cdots, \theta_l\right) = K_i g\left(D; \theta_1, \theta_2, \theta_3, \cdots, \theta_l\right) \\
&\cdots \\
&\pi\left(\theta_l \middle| D, \theta_1, \theta_2, \cdots, \theta_{l-1}\right) = K_l g\left(D; \theta_1, \theta_2, \theta_3, \cdots, \theta_l\right)
\end{aligned} \tag{173}$$

where the factor K_i is the normalized factor which is decided as we regard the parameter θ_i as a variable while the others as constant numbers. However, we do not care about the value of it when we evaluate the ratio. We then apply the same procedures.

4.4. Hierarchical Bayesian Theory

The shape of the posterior distribution is limited to the function we use for likelihood and priori distributions. We can improve the flexibility for the shape using a hierarchical Bayesian theory.

We consider a case where 20 members perform an examination as shown in Figure 12. The examination consists of 10 problems, and the members select an answer in each problem. The score of each member is the number of the correct answer, which is denoted as x_i for a member i. Therefore, the value of x_i is then between 0 and 10.

Figure 12 shows the assumed obtained data. The average μ_0 and the standard deviation σ_0 of the examination are given by

$$\mu_0 = 5.85 \tag{174}$$

$$\sigma_0 = 3.81 \tag{175}$$

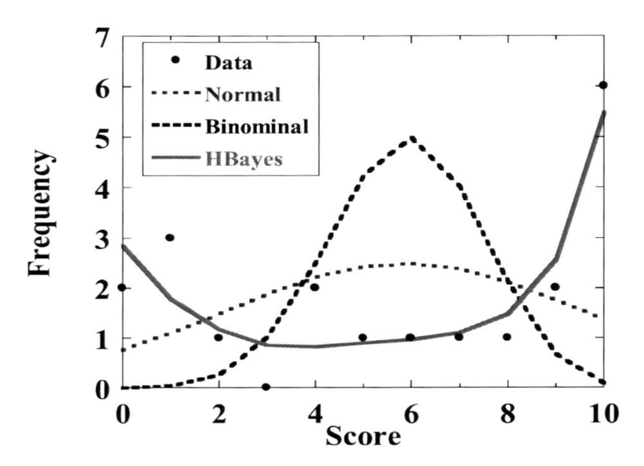

Figure 12. Score distribution and expected value.

If we assume that the member number of the scores follows the normal distribution, and the member number of score k is given by

$$N(k) \propto \frac{1}{\sqrt{2\pi}\sigma_0} \exp\left(-\frac{(k-\mu_0)^2}{2\sigma_0{}^2} \right) \tag{176}$$

This can be modified as

$$N(k) = 20 \times \frac{\exp\left(-\frac{(k-\mu_0)^2}{2\sigma_0{}^2} \right)}{\sum_{i=0}^{10} \exp\left(-\frac{(i-\mu_0)^2}{2\sigma_0{}^2} \right)} \tag{177}$$

The score distribution is far from the bell shape ones, which cannot be reproduced with the normal distribution as shown in Figure 12.

We try to explain the distribution with a hierarchical Bayes theory using a binominal distribution.

The probability that a member obtain a correct answer for each problem is assumed to be θ, and we have a corresponding the likelihood function given by

$$\begin{aligned} f(D|\theta) &\propto {}_nC_{x_1}\theta^{x_1}(1-\theta)^{n-x_1}\ {}_nC_{x_2}\theta^{x_2}(1-\theta)^{n-x_2}\cdots{}_nC_{x_{20}}\theta^{x_{20}}(1-\theta)^{n-x_{20}} \\ &\propto \theta^X(1-\theta)^{10\times 20-X} \end{aligned} \tag{178}$$

where

$$\begin{aligned} X &= \sum_{i=1}^{20} x_i \\ &= 117 \end{aligned} \tag{179}$$

The peak value of the likelihood function can be evaluated from

$$\begin{aligned} \frac{\partial f(D|\theta)}{\partial\theta} &\propto \frac{\partial\left[\theta^X(1-\theta)^{200-X}\right]}{\partial\theta} \\ &= \theta^{X-1}(1-\theta)^{200-X-1}\left[X(1-\theta)-(200-X)\theta\right] \\ &= 0 \end{aligned} \tag{180}$$

We therefore obtain the peak value of the likelihood function at

$$\begin{aligned} \theta_0 &= \frac{X}{200} \\ &= 0.585 \end{aligned} \tag{181}$$

The expected number for score k is then given by

$$N(k) = 20\times {}_{10}C_k\theta_0^k(1-\theta_0)^{10-k} \tag{182}$$

Figure 12 compares the data with the evaluated values. The expected values deviate significantly. It is obvious that we apply the same probability θ_0 to any member. This treatment is valid if it is such kind of coin toss with the similar coin. This treatment cannot explain the data such as the figure.

We assume that the peculiar shape of the distribution of Figure 12 comes from the various skill of the person. How can we include the different skills?

We regard that each member comes from his own group with different macro parameters of θ, which is shown schematically in Figure 13.

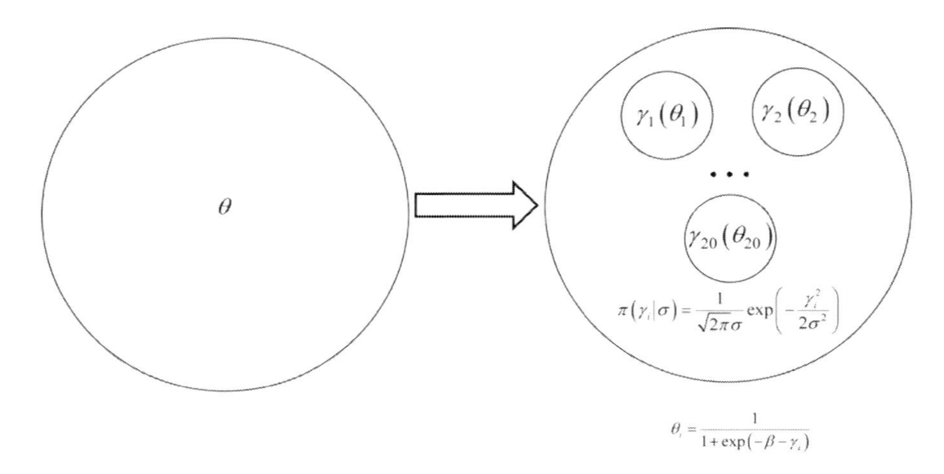

Figure 13. Schematic treatment for a hierarchical Bayes theory. We regard that each data comes from different group and each group is controlled by the priori distribution.

We can add the variety of the shape of the distribution by changing θ in each trial. Therefore, we modify the likelihood function as

$$
\begin{aligned}
& f\left(D|\theta_1,\theta_2,\cdots,\theta_{20}\right) \\
& \propto {}_nC_{x_1}\theta_1^{x_1}\left(1-\theta_1\right)^{n-x_1}\ {}_nC_{x_2}\theta_2^{x_2}\left(1-\theta_2\right)^{n-x_2}\cdots{}_nC_{x_{20}}\theta_5^{x_{20}}\left(1-\theta_{20}\right)^{n-x_{20}} \\
& \propto \theta_1^{x_1}\left(1-\theta_1\right)^{n-x_1}\ \theta_2^{x_2}\left(1-\theta_2\right)^{n-x_2}\cdots\theta_5^{x_{20}}\left(1-\theta_{20}\right)^{n-x_{20}}
\end{aligned}
\tag{183}
$$

It should be noted that we now have undecided parameter number of 20 from 1. We can do anything with this method. However, we add some constraint to the functions. The distribution of θ_i is controlled by the priori distribution.

A parameter θ_i defined in the range of 0 and 1, and it is frequently converted as

$$
\ln\left(\frac{\theta_i}{1-\theta_i}\right) = \beta + \gamma_i \tag{184}
$$

We then have

$$\theta_i = \frac{1}{1+\exp(-\beta-\gamma_i)} \tag{185}$$

Figure 14 shows the relationship between γ and θ with various β. θ increases monotonically with increasing γ and decreases with increasing β. Note that the definition range for γ is $[-\infty, \infty]$, which is available for a normal distribution.

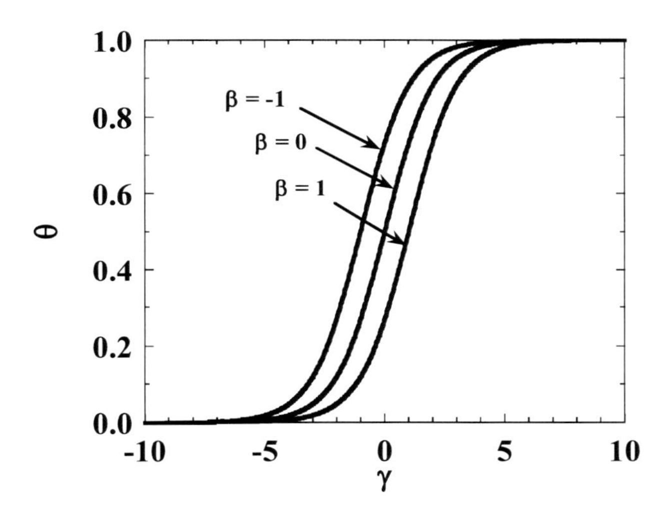

Figure 14. Relationship between γ and θ with various β.

We assume the normal distribution for γ_i as

$$\pi(\gamma_i|\sigma) = \frac{1}{\sqrt{2\pi}\sigma}\exp\left(-\frac{\gamma_i^2}{2\sigma^2}\right) \tag{186}$$

where σ is also an undecided parameter. We introduced a parameter β and it is assumed to be constant although the value is not decided at present. Therefore, we do not define probability distribution for β.

Therefore, we assume the priori distribution for γ_i as

$$\frac{1}{\sqrt{2\pi}\sigma}\exp\left(-\frac{\gamma_1^2}{2\sigma^2}\right)\frac{1}{\sqrt{2\pi}\sigma}\exp\left(-\frac{\gamma_2^2}{2\sigma^2}\right)\cdots\frac{1}{\sqrt{2\pi}\sigma}\exp\left(-\frac{\gamma_{20}^2}{2\sigma^2}\right) \tag{187}$$

The posterior distribution is given by

$$
\begin{aligned}
&{}_{10}C_{x_1}\theta_1^{x_1}\left(1-\theta_1\right)^{n-x_1}\frac{1}{\sqrt{2\pi}\sigma}\exp\left(-\frac{\gamma_1^2}{2\sigma^2}\right)\\
&\times{}_{10}C_{x_2}\theta_2^{x_2}\left(1-\theta_2\right)^{n-x_2}\frac{1}{\sqrt{2\pi}\sigma}\exp\left(-\frac{\gamma_2^2}{2\sigma^2}\right)\\
&\cdots\\
&\times{}_{10}C_{x_{20}}\theta_{20}^{x_{20}}\left(1-\theta_{20}\right)^{n-x_{20}}\frac{1}{\sqrt{2\pi}\sigma}\exp\left(-\frac{\gamma_{20}^2}{2\sigma^2}\right)
\end{aligned}
\tag{188}
$$

The probability associated with the score k is evaluated as

$$
\begin{aligned}
P(k) &= {}_{10}C_k\theta^k\left(1-\theta\right)^{n-k}\frac{1}{\sqrt{2\pi}\sigma}\exp\left(-\frac{\gamma^2}{2\sigma^2}\right)\\
&= {}_{10}C_k\left(\frac{1}{1+e^{-\beta-\gamma}}\right)^k\left(1-\frac{1}{1+e^{-\beta-\gamma}}\right)^{10-k}\frac{1}{\sqrt{2\pi}\sigma}\exp\left(-\frac{\gamma^2}{2\sigma^2}\right)
\end{aligned}
\tag{189}
$$

Figure 15 (a) shows the dependence of $P(k)$ on γ with two standard deviations of σ. The peak position of γ increases with increasing k, and the profile becomes broad with increasing σ.

Figure 15 (b) shows the dependence of $P(k)$ on γ with various β. The profile is symmetrical with respect to k with $\beta = 0$, and the probability with higher k is larger with positive β, and vice versa.

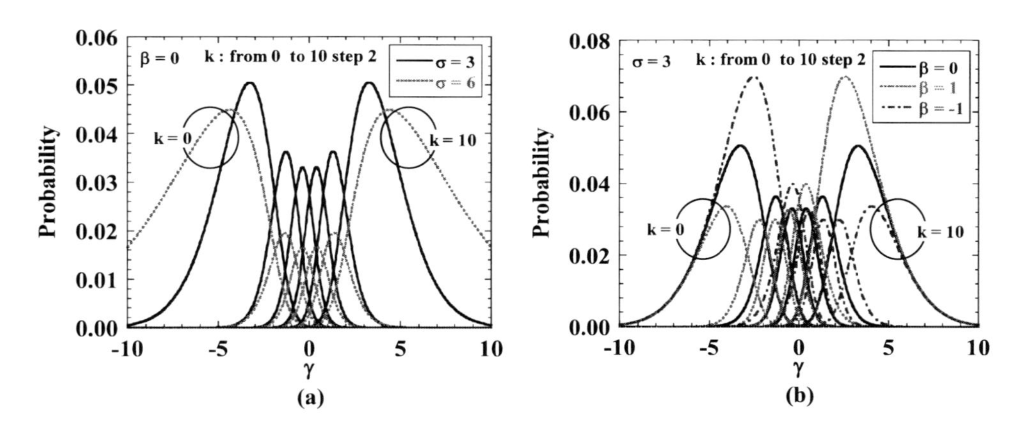

Figure 15. Dependence of probability for number k on γ. (a) Two standard deviations σ. (b) Various β.

Neglecting the factors which are not related to the parameters, we obtain the function given by

$$\begin{aligned} L(\sigma,\beta) &= \prod_{i=1}^{20} L_i \\ &= \prod_{i=1}^{20} \theta_i^{x_i} (1-\theta_i)^{n-x_i} \frac{1}{\sigma}\exp\left(-\frac{\gamma_1^2}{2\sigma^2}\right) \\ &= \prod_{i=1}^{20} \left[1+\exp(-\beta-\gamma_i)\right]^{-x_i} \left[1+\exp(\beta+\gamma_i)\right]^{-(n-x_i)} \frac{1}{\sigma}\exp\left(-\frac{\gamma_1^2}{2\sigma^2}\right) \end{aligned} \tag{190}$$

where

$$\begin{aligned} L_i(\gamma_i,\sigma,\beta) &= \theta_i^{x_i} (1-\theta_i)^{n-x_i} \frac{1}{\sigma}\exp\left(-\frac{\gamma_1^2}{2\sigma^2}\right) \\ &= \left[1+\exp(-\beta-\gamma_i)\right]^{-x_i} \left[1+\exp(\beta+\gamma_i)\right]^{-(n-x_i)} \frac{1}{\sigma}\exp\left(-\frac{\gamma_1^2}{2\sigma^2}\right) \end{aligned} \tag{191}$$

We first decide parameters that do not depend on member, which are β and σ. We eliminate parameters of γ_i as

$$\begin{aligned} &\int_{-\infty}^{\infty} L_1(\gamma_1,\beta,\sigma) d\gamma_1 \int_{-\infty}^{\infty} L_2(\gamma_2,\beta,\sigma) d\gamma_2 \cdots \int_{-\infty}^{\infty} L_{20}(\gamma_{20},\beta,\sigma) d\gamma_{20} \\ &= Q_1(\beta,\sigma) Q_2(\beta,\sigma) \cdots Q_{20}(\beta,\sigma) \end{aligned} \tag{192}$$

We perform the logarithm for the equation given by

$$G = \ln\left[Q_1(\beta,\sigma)\right] + \ln\left[Q_2(\beta,\sigma)\right] + \cdots + \ln\left[Q_{20}(\beta,\sigma)\right] \tag{193}$$

We can obtain the values for β and σ that maximize G using a solver in a software such as Excel in Micro Soft, and they are denoded as β_0 and σ_0. In this case, we obtain

$$\begin{cases} \beta_0 = 0.86 \\ \sigma_0 = 3.08 \end{cases} \tag{194}$$

We can then perform the MHMC method and obtain the parameter values of γ_i as shown in Table 4 where we evaluated 5000 cycle times data and select the last 4000 data. The expected score values are evaluated as

$$N(k) = 20 \times \frac{1}{20} \sum_{i=1}^{20} {}_{10}C_k \theta_i^k \left(1-\theta_i\right)^{10-k} \tag{195}$$

which is shown in Figure 12. The expected values well reproduces the data.

Table 4. Evaluated parameter values

Data ID	1	2	3	4	5	6	7	8	9	10	11	12	13	14	15	16	17	18	19	20
γ	-3.22	-4.87	3.80	-1.22	3.07	1.99	3.64	-0.28	-1.30	2.42	-2.76	1.32	-3.93	-0.97	4.92	-0.10	-2.80	2.57	-2.06	0.59
θ	0.09	0.02	0.99	0.41	0.98	0.95	0.99	0.64	0.39	0.96	0.13	0.90	0.04	0.47	1.00	0.68	0.13	0.97	0.23	0.81
Score	1	0	10	4	10	10	10	6	4	10	1	9	0	5	10	7	1	9	2	8

SUMMARY

To summarize the results in this chapter—

We obtain data x, and the population parameter is denoted as θ. We modify the Bayes' theory as

$$P(H|D) = \frac{P(D|H)P(H)}{P(D)} \rightarrow \pi(\theta|D) = \frac{f(D|\theta)\pi(\theta)}{P(D)} = Kf(D|\theta)\pi(\theta)$$

where K is a normalized factor. We call $\pi(\theta)$ as the priori probability distribution which depends on a macroscopic parameter or constant. $f(D|\theta)$ is the likelihood probability where we have data D under the macroscopic value of θ. $\pi(\theta|D)$ is the posterior probability distribution.

There are four convenient combinations for $\pi(\theta)$, $f(D|\theta)$, which is called as conjugate functions, and $\pi(\theta|D)$ are the as follows.

Beta/Binominal (Beta)/Beta Distribution

We assume the probability where we obtain x target events among n trials, and the corresponding likelihood function is given by

$$f(D|\theta) = \frac{\theta^{\alpha_L - 1}(1-\theta)^{\beta_L - 1}}{\Gamma(\alpha_L, \beta_L)}$$

where

$$\alpha_L = x + 1$$

$$\beta_L = n - x + 1$$

The priori probability distribution for this case is given by

$$\pi(\theta) = B_e(\alpha_0, \beta_0) = \frac{1}{B(\alpha_0, \beta_0)} \theta^{\alpha_0 - 1}(1-\theta)^{\beta_0 - 1}$$

The average and variance for the priori distributions are related to the parameters as

$$\alpha_0 = \frac{\mu_0^2}{\sigma_0^2}(1-\mu_0) - \mu_0$$

$$\beta_0 = \frac{1-\mu_0}{\mu_0}\left[\frac{\mu_0^2}{\sigma_0^2}(1-\mu_0) - \mu_0\right]$$

The posterior function is given by

$$\pi(\theta|D) = \frac{\theta^{\alpha^* - 1}(1-\theta)^{\beta^* - 1}}{B(\alpha^*, \beta^*)}$$

where

$$\alpha^* = \alpha_0 + \alpha_L - 1$$

$$\beta^* = \beta_0 + \beta_L - 1$$

The average and variance of θ are denoted as μ_θ and σ_θ^2 are given by

$$\mu_\theta = \frac{\alpha^*}{\alpha^* + \beta^*}$$

$$\sigma_\theta^2 = \frac{\alpha^* \beta^*}{\left(\alpha^* + \beta^* + 1\right)\left(\alpha^* + \beta^*\right)^2}$$

Dirichlet/Multinomial (Dirichlet)/Dirichlet Distribution

In general, we need to consider multiple events of more than two kinds. Let us consider m kinds of events and the corresponding probability for $P\left(X_1 = x_1, X_2 = x_2, \cdots, X_m = x_m\right)$ is given by

$$f\left(x_1, x_2, \cdots, x_m\right) = \frac{n!}{x_1! x_2! \cdots x_m!} p^{x_1} p^{x_2} \cdots p^{x_m}$$

where

$$p_1 + p_2 + \cdots + p_m = 1$$

$$x_1 + x_2 + \cdots + x_m = n$$

The corresponding likelihood function is given by

$$f\left(\theta_1, \theta_2, \cdots, \theta_m \middle| D\right) = \frac{\Gamma(\alpha)}{\Gamma(\alpha_1)\Gamma(\alpha_2)\cdots\Gamma(\alpha_m)} \theta_1^{\alpha_1 - 1} \theta_2^{\alpha_2 - 1} \cdots \theta_m^{\alpha_m - 1}$$

where

$$\alpha_i = x_i + 1$$

The priori probability distribution for this case is given by

$$\pi\left(\theta_1,\theta_2,\cdots,\theta_m\right)=\frac{\Gamma\left(\alpha_0\right)}{\Gamma\left(\alpha_{10}\right)\Gamma\left(\alpha_{20}\right)\cdots\Gamma\left(\alpha_{m0}\right)}\theta_1^{\alpha_{10}-1}\theta_2^{\alpha_{20}-1}\cdots\theta_m^{\alpha_{m0}-1}$$

Finally, we obtain the posterior distribution as

$$\pi\left(\theta_1,\theta_2,\cdots,\theta_m\middle|D\right)=\frac{\Gamma\left(\alpha^*\right)}{\Gamma\left(\alpha_1^*\right)\Gamma\left(\alpha_2^*\right)\cdots\Gamma\left(\alpha_m^*\right)}\theta_1^{\alpha_1^*-1}\theta_2^{\alpha_2^*-1}\cdots\theta_m^{\alpha_m^*-1}$$

where

$$\alpha_i^* = \alpha_i + \alpha_{i0} - 1$$

Gamma/Poisson(Gamma)/Gamma Distribution

We assume that we obtain x times target events among n, and the average number we obtain is given by

$$\theta = np$$

where p is the probability that we obtain a target event in each trial and assumed to be small.

The corresponding likelihood function is given by

$$f\left(D\middle|\theta\right)=\frac{\theta^{n_L-1}e^{-\theta}}{\Gamma\left(n_L\right)}$$

where

$$n_L = x+1$$

The priori distribution $\pi(\theta)$ is given by

$$\pi\left(\theta\right)=\frac{\lambda_0^{n_0}\theta^{n-1}e^{-\lambda_0\theta}}{\Gamma\left(n_0\right)}$$

The average and variance for the priori distributions are related to the parameters as

$$\lambda_0 = \frac{\mu_0}{\sigma_0^2}$$

$$n_0 = \frac{\mu_0^2}{\sigma_0^2}$$

The posterior distribution is given by

$$\pi(\theta|D) = \frac{\theta^{n^*-1} e^{-\lambda^*\theta}}{\Gamma(n^*)}$$

where

$$n^* = n_0 + n_L - 1$$

$$\lambda^* = \lambda_0 + 1$$

Normal/Normal/Normal Distribution

If we obtained n data from a set as $x_1, x_2, \cdots, x_n$, the corresponding likelihood function is given by

$$f(D|\theta) = \frac{1}{\sqrt{2\pi\sigma_L^2}} \exp\left[-\frac{(\theta-\mu_L)^2}{2\sigma_L^2}\right]$$

where

$$\sigma_L^2 = \frac{\sigma^2}{n}$$

$$\mu_L = \bar{x} = \frac{1}{n}\sum_{i=1}^{n} x_i$$

The priori distribution is given by

$$\pi(\theta)=\frac{1}{\sqrt{2\pi}\sigma_0}\exp\left[-\frac{(\theta-\mu_0)^2}{2\sigma_0^2}\right]$$

The corresponding posterior distribution is given by

$$\pi(\theta|D)=\frac{1}{\sqrt{2\pi}\sigma^*}\exp\left[-\frac{(\theta-\mu^*)^2}{2\sigma^{*2}}\right]$$

where

$$\frac{1}{\sigma^{*2}}=\frac{1}{\sigma_0^2}+\frac{1}{\sigma_L^2}$$

$$\mu^*=\left(\frac{\mu_0}{\sigma_0^2}+\frac{\mu_L}{\sigma_L^2}\right)\sigma^{*2}$$

Inverse Gamma/Normal (Inverse Gamma)/Inverse Gamma Distribution

We extract a data set from a set as $x_1,x_2,\cdots,x_n$, and the corresponding likelihood distribution is given by

$$f(D|\theta)=\frac{\theta^{-n_L-1}\exp\left(-\frac{\lambda_L}{\theta}\right)}{\Gamma(n_L)}$$

where

$$n_L=\frac{n}{2}-1$$

$$\lambda_L=\frac{1}{2}\left\{n\left[(\mu-\bar{x})^2-\bar{x}^2\right]+\sum_{i=1}^{n}x_i^2\right\}$$

The priori distribution is given by

$$\pi(\theta) = \frac{\lambda_0^{n_0} \theta^{-n_0-1} Exp\left(-\frac{\lambda_0}{\theta}\right)}{\Gamma(n_0)}$$

The average and variance for the priori distributions are related to the parameters as

$$n_0 = 2 + \frac{\mu_0^2}{\sigma_0^2}$$

$$\lambda_0 = \mu_0\left(1 + \frac{\mu_0^2}{\sigma_0^2}\right)$$

The posterior distribution is given by

$$\pi(\theta|D) = \frac{\theta^{-n^*-1} \exp\left(-\frac{\lambda^*}{\theta}\right)}{\Gamma(n^*)}$$

where

$$n^* = n_0 + n_L + 1$$

$$\lambda^* = \lambda_0 + \lambda_L$$

We can easily extend these procedures to obtain many macro parameters of a population set. We need to obtain posterior l distribution functions with l parameters, which is denoted as

$$\pi(\theta_1|D, \theta_2, \theta_3, \cdots, \theta_l)$$
$$\pi(\theta_2|D, \theta_1, \theta_3, \cdots, \theta_l)$$
$$\cdots$$
$$\pi(\theta_i|D, \theta_1, \theta_2, \cdots, \theta_{i-1}, \theta_{i+1}, \cdots, \theta_l)$$
$$\cdots$$
$$\pi(\theta_l|D, \theta_1, \theta_2, \cdots, \theta_{l-1})$$

In the Gibbs sampling method, the parameter values are updated by generating corresponding random numbers as

$$\theta_i^{(k+1)} = rand\left[Inv\_\pi\left(\theta_i \middle| D, \theta_1^{(k+1)}, \theta_2^{(k+1)}, \cdots, \theta_{i-1}^{(k+1)}, \theta_{i+1}^{(k)}, \cdots, \theta_l^{(k)}\right), \theta_i \right]$$

In the MH method, the parameter values are updated as follows.

We generate θ_{it} as

$$\theta_{it} = \theta_i^{(k)} + rand\left[Inv\_N\left[0, \varepsilon_{\theta_i}\right]\right]$$

We then evaluate the ratio

$$p_{\theta_i} = \frac{\pi\left(\theta_{it} \middle| D, \theta_1^{(k)}, \theta_2^{(k)}, \cdots, \theta_{i-1}^{(k)}, \theta_{i+1}^{(k)}, \cdots, \theta_l^{(k)}\right)}{\pi\left(\theta_i^{(k)} \middle| D, \theta_1^{(k)}, \theta_2^{(k)}, \cdots, \theta_{i-1}^{(k)}, \theta_{i+1}^{(k)}, \cdots, \theta_l^{(k)}\right)}$$

We also generate uniform random number $q^{(k+1)}$ as

$$q^{(k+1)} = rand(\ )$$

We then update the parameter as

$$\theta_i^{(k+1)} = \begin{cases} \theta_{it} & for\ p_{\theta_i} \geq 1\ or\ p_{\theta_i} \geq q^{(k+1)} \\ \theta_i^{(k)} & for\ p_{\theta_i} < q^{(k+1)} \end{cases}$$

We can extend this theory to use various parameters in a likelihood function which is controlled with priori function, which is called as a hierarchical Bayes' theory.

Chapter 8

ANALYSIS OF VARIANCES

ABSTRACT

We sometimes want to improve a certain objective value by trying various treatments. For example, we treat various kinds of fertilizer to improve harvest. We probably judge the effectiveness of the treatment by comparing corresponding average of the objective values of amount of harvest. However, there should be a scattering in the data values in each treatment. Even when we have a better average data, some data in the best treatment are less than the one in the other treatment. We show how we decide the effectiveness of the treatment considering the data scattering.

Keywords: analysis of variance, correlation ratio, interaction, studentized range distribution

1. INTRODUCTION

In agriculture, we want to harvest grain much, and try various treatments and want to clarify effectiveness of various treatments. One example factor is kinds of fertilizers. We denote the factor as A and kinds of fertilizers as A_j. We want to know the effectiveness of various kinds of fertilizers. We obtain various harvest data for each fertilizer. Although the average values have clear difference among the fertilizers, they are usually scattered in each fertilizers' group. Analysis of variance is developed to decide validity of the effectiveness. Further, we need to consider more factors for the harvest such as temperature and humidity. Therefore, we need to include various factors in the analysis of the variances. We start with one factor, and then extend two factors.

2. Correlation Ratio (One Way Analysis)

We study the dependence of three kinds of fertilizer A_1, A_2, and A_3 on a grain harvest. We call the kinds of fertilizer as levels in general. We use 1 kg seed for each and harvest grain and check the weight with unit of kg. The data are shown in the Table 1.

Table 1. Dependence of harvest on kinds of fertilizer

	A:Fertilizer		
	A1	A2	A3
	5.44	7.33	9.44
	7.66	6.32	7.33
	7.55	6.76	7.66
	4.33	5.33	6.43
	5.33	5.89	7.45
	4.55	6.33	6.86
	6.33	5.44	9.78
	5.98	8.54	6.34
Average	5.90	6.49	7.66

The data number for each level are given by

$$n_{A_1} = 8, n_{A_2} = 8, n_{A_3} = 8 \tag{1}$$

$$n = n_{A_1} + n_{A_2} + n_{A_3} = 24 \tag{2}$$

The corresponding averages are given by

$$\mu_{A_1} = \frac{1}{n_{A_1}} \sum_{i=1}^{n_{A_1}} x_{iA_1} = 5.90 \tag{3}$$

$$\mu_{A_2} = \frac{1}{n_{A_2}} \sum_{i=1}^{n_{A_2}} x_{iA_2} = 6.49 \tag{4}$$

$$\mu_{A_3} = \frac{1}{n_{A_3}} \sum_{i=1}^{n_{A_3}} x_{iA_3} = 7.66 \tag{5}$$

The total average is given by

$$\mu = \frac{1}{n_{A_1} + n_{A_2} + n_{A_3}} \left(\sum_{i=1}^{n_{A_1}} x_{iA_1} + \sum_{i=1}^{n_{A_2}} x_{iA_2} + \sum_{i=1}^{n_{A_3}} x_{iA_3} \right) = 6.68 \tag{6}$$

The average harvested weight was 5.90 kg for fertilizer A_1, 6.49 kg for fertilizer A_2, and 7.66 kg for fertilizer A_3. It seems that fertilizer A_3 is most effective. Fertilizer A_1 seems to be less effective. However, there is a datum for fertilizer A_1 that is larger than the one for fertilizer A_3. How can we judge the effectiveness of levels statistically?

We judge the variance associated with the level versus the variance associated with each data within each level. If the former one is larger than the latter one, we judge the level change is effective, and vice versa. Let us express the above discussion quantitatively.

We evaluate the variance associated with the effectiveness of the level, which is denoted as S_{ex}^2, and is given by

$$S_{ex}^2 = \frac{n_{A_1}\left(\mu_{A_1} - \mu\right)^2 + n_{A_2}\left(\mu_{A_2} - \mu\right)^2 + n_{A_3}\left(\mu_{A_3} - \mu\right)^2}{n_{A_1} + n_{A_2} + n_{A_3}} = 0.53 \tag{7}$$

We then consider the variance associated with the group in each level. It is denoted as S_{in}^2, and is given by

$$S_{in}^2 = \frac{\sum_{i=1}^{n_{A_1}}\left(x_{iA_1} - \mu_{A_1}\right)^2 + \sum_{i=1}^{n_{A_2}}\left(x_{iA_2} - \mu_{A_2}\right)^2 + \sum_{i=1}^{n_{A_3}}\left(x_{iA_3} - \mu_{A_3}\right)^2}{n_{A_1} + n_{A_2} + n_{A_3}} = 1.27 \tag{8}$$

The correlation ratio η^2 is given by

$$\eta^2 = \frac{S_{ex}^2}{S_{in}^2 + S_{ex}^2} = 0.30 \tag{9}$$

This indicates the significance of the dependence. If it is large, the effectiveness of changing fertilizer is large. It is also obvious that the correlation ratio is less than 1.

However, we do not use this correlation ratio for the final effectiveness judge. The procedure is shown in the next section.

3. TESTING OF THE DEPENDENCE

We consider the freedom of S_{ex}^2. We denote it as ϕ_{ex}. The variable in the variance is three of $\mu_{A_1}, \mu_{A_2}, \mu_{A_3}$. However, it should satisfy Eq. (6). Therefore, the freedom is the number of fertilizer kinds minus 1. Therefore the freedom ϕ_{ex} is given by

$$\phi_{ex} = p - 1 = 3 - 1 = 2 \tag{10}$$

where p is the number of fertilizer kinds and is 3 in this case. Therefore, the unbiased variance s_{ex}^2 is given by

$$s_{ex}^2 = \frac{n}{\phi_{ex}} S_{ex}^2 = 6.45$$

We consider the freedom of S_{in}^2. We denote it as ϕ_{in}. Each group has $n_{A_1}, n_{A_2}, n_{A_3}$ data, but each group has their sample averages and the freedom is decreased by 1 for each group. Therefore, the total freedom is given by

$$\begin{aligned}\phi_{in} &= (n_A - 1) + (n_B - 1) + (n_C - 1) \\ &= n - p \\ &= 24 - 3 \\ &= 21\end{aligned} \tag{11}$$

The corresponding unbiased variance is given by s_{in}^2

$$s_{in}^2 = \frac{n}{\phi_{in}} S_{in}^2 = 1.44 \tag{12}$$

Finally, the ratio of the unbiased variance is denoted by F and is given by

$$F = \frac{s_{ex}^2}{s_{in}^2} = 4.46 \tag{13}$$

This follows the F distribution with a freedom of $\left(\phi_{ex},\phi_{in}\right)$. The P point for the F distribution with the predictive probability of 95% is denoted as $F_C\left(\phi_{ex},\phi_{in},0.95\right)$, and is

$$F_C\left(\phi_{ex},\phi_{in},0.95\right)=3.47 \tag{14}$$

which is shown in Figure 1. Therefore, $F > F_C$, and hence the dependence is valid.

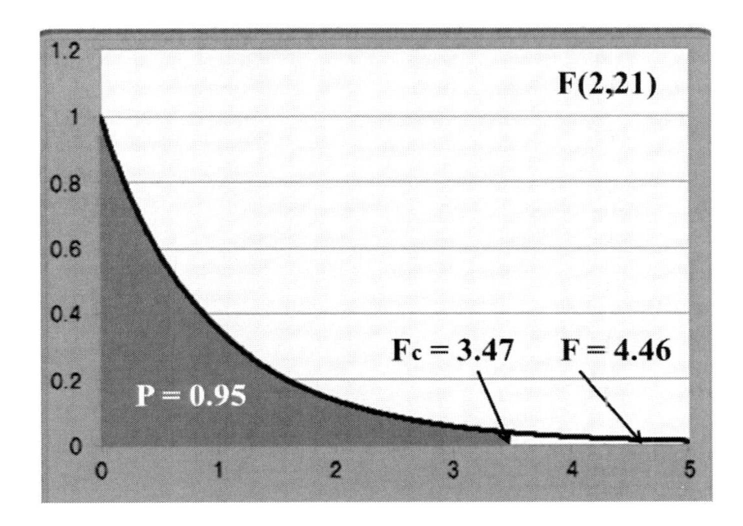

Figure 1. F distribution testing.

4. The Relationship between Various Variances

We treat two kinds of variances in the previous section. We can also define the total variance with respect to the total average. We investigate the relationship between the variances. We assume that the level number is three and denoted as A, B, and C, and each level has the same data number of n.

The total variance S^2 is constituted by summing up the deviation of each data with respect to the total average, and is given by

$$S_{tot}^2=\frac{\sum_{i=1}^{n_{A_1}}\left(x_{iA_1}-\mu\right)^2+\sum_{i=1}^{n_{A_2}}\left(x_{iA_2}-\mu\right)^2+\sum_{i=1}^{n_{A_3}}\left(x_{iA_3}-\mu\right)^2}{n} \tag{15}$$

Let us consider the first term of numerator of Eq. (15), which is modified as

$$\sum_{i=1}^{n_A}\left(x_{iA_1}-\mu\right)^2=\sum_{i=1}^{n_{A_1}}\left(x_{iA_1}-\mu_{A_1}+\mu_{A_1}-\mu\right)^2$$
$$=\sum_{i=1}^{n_{A_1}}\left(x_{iA_1}-\mu_{A_1}\right)^2+2\sum_{i=1}^{n_{A_1}}\left(x_{iA_1}-\mu_{A_1}\right)\left(\mu_{A_1}-\mu\right)+\sum_{i=1}^{n_{A_1}}\left(\mu_{A_1}-\mu\right)^2$$
$$=\sum_{i=1}^{n_{A_1}}\left(x_{iA_1}-\mu_{A_1}\right)^2+n_{A_1}\left(\mu_{A_1}-\mu\right)^2 \tag{16}$$

Similar analysis is performed for the other terms, and we obtain

$$S_{tot}^2=\frac{\sum_{i=1}^{n_{A_1}}\left(x_{iA_1}-\mu_{A_1}\right)^2+\sum_{i=1}^{n_{A_2}}\left(x_{iA_2}-\mu_{A_2}\right)^2+\sum_{i=1}^{n_{A_3}}\left(x_{iA_3}-\mu_{A_3}\right)^2}{n}$$
$$+\frac{n_{A_1}\left(\mu_{A_1}-\mu\right)^2+n_{A_2}\left(\mu_{A_2}-\mu\right)^2+n_{A_3}\left(\mu_{A_3}-\mu\right)^2}{n}$$
$$=S_{ex}^2+S_{in}^2 \tag{17}$$

Therefore, the total variance is the sum of the variance associated with various levels and one associated with each level. Therefore, the correlation ratio is expressed by

$$\eta^2=\frac{S_{ex}^2}{S_{in}^2+S_{ex}^2}=\frac{S_{ex}^2}{S_{tot}^2} \tag{18}$$

5. MULTIPLE COMPARISONS

In the analysis comparisons, we test whether whole averages are the same or not. When we judge whole averages are not the same, we want to know which one is different. One way to test it is to select one pair and judge the difference as shown in the previous section. We test each pair setting a prediction probability. The total results are not related to the setting prediction probability since they are not independent. We want to test them all pairs with a certain prediction probability. Turkey-Kramer multiple comparisons procedure is frequently used and we introduce it here.

We compare the average of level i and j, and obtain the absolute value of the difference of $\left|\mu_{A_i}-\mu_{A_j}\right|$, and form a normalized value given by

$$z_{ij} = \frac{\left|\mu_{A_i} - \mu_{A_j}\right|}{\sqrt{\frac{s_{in}^2}{2}\left(\frac{1}{n_{A_i}} + \frac{1}{n_{Aj}}\right)}} \tag{19}$$

We compare the normalized value z to the studentized range distribution critical value of $q(p, n-p, P)$, where p is the level number, and n is the total sample number. The corresponding table for $q(r, n-p, P)$ with $P = 0.95$ and $P = 0.99$ are shown in Table 2.

The other simple way to evaluate the difference is the one with

$$z_i = \frac{\mu_{A_i} - \mu}{\sqrt{\frac{s_{in}^2}{2}\left(\frac{1}{n_{A_i}} + \frac{1}{n}\right)}} \tag{20}$$

We may be able to compare this normalized value to $q(p, n-p, P)$.

We need to interpolate the value of $q(r, n-r, P)$ from the table. Especially, the interpolation associated with $n-r$ is important.

For simplicity, we set

$$k = n - p \tag{21}$$

The row number before the final is 120 in the table.

If k is less than 120, we can easily interpolate the value of $q(p, k, P)$ as

$$q(p,k,P) = q(p,i,P) + \frac{q(p,k,P) - q(p,i,P)}{q(p,j,P) - q(p,i,P)}\left[q(p,j,P) - q(p,i,P)\right] \tag{22}$$

Where i, j are the both sides of the row number that holds

$$i \le k < j \tag{23}$$

The difficulty exists in that the final row number is infinity and the number before the final is 120 in general. Therefore, we need to study how to interpolate where $k \ge 120$.

We introduce variables as

$$\Delta = \frac{k-120}{120} = \ln\left(\frac{1+\varsigma}{1-\varsigma}\right) \tag{24}$$

Table 2. Studentized range distribution table for prediction probabilities of 0.95 and 0.99

One sided P = 0.95

		Freedom of denominator															
		2	3	4	5	6	7	8	9	10	11	12	13	14	15	16	17
Freedom of numerator	1	17.97	26.98	32.82	37.08	40.41	43.12	45.40	43.36	49.07	59.59	51.96	53.20	54.33	55.36	56.32	57.22
	2	6.09	8.33	9.80	10.88	11.74	12.44	13.03	13.54	13.99	14.39	14.75	15.08	15.38	15.65	15.91	16.14
	3	4.50	5.91	6.83	7.50	8.04	8.48	8.85	9.18	9.46	9.72	9.95	10.15	10.35	10.53	10.61	10.84
	4	3.93	5.04	5.76	6.29	6.71	7.05	7.35	7.60	7.83	8.03	8.21	8.37	8.53	8.66	8.79	8.91
	5	3.64	4.60	5.22	5.67	6.03	6.33	6.58	6.80	7.00	7.17	7.32	7.47	7.60	7.72	7.83	7.93
	6	3.46	4.34	4.90	5.31	5.63	5.90	6.12	6.32	6.49	6.65	6.79	6.92	7.03	7.14	7.24	7.34
	7	3.34	4.17	4.68	5.06	5.36	5.61	5.82	6.00	6.16	6.30	6.43	6.55	6.66	6.76	6.85	6.94
	8	3.26	4.04	4.53	4.89	5.17	5.40	5.60	5.77	5.92	6.05	6.18	6.29	6.39	6.48	6.57	6.65
	9	3.20	3.95	4.42	4.76	5.02	5.24	5.43	5.60	5.74	5.87	5.98	6.09	6.19	6.28	6.36	6.44
	10	3.15	3.88	4.33	4.65	4.91	5.12	5.31	5.46	5.60	5.72	5.83	5.93	6.03	6.11	6.20	6.27
	11	3.11	3.82	4.26	4.57	4.82	5.03	5.20	5.35	5.49	5.61	5.71	5.81	5.90	5.98	6.06	6.13
	12	3.08	3.77	4.20	4.51	4.75	4.95	5.12	5.27	5.40	5.51	5.62	5.71	5.80	5.88	5.95	6.02
	13	3.06	3.74	4.15	4.45	4.69	4.89	5.05	5.19	5.32	5.43	5.53	5.63	5.71	5.79	5.86	5.93
	14	3.03	3.70	4.11	4.41	4.64	4.83	4.99	5.13	5.25	5.36	5.46	5.55	5.64	5.71	5.79	5.85
	15	3.01	3.67	4.08	4.37	4.60	4.78	4.94	5.08	5.20	5.31	5.40	5.49	5.57	5.65	5.72	5.79
	16	3.00	3.65	4.05	4.33	4.56	4.74	4.90	5.03	5.15	5.26	5.35	5.44	5.52	5.59	5.66	5.73
	17	2.98	3.63	4.02	4.30	4.52	4.71	4.86	4.99	5.11	5.21	5.31	5.39	5.47	5.54	5.61	5.68
	18	2.97	3.61	4.00	4.28	4.50	4.67	4.82	4.96	5.07	5.17	5.27	5.35	5.43	5.50	5.57	5.63
	19	2.96	3.59	3.98	4.25	4.47	4.65	4.79	4.92	5.04	4.14	5.23	5.32	5.39	5.46	5.53	5.59
	20	2.95	3.58	3.96	4.23	4.45	4.62	4.77	4.90	5.01	5.11	5.20	5.28	5.36	5.43	5.49	5.55
	24	2.92	3.53	3.90	4.17	4.37	4.54	4.68	4.81	4.92	5.01	5.10	5.18	5.25	5.32	5.38	5.44
	30	2.89	3.49	3.85	4.10	4.30	4.46	4.60	4.72	4.82	4.92	5.00	5.08	5.15	5.21	5.27	5.33
	40	2.86	3.44	3.79	4.04	4.23	4.39	4.52	4.64	4.74	4.82	4.90	4.98	5.04	5.11	5.16	5.22
	60	2.83	3.40	3.74	3.98	4.16	4.31	4.44	4.55	4.65	4.73	4.81	4.88	4.94	5.00	5.06	5.11
	120	2.80	3.36	3.69	3.92	4.10	4.24	4.36	4.47	4.56	4.64	4.71	4.78	4.84	4.90	4.95	5.00
	∞	2.77	3.31	3.63	3.86	4.03	4.17	4.29	4.39	4.47	4.55	4.62	4.68	4.74	4.80	4.85	4.89

One sided P = 0.99

		Freedom of denominator															
		2	3	4	5	6	7	8	9	10	11	12	13	14	15	16	17
Freedom of numerator	1	90.03	135.00	164.30	185.60	202.20	215.80	227.20	237.00	245.60	253.20	260.00	266.20	271.80	277.00	281.80	286.30
	2	14.04	19.02	22.29	24.72	26.63	28.20	29.53	30.68	31.69	32.59	33.40	34.13	34.81	35.43	36.00	36.53
	3	8.26	10.62	12.17	13.33	14.24	15.00	15.64	16.20	16.69	17.13	17.53	17.89	18.22	18.52	18.81	19.07
	4	6.51	8.12	9.17	9.96	10.58	11.10	11.55	11.93	12.27	12.57	12.84	13.09	13.32	13.53	13.73	13.91
	5	5.70	6.98	7.80	8.42	8.91	9.32	9.67	9.97	10.24	10.48	10.70	10.89	11.08	11.24	11.40	11.55
	6	5.24	6.33	7.03	7.56	7.97	8.32	8.61	8.87	9.10	9.30	9.49	9.65	9.81	9.95	10.08	10.21
	7	4.95	5.92	6.54	7.01	7.37	7.68	7.94	8.17	8.37	8.55	8.71	8.86	9.00	9.12	9.24	9.35
	8	4.75	5.64	6.20	6.63	6.96	7.24	7.47	7.68	7.86	8.03	8.18	8.31	8.44	8.55	8.66	8.76
	9	4.60	5.43	5.96	6.35	6.66	6.92	7.13	7.32	7.50	7.65	7.78	7.91	8.03	8.13	8.23	8.33
	10	4.48	5.27	5.77	6.14	6.43	6.67	6.87	7.06	7.21	7.36	7.49	7.60	7.71	7.81	7.91	7.99
	11	4.39	5.15	5.62	5.97	6.25	6.48	6.67	6.84	6.99	7.13	7.25	7.36	7.47	7.56	7.65	7.73
	12	4.32	5.04	5.50	5.84	6.10	6.32	6.51	6.67	6.81	6.94	7.06	7.17	7.26	7.36	7.44	7.52
	13	4.26	4.96	5.40	5.73	5.98	6.19	6.37	6.53	6.67	6.79	6.90	7.01	7.10	7.19	7.27	7.35
	14	4.21	4.90	5.32	5.63	5.88	6.09	6.26	6.41	6.54	6.66	6.77	6.87	6.96	7.05	7.13	7.20
	15	4.17	4.84	5.25	5.56	5.80	5.99	6.16	6.31	6.44	6.56	6.66	6.76	6.85	6.93	7.00	7.07
	16	4.13	4.79	5.19	5.49	5.72	5.92	6.08	6.22	6.35	6.46	6.56	6.66	6.74	6.82	6.90	6.97
	17	4.10	4.74	5.14	5.43	5.66	5.85	6.01	6.15	6.27	6.38	6.48	6.57	6.66	6.73	6.81	6.87
	18	4.07	4.70	5.09	5.38	5.60	5.79	5.94	6.08	6.20	6.31	6.41	6.50	6.58	6.66	6.73	6.79
	19	4.05	4.67	5.05	5.33	5.55	5.74	5.89	6.02	6.14	6.25	6.34	6.43	6.51	6.59	6.65	6.72
	20	4.02	4.64	5.02	5.29	5.51	5.69	5.84	5.97	6.09	6.19	6.29	6.37	6.45	6.52	6.59	6.65
	24	3.96	4.55	4.91	5.17	5.37	5.54	5.69	5.81	5.92	6.02	6.11	6.19	6.26	6.33	6.39	6.45
	30	3.89	4.46	4.80	5.05	5.24	5.40	5.54	5.65	5.76	5.85	5.93	6.01	6.08	6.14	6.20	6.26
	40	3.83	4.37	4.70	4.93	5.11	5.27	5.39	5.50	5.60	5.69	5.76	5.84	5.90	5.96	6.02	6.07
	60	3.76	4.28	4.60	4.82	4.99	5.13	5.25	5.36	5.45	5.53	5.60	5.67	5.73	5.79	5.84	5.89
	120	3.70	4.20	4.50	4.71	4.87	5.01	5.12	5.21	5.30	5.38	5.44	5.51	5.56	5.61	5.66	5.71
	∞	3.64	4.12	4.40	4.60	4.76	4.88	4.99	5.08	5.16	5.23	5.29	5.35	5.40	5.45	5.49	5.54

The ς is related to the value of $q(p,k,P)$ as

$$\varsigma = \frac{q(p,k,P) - q(p,120P)}{q(p,\infty,P) - q(p,120,P)} \tag{25}$$

We can evaluate Δ from the thirst relationship as

$$\Delta = \frac{k-120}{120} \tag{26}$$

We can then obtain

$$\Delta = \ln\left(\frac{1+\varsigma}{1-\varsigma}\right) \tag{27}$$

Solving this equation with respect to ς, we obtain

$$\varsigma = \frac{e^{\Delta}-1}{e^{\Delta}+1} \tag{28}$$

Comparing Eq. (28) with Eq. (25), we obtain an interpolate value as

$$q(p,k,P) = q(p,120,P) + \frac{e^{\Delta}-1}{e^{\Delta}+1}\left[q(p,\infty,P) - q(p,120,P)\right] \tag{29}$$

6. Analysis of Variance (Two Way Analysis with Non-Repeated Data)

We consider two factors of A and B. For example, factor A corresponds to the kinds of fertilizers, and factor B corresponds to the temperature. We assume the factor A has four levels of A_1, A_2, A_3, A_4, denoted as n_A, and the factor B has three levels of B_1, B_2, B_3 denoted as n_B, where $n_A = 4$ and $n_B = 3$. We assume that the data is only one, that is, non-repeated data as shown in Table 3. The total data number n is given by

$$n = n_A \times n_B \tag{30}$$

In this case, each data x_{ij} can be expressed by the deviation from the average, and is given by

$$x_{ij} = \mu + \left(\mu_{A_j} - \mu\right) + \left(\mu_{B_i} - \mu\right) + e_{ij} \tag{31}$$

This can be modified as

$$x_{ij} - \mu = \left(\mu_{A_j} - \mu\right) + \left(\mu_{B_i} - \mu\right) + e_{ij} \tag{32}$$

The corresponding freedom relationship is given by

$$\phi_{tot} = \phi_A + \phi_B + \phi_e \tag{33}$$

where

$$\begin{aligned} \phi_{tot} &= n - 1 \\ &= 12 - 1 \\ &= 11 \end{aligned} \tag{34}$$

$$\begin{aligned} \phi_A &= n_A - 1 \\ &= 4 - 1 \\ &= 3 \end{aligned} \tag{35}$$

$$\begin{aligned} \phi_B &= n_B - 1 \\ &= 3 - 1 \\ &= 2 \end{aligned} \tag{36}$$

We can obtain the freedom associated with the error and it is given by

$$\begin{aligned} \phi_e &= \phi_{tot} - \left(\phi_A + \phi_B\right) \\ &= \left(n - 1\right) - \left(n_A - 1 + n_B - 1\right) \\ &= n - \left(n_A + n_B\right) + 1 \\ &= 12 - \left(4 + 3\right) + 1 \\ &= 6 \end{aligned} \tag{37}$$

The discussion is almost the same as that for one-way analysis.

We evaluate the total average μ and evaluate the average associated with factor A denoted as μ_{A_j} where $j=1,2,3,4$ and factor B denoted as μ_{B_i} where $i=1,2,3$, which are shown in Table 3.

Table 3. Two-factors cross data table with no repeat

		A					
		A_1	A_2	A_3	A_4	μ_{Bi}	$D\mu_{Bi}$
B	B_1	7.12	8.33	9.12	11.21	8.95	-1.20
	B_2	8.63	8.15	10.98	15.43	10.80	0.65
	B_3	7.99	8.55	11.79	14.45	10.70	0.55
	μ_{Aj}	7.91	8.34	10.63	13.70		
	$D\mu_{Aj}$	-2.23	-1.80	0.48	3.55		

Total average	10.15

We consider the effect of a factor A where 4 levels exist. We can evaluate the associated variance as

$$\begin{aligned} S_A^2 &= \frac{n_B \sum_{j=1}^{n_A} \left(\mu_{A_j} - \mu\right)^2}{n} \\ &= \frac{1}{n_A} \sum_{j=1}^{n_A} \left(\mu_{A_j} - \mu\right)^2 \\ &= \frac{1}{4}\left[(-2.23)^2 + (-1.80)^2 + (0.48)^2 + (3.55)^2\right] \\ &= 5.26 \end{aligned} \tag{38}$$

The freedom is level number 4 -1 =3. Therefore, the unbiased variance is given by

$$\begin{aligned} s_A^2 &= \frac{n}{\phi_A} S_A^2 \\ &= \frac{12}{3} \times 5.26 \\ &= 21.08 \end{aligned} \tag{39}$$

We next evaluate the effect of a factor B where three level s exist. We evaluate three deviations of each level average with respect to the total average, given by

$$\Delta\mu_{B_i} = \mu_{B_i} - \mu \tag{40}$$

The corresponding variance can be evaluated as

$$
\begin{aligned}
S_B^2 &= \frac{n_A \sum_{i=1}^{n_{B_i}} \left(\mu_{B_i} - \mu\right)^2}{n} \\
&= \frac{1}{n_B} \sum_{i=1}^{n_{B_i}} \left(\mu_{B_i} - \mu\right)^2 \\
&= \frac{1}{3}\left[(-1.20)^2 + (0.65)^2 + (0.55)^2\right] \\
&= 0.72
\end{aligned}
\tag{41}
$$

The unbiased variance s_B^2 is then given by

$$
\begin{aligned}
s_B^2 &= \frac{n}{\phi_B} S_B^2 \\
&= \frac{12}{2} \times 0.72 \\
&= 4.34
\end{aligned}
\tag{42}
$$

The error is given by

$$
e_{ij} = x_{ij} - \mu - \left[\left(\mu_{A_j} - \mu\right) + \left(\mu_{B_i} - \mu\right)\right]
\tag{43}
$$

which is shown in Table 4.

Table 4. Error data

		A			
		A_1	A_2	A_3	A_4
B	B_1	0.41	1.19	-0.31	-1.29
	B_2	0.06	-0.85	-0.30	1.08
	B_3	-0.47	-0.34	0.61	0.20

The corresponding variance S_e^2 is given by

$$S_e^2 = \frac{1}{n}\sum_{i=1}^{3}\sum_{j=1}^{4} e_{ij}^2 \\ = 0.50 \tag{44}$$

The freedom of e is given by

$$\begin{aligned} \phi_e &= \phi_{tot} - \left(\phi_{aLex} + \phi_{bLex}\right) \\ &= (n-1) - (n_a - 1 + n_b - 1) \\ &= n - (n_a + n_b) + 1 \\ &= 12 - (4+3) + 1 \\ &= 6 \end{aligned} \tag{45}$$

Therefore, the unbiased variance is given by

$$\begin{aligned} s_e^2 &= \frac{n}{\phi_e} S_e^2 \\ &= \frac{12}{6} \times 0.50 \\ &= 1.01 \end{aligned} \tag{46}$$

The F associated with a factor A is given by

$$F_A = \frac{s_A^2}{s_e^2} = \frac{21.08}{1.01} = 20.87 \tag{47}$$

The critical value for F_{Ac} is given by

$$F_{Ac}(3, 6, 0.95) = 4.76 \tag{48}$$

Therefore, F_A is larger than F_{Ac}, and hence there is dependence of a factor B. The F value associate with the factor B is given by

$$F_B = \frac{s_B^2}{s_e^2} = \frac{4.34}{1.01} = 4.29 \tag{49}$$

The critical value for F is given by

$$F_{Bc}(2,6,0.95) = 5.14 \tag{50}$$

Therefore, F is smaller than F_{Bc}, and hence there is no clear dependence on the factor B.

7. Analysis of Variance (Two Way Analysis with Repeated Data)

We consider two factors of A and B, and the factor A has four levels of A_1, A_2, A_3, A_4 denoted as n_A, and the factor B has three levels of B_1, B_2, B_3 denoted as n_B, where $n_A = 4$ and $n_B = 3$. In the previous section, we assumed a one set data. That is, there is only one data for the level combination of $A_j B_i$. We assume that the data are two sets, that is, repeated data as shown in Table 5.

There are two data for $A_j B_i$ and are denoted as x_{ij_s} and $s = 1,2$.

The total data number n is given by

$$\begin{aligned} n &= n_A \times n_B \times n_s \\ &= 3 \times 4 \times 2 \\ &= 24 \end{aligned} \tag{51}$$

The total average μ is evaluated as

$$\begin{aligned} \mu &= \frac{1}{n} \sum_{s=1}^{n_s} \sum_{j=1}^{n_A} \sum_{i=1}^{n_B} x_{ij\_s} \\ &= 10.07 \end{aligned} \tag{52}$$

We can then evaluate each data deviation Δx_{ij_s} from the total average, and it is given by

$$\Delta x_{ij\_s} = x_{ij\_s} - \mu \tag{53}$$

which is shown in Table 6.

Table 5. Two-factors cross data table with repeat of data 1 and 2

Data 1		A			
		A_1	A_2	A_3	A_4
B	B_1	7.55	8.98	8.55	11.02
	B_2	8.12	6.96	11.89	11.89
	B_3	8.97	10.95	12.78	16.55

Total mean	10.07

Data 2		A			
		A_1	A_2	A_3	A_4
B	B_1	6.11	8.78	8.44	9.89
	B_2	9.01	9.65	9.86	13.22
	B_3	8.79	9.21	12.01	12.47

Table 6. Deviation of each data with respect to the total average

Data 1 deviation		A			
		A_1	A_2	A_3	A_4
B	B_1	-2.52	-1.09	-1.52	0.95
	B_2	-1.95	-3.11	1.82	1.82
	B_3	-1.10	0.88	2.71	6.48

Data 2 deviation		A			
		A_1	A_2	A_3	A_4
B	B_1	-3.96	-1.29	-1.63	-0.18
	B_2	-1.06	-0.42	-0.21	3.15
	B_3	-1.28	-0.86	1.94	2.40

We form a table averaged the data 1 and 2, and it is given by

$$\bar{x}_{ij} = \frac{1}{2}\left(x_{ij\_1} + x_{ij\_1}\right) \tag{54}$$

which is shown in Table 7. We regard that we have two sets of the same data.

Table 7. Average of data 1 and 2

Average of data 1,2		A					
		A_1	A_2	A_3	A_4	μ_{Bi}	$\Delta\mu_{Bi}$
B	B_1	6.83	8.88	8.50	10.46	8.67	-1.40
	B_2	8.57	8.31	10.88	12.56	10.08	0.01
	B_3	8.88	10.08	12.40	14.51	11.47	1.40
	μ_{Aj}	8.09	9.09	10.59	12.51		
	$D\mu_{Aj}$	-1.98	-0.98	0.52	2.44		

The averages are evaluated as

$$\mu_{Aj} = \frac{1}{n_B}\sum_{i=1}^{n_B}\bar{x}_{ij} \tag{55}$$

$$\mu_{Bi} = \frac{1}{n_A}\sum_{j=1}^{n_A}\bar{x}_{ij} \tag{56}$$

and the average deviation can be evaluated as

$$\Delta\mu_{Aj} = \mu_{Aj} - \mu \tag{57}$$

$$\Delta\mu_{Bi} = \mu_{Bi} - \mu \tag{58}$$

Let us consider the effectiveness of a factor A. The related sum of square S_{Aex}^2 is given by

$$\begin{aligned} S_{Aex}^2 &= \frac{\left[\sum_{i=1}^{n_A}\left(\mu_{A_i} - \mu\right)^2\right] n_B \times 2}{n} \\ &= \frac{\left[\sum_{i=1}^{n_A}\left(\mu_{A_i} - \mu\right)^2\right] n_B \times 2}{2n_A n_B} \\ &= \frac{\sum_{i=1}^{n_A}\left(\mu_{A_i} - \mu\right)^2}{n_A} \\ &= \frac{(-1.98)^2 + (-0.98)^2 + (0.52)^2 + (2.44)^2}{4} \\ &= 2.77 \end{aligned} \tag{59}$$

The related freedom ϕ_{Aex} is the level number of 4 minus 1, and is given by

$$\begin{aligned} \phi_{Aex} &= n_A - 1 \\ &= 4 - 1 \\ &= 3 \end{aligned} \tag{60}$$

Therefore, the corresponding unbiased variance s_{Aex}^2 is given by

$$\begin{aligned} s_{Aex}^2 &= \frac{n}{\phi_{Aex}} S_{Aex}^2 \\ &= \frac{24}{3} \times 2.77 \\ &= 22.17 \end{aligned} \tag{61}$$

Let us consider the effectiveness of the factor B. The related sum of square S_{Bex}^2 is given by

$$S_{Bex}^2 = \frac{\left[\sum_{i=1}^{n_B}\left(\mu_{B_i} - \mu\right)^2\right] n_A \times 2}{n}$$
$$= \frac{\left[\sum_{i=1}^{n_B}\left(\mu_{B_i} - \mu\right)^2\right] n_A \times 2}{2n_A n_B}$$
$$= \frac{\sum_{i=1}^{n_B}\left(\mu_{B_i} - \mu\right)^2}{n_B}$$
$$= \frac{(-1.40)^2 + (0.01)^2 + (1.40)^2}{3}$$
$$= 1.31 \tag{62}$$

The related freedom ϕ_{Bex} is the level number of 3 minus 1, and is given by

$$\phi_{Bex} = n_B - 1$$
$$= 3 - 1$$
$$= 2 \tag{63}$$

Therefore, the corresponding unbiased variance s_{Bex}^2 is given by

$$s_{Bex}^2 = \frac{n}{\phi_{bLex}} S_{Bex}^2$$
$$= \frac{24}{2} \times 1.31$$
$$= 15.69 \tag{64}$$

Table 8. Deviation of the data with respect the average for each level

Data 1		A			
		A_1	A_2	A_3	A_4
B	B_1	0.72	0.10	0.05	0.57
	B_2	-0.45	-1.35	1.02	-0.66
	B_3	0.09	0.87	0.39	2.04

Data 2		A			
		A_1	A_2	A_3	A_4
B	B_1	-0.72	-0.10	-0.06	-0.57
	B_2	0.45	1.35	-1.02	0.67
	B_3	-0.09	-0.87	-0.39	-2.04

We obtain two sets of data and the both data for the same levels of $A_j B_i$ are different in general. This difference is called as a pure error and is given by

$$e_{pure\_ij\_s} = x_{ij\_s} - \bar{x}_{ij} \tag{65}$$

It should be noted that

$$e_{pure\_ij\_2} = -e_{pure\_ij\_1} \tag{66}$$

which is shown in Table 8.

We can evaluate the variance associated with the difference $S^2_{e\_pure}$ and is given by

$$\begin{aligned} S^2_{e\_pure} &= \frac{1}{n}\left[\sum_j^{n_B}\sum_i^{n_A}\left(e_{pure\_ij\_1}\right)^2\right]\times 2 \\ &= \frac{1}{24}\left\{\left[(0.72)^2+(0.10)^2+(0.05)^2+(0.57)^2\right]\right. \\ &\quad +\left[(-0.45)^2+(-1.35)^2+(1.02)^2+(-0.66)^2\right] \\ &\quad \left.+\left[(0.09)^2+(0.87)^2+(0.39)^2+(2.04)^2\right]\right\}\times 2 \\ &= 0.78 \end{aligned} \tag{67}$$

Let us consider the corresponding freedom ϕ_{e_pure}. The data number is given by

$$n = n_a \times n_b \times 2 \tag{68}$$

We use averages of $n_a \times n_b$ kinds. Therefore, we have a freedom of

$$\begin{aligned} \phi_{e\_pure} &= n_a \times n_b \times 2 - n_a \times n_b \\ &= n_a \times n_b \times (2-1) \\ &= 4\times 3 \\ &= 12 \end{aligned} \tag{69}$$

Therefore, the corresponding unbiased variation $s^2_{e\_pure}$ is given by

$$s^2_{e\_pure} = \frac{n}{\phi_{e\_pure}} S^2_{e\_pure} = \frac{24}{12} \times 0.78 = 1.57 \tag{70}$$

The deviation of each data from the total average e_{ij_s} is given by

$$e_{ij\_s} = x_{ij\_s} - \mu - \left[\left(\mu_{A_j} - \mu \right) + \left(\mu_{B_i} - \mu \right) \right] \tag{71}$$

What is the difference between the deviation and the pure error plus effectiveness of factors A and B means? If the effectiveness of the factors A and B is independent, the error is equal to the pure error. Therefore, the difference is related to the interaction between the two factors. The difference is called with an interaction, and is given by

$$\begin{aligned} e_{\text{interact}\_ij\_s} &= e_{ij\_s} - e_{pure\_ij\_s} \\ &= \left\{ x_{ij\_s} - \mu - \left[\left(\mu_{A_j} - \mu \right) + \left(\mu_{B_i} - \mu \right) \right] \right\} - \left(x_{ij\_s} - \bar{x}_{ij} \right) \\ &= \bar{x}_{ij} - \mu - \left[\left(\mu_{A_j} - \mu \right) + \left(\mu_{B_i} - \mu \right) \right] \end{aligned} \tag{72}$$

Note that this does not depend on s. This is evaluated as shown in Table 9. The corresponding variance S^2_{interact} is given by

$$S^2_{\text{interact}} = \frac{2 \times \left\{ (0.14)^2 + (1.20)^2 + \cdots + (0.41)^2 + (0.61)^2 \right\}}{24} = 0.36 \tag{73}$$

Let us consider the related freedom ϕ_{interact}. We obtain the equations associated with a freedom given by

$$\phi_{tot} = \phi_{aLex} + \phi_{bLex} + \phi_{\text{interact}} + \phi_{e\_pure} \tag{74}$$

Therefore, we obtain

$$\begin{aligned}\phi_{\text{interact}} &= \phi_{tot} - \left(\phi_{Aex} + \phi_{Bex} + \phi_{e\_pure}\right) \\ &= n - 1 - \left(n_A - 1 + n_B - 1 + n_A \times n_B \times \left(n_s - 1\right)\right) \\ &= 23 - \left(3 + 2 + 12\right) \\ &= 6\end{aligned} \tag{75}$$

The corresponding unbiased variance s_{interact}^2 is given by

$$\begin{aligned}s_{\text{interact}}^2 &= \frac{n}{\phi_{\text{interact}}} S_{\text{interact}}^2 \\ &= \frac{24}{6} \times 0.36 \\ &= 1.45\end{aligned} \tag{76}$$

Table 9. Interactions

Data 1		A			
		A$_1$	A$_2$	A$_3$	A$_4$
B	B$_1$	0.14	1.20	-0.69	-0.65
	B$_2$	0.47	-0.79	0.28	0.04
	B$_3$	-0.61	-0.41	0.41	0.61

Data 2		A			
		A$_1$	A$_2$	A$_3$	A$_4$
B	B$_1$	-1.84	0.22	-0.17	1.79
	B$_2$	-1.51	-1.77	0.80	2.48
	B$_3$	-2.59	-1.39	0.93	3.04

Effectiveness of each factor can be evaluated with corresponding F values.
The effectiveness of the factor A can be evaluated as

$$F_A = \frac{s_{Aex}^2}{s_{e\_pure}^2} = 14.14 \tag{77}$$

The critical F value F_{ac} is given by

$$\begin{aligned}F_{Ac} &= F\left(\phi_{Aex}, \phi_e, P\right) \\ &= F\left(3, 12, 0.95\right) \\ &= 3.49\end{aligned} \tag{78}$$

Therefore, the factor A is effective.
The effectiveness of the factor B can be evaluated as

$$F_B = \frac{s_{Bex}^2}{s_{e\_pure}^2} = 10.01 \tag{79}$$

The critical F value F_{Bc} is given by

$$\begin{aligned} F_{Bc} &= F\left(\phi_{Bex}, \phi_e, P\right) \\ &= F\left(2,12,0.95\right) \\ &= 3.89 \end{aligned} \tag{80}$$

Therefore, the factor B is effective.
The effectiveness of the interaction can be evaluated as

$$F_{\text{interact}} = \frac{s_{\text{interact}}^2}{s_{e\_pure}^2} = 0.92 \tag{81}$$

The critical F value $F_{\text{interactc}}$ is given by

$$\begin{aligned} F_{\text{interactc}} &= F\left(\phi_{\text{interact}}, \phi_e, P\right) \\ &= F\left(6,6,0.95\right) \\ &= 3.00 \end{aligned} \tag{82}$$

Therefore, the interaction is not effective.

SUMMARY

This chapter is summarized in the following.

One Way Analysis

The effectiveness of the level is evaluated with S_{ex}^2 given by

$$S_{ex}^2 = \frac{n_{A_1}\left(\mu_{A_1} - \mu\right)^2 + n_{A_2}\left(\mu_{A_2} - \mu\right)^2 + n_{A_3}\left(\mu_{A_3} - \mu\right)^2}{n_{A_1} + n_{A_2} + n_{A_3}}$$

The scattering of the data is expressed with S_{in}^2, and it is given by

$$S_{in}^2 = \frac{\sum_{i=1}^{n_{A_1}} \left(x_{iA_1} - \mu_{A_1}\right)^2 + \sum_{i=1}^{n_{A_2}} \left(x_{iA_2} - \mu_{A_2}\right)^2 + \sum_{i=1}^{n_{A_3}} \left(x_{iA_3} - \mu_{A_3}\right)^2}{n_{A_1} + n_{A_2} + n_{A_3}}$$

The correlation ratio η^2 is given by

$$\eta^2 = \frac{S_{ex}^2}{S_{in}^2 + S_{ex}^2}$$

This is between 0 and 1, and the effectiveness of the factor can be regarded as significant with larger η^2.

We form an unbiased variable as

$$s_{ex}^2 = \frac{n}{\phi_{ex}} S_{ex}^2$$

where

$$\phi_{ex} = p$$

p is the level number.

$$s_{in}^2 = \frac{n}{\phi_{in}} S_{in}^2$$

where

$$\phi_{in} = n - p$$

Finally, the ratio of the unbiased variance is denoted by F and it is given by

$$F = \frac{s_{ex}^2}{s_{in}^2}$$

This follows the F distribution with a freedom of $\left(\phi_{ex}, \phi_{in}\right)$. The P point for the F distribution is denoted as $F_C\left(\phi_{ex}, \phi_{in}, P\right)$.

If $F > F_c$ we regard that the factor is valid, and vice versa.

The effectiveness between each level can be evaluated as

$$\frac{\left|\mu_{A_i} - \mu_{A_j}\right|}{\sqrt{\frac{s_{in}^2}{2}\left(\frac{1}{n_{A_i}} + \frac{1}{n_{Aj}}\right)}}$$

If this value is larger than the studentized range distribution table value of $q(r, n-r, P)$, we judge that the difference is effective.

The other simple way to evaluate the difference is the one with

$$z_i = \frac{\mu_{A_i} - \mu}{\sqrt{\frac{s_{in}^2}{2}\left(\frac{1}{n_{A_i}} + \frac{1}{n}\right)}}$$

We may be able to compare absolute value of this with z_p for a normal distribution.

Two way analysis without repeated data

We consider two factors of A : A_1, A_2, A_3, A_4 and B : B_1, B_2, B_3 , where $n_A = 4$ and $n_B = 3$. The total data number n is given by

$$n = n_A \times n_B$$

In this case, each data x_{ij} can be expressed by the deviation from the average, and is given by

$$x_{ij} = \mu + \left(\mu_{A_j} - \mu\right) + \left(\mu_{B_i} - \mu\right) + e_{ij}$$

The various variances are given by

$$S_{Aex}^2 = \frac{\sum_{i=1}^{n_A}\left(\mu_{A_i} - \mu\right)^2}{n_A}$$

$$S_{Bex}^2 = \frac{\sum_{i=1}^{n_B} \left(\mu_{B_i} - \mu \right)^2}{n_B}$$

$$S_e^2 = \frac{1}{n} \sum_{i=1}^{n_B} \sum_{j=1}^{n_A} e_{ij}^2$$

The various freedoms are given by

$$\phi_{tot} = n - 1$$

$$\phi_A = n_A - 1$$

$$\phi_B = n_B - 1$$

The freedom associated with error is given by

$$\begin{aligned} \phi_e &= \phi_{tot} - \left(\phi_A + \phi_B \right) \\ &= \left(n - 1 \right) - \left(n_A - 1 + n_B - 1 \right) \\ &= n - \left(n_A + n_B \right) + 1 \end{aligned}$$

Therefore, the unbiased variances are given by

$$s_A^2 = \frac{n}{\phi_A} S_A^2$$

$$s_B^2 = \frac{n}{\phi_B} S_B^2$$

$$s_e^2 = \frac{n}{\phi_e} S_e^2$$

The F associated with the factor A is given by

$$F_A = \frac{s_A^2}{s_e^2}$$

This is compared with the F critical value of $F_{Ac}\left(\phi_A, \phi_c, P\right)$

$$F_B = \frac{s_B^2}{s_e^2}$$

This is compared with the F critical value of $F_{Bc}\left(\phi_B, \phi_c, P\right)$

- Two way analysis with repeated data

We consider two factors of A: $A_1, A_2, \cdots, A_{n_A}$ and $B: B_1, B_2, \cdots, B_{n_B}$, and we have n_s set.

The total data number n is given by

$$n = n_A \times n_B \times n_s$$

The total average is given by

$$\mu = \frac{1}{n}\sum_{s=1}^{n_s}\sum_{j=1}^{n_A}\sum_{i=1}^{n_B} x_{ij\_s}$$

Each data deviation Δx_{ij_s} from the total average is given by

$$\Delta x_{ij\_s} = x_{ij\_s} - \mu$$

The average data for each level is given by

$$\bar{x}_{ij} = \frac{1}{2}\left(x_{ij\_1} + x_{ij\_1}\right)$$

The averages are evaluated as

$$\mu_{Aj} = \frac{1}{n_B}\sum_{i=1}^{n_B}\bar{x}_{ij}$$

$$\mu_{Bi} = \frac{1}{n_A} \sum_{j=1}^{n_A} \bar{x}_{ij}$$

and the average deviations can be evaluated as

$$\Delta\mu_{Aj} = \mu_{Aj} - \mu$$

$$\Delta\mu_{Bi} = \mu_{Bi} - \mu$$

$$S_{Aex}^2 = \frac{\sum_{i=1}^{n_A} \left(\mu_{A_i} - \mu \right)^2}{n_A}$$

$$\phi_{Aex} = n_A - 1$$

Therefore, the corresponding unbiased variance s_{Aex}^2 is given by

$$s_{Aex}^2 = \frac{n}{\phi_{Aex}} S_{Aex}^2$$

$$S_{Bex}^2 = \frac{\sum_{i=1}^{n_B} \left(\mu_{B_i} - \mu \right)^2}{n_B}$$

$$\phi_{Bex} = n_B - 1$$

The corresponding unbiased variance s_{Bex}^2 is given by

$$s_{Bex}^{(2)} = \frac{n}{\phi_{bLex}} S_{Bex}^{(2)}$$

A pure error is given by

$$e_{pure\_ij\_s} = x_{ij\_s} - \bar{x}_{ij}$$

$$S^2_{e\_pure} = \frac{1}{n_A n_B n_s}\left[\sum_j^{n_B}\sum_i^{n_A}\sum_s^{n_s}\left(e_{pure\_ij\_s}\right)^2\right]$$

The deviation of each data from the total average e_{ij_s} is given by

$$e_{ij\_s} = x_{ij\_s} - \mu - \left[\left(\mu_{A_j} - \mu\right) + \left(\mu_{B_i} - \mu\right)\right]$$

The difference associated with interaction is given by

$$\begin{aligned} e_{\text{interact}\_ij\_s} &= e_{ij\_s} - e_{pure\_ij\_s} \\ &= \bar{x}_{ij} - \mu - \left[\left(\mu_{A_j} - \mu\right) + \left(\mu_{B_i} - \mu\right)\right] \end{aligned}$$

$$\begin{aligned} S^2_{\text{interact}} &= \frac{1}{n_A n_B n_s}\sum_j^{n_A}\sum_i^{n_B}\sum_s^{n_s}\left(e_{\text{interact}\_ij\_s}\right)^2 \\ &= \frac{1}{n_A n_B}\sum_j^{n_A}\sum_i^{n_B}\left\{\bar{x}_{ij} - \mu - \left[\left(\mu_{A_j} - \mu\right) + \left(\mu_{B_i} - \mu\right)\right]\right\}^2 \end{aligned}$$

$$\begin{aligned} \phi_{\text{interact}} &= \phi_{tot} - \left(\phi_{Aex} + \phi_{Bex} + \phi_{e\_pure}\right) \\ &= n - 1 - \left(n_A - 1 + n_B - 1 + n_A \times n_B \times \left(n_s - 1\right)\right) \end{aligned}$$

$$s^2_{\text{interact}} = \frac{n}{\phi_{\text{interact}}} S^2_{\text{interact}}$$

$$F_A = \frac{s^2_{Aex}}{s^2_{e\_pure}}$$

The critical F value F_{ac} is given by

$$F_{Ac} = F\left(\phi_{Aex}, \phi_e, P\right)$$

Therefore, the factor A is effective.
The effectiveness of the factor B can be evaluated as

$$F_B = \frac{s_{Bex}^2}{s_{e\_pure}^2}$$

The critical F value F_{Bc} is given by

$$F_{Bc} = F\left(\phi_{Bex}, \phi_e, P\right)$$

Therefore, the factor B is effective.
The effectiveness of the interaction can be evaluated as

$$F_{\text{interact}} = \frac{s_{\text{interact}}^2}{s_{e\_pure}^2}$$

The critical F value $F_{\text{interactc}}$ is given by

$$F_{\text{interactc}} = F\left(\phi_{\text{interact}}, \phi_e, P\right)$$

Chapter 9

Analysis of Variances Using an Orthogonal Table

Abstract

We show a variable analysis using an orthogonal table which enables us to treat factors more than two. Although the level number for each factor is limited to two, many factors can be assigned to columns in the orthogonal table and we can also evaluate interaction of targeted combination, and obtain easily the effectiveness of each factor.

Keywords: analysis of variance, orthogonal table, interaction

1. Introduction

We studied variable analysis with two factors in the previous chapter. However, we have a case to treat factors more than two. For example, when we treat 4 factors with two levels of each, we need the data number at least of $2^4 = 16$. The factor number a with level number m requires the minimum data number of a^m. The number of levels we should try to obtain a result increases significantly with increasing a and m. Therefore, we need some procedure to treat such many factors, which is called as experimental design using an orthogonal table. We focus on the two levels with many factors in this chapter.

2. TWO FACTOR ANALYSIS

We start with two factors where each level number is two.

The factor is donated as A and B, and the levels for each factors are denoted as A_1, A_2 and B_1, B_2. The corresponding data are shown in Table 1. The related data for given factors are denoted as y_{ij}, where $i, j = 1, 2$.

This table is converted as a data list for a given combination of factors shown in Table 2.

The combination is also expressed as the two columns as shown in Table 3.

Finally, we convert the notation 1 to -1, and 2 to 1, and obtain Table 4. We regard each column as vector, and the dot product of the two vectors is given by

$$\begin{pmatrix} -1 & -1 & 1 & 1 \end{pmatrix} \begin{pmatrix} -1 \\ 1 \\ -1 \\ 1 \end{pmatrix} = 1 - 1 - 1 + 1 = 0 \tag{1}$$

Therefore, the two vectors are orthogonal to each other, which is the reason why the table is called as orthogonal.

Table 1. Factors, levels, and related data

		B	
		B_1	B_2
A	A_1	y_{11}	y_{12}
	A_2	y_{21}	y_{22}

Table 2. Combination of factors, and related data

Combination	Data
A_1B_1	y_{11}
A_1B_2	y_{12}
A_2B_1	y_{21}
A_2B_2	y_{22}

Table 3. Two column expression for factors, and related data

A	B	Data
1	1	y_{11}
1	2	y_{12}
2	1	y_{21}
2	2	y_{22}

Table 4. Two column expression for factors, and related data

A	B	Data
-1	-1	y_{11}
-1	1	y_{12}
1	-1	y_{21}
1	1	y_{22}

We assume that the data y_{ij} is composed of many factors: the contribution of a factor A_i denoted as α_i, a factor B_j denoted as β_j, interaction between α_i and β_j denoted as $(\alpha\beta)_{ij}$, an independent constant number denoted as μ, and an error denoted as e_{ij}. y_{ij} is hence expressed by

$$y_{ij} = \mu + \alpha_i + \beta_j + (\alpha\beta)_{ij} + e_{ij} \tag{2}$$

The number of y_{ij} is denoted as n, and in this case, it is given by

$$n = 2^2 = 4 \tag{3}$$

We impose the restrictions to the variables as

$$\alpha_1 + \alpha_2 = 0 \tag{4}$$

$$\beta_1 + \beta_2 = 0 \tag{5}$$

$$(\alpha\beta)_{11} + (\alpha\beta)_{12} = 0 \tag{6}$$

$$(\alpha\beta)_{21} + (\alpha\beta)_{22} = 0 \tag{7}$$

$$(\alpha\beta)_{11} + (\alpha\beta)_{21} = 0 \tag{8}$$

$$(\alpha\beta)_{12} + (\alpha\beta)_{22} = 0 \tag{9}$$

The above treatment does not vanish generality. Therefore, we decrease the variables number as

$$-\alpha_1 = \alpha_2 \equiv \alpha \tag{10}$$

$$-\beta_1 = \beta_2 \equiv \beta \tag{11}$$

$$(\alpha\beta)_{11} = -(\alpha\beta)_{21} = -(\alpha\beta)_{21} = (\alpha\beta)_{22} = (\alpha\beta) \tag{12}$$

Each data is then given by

$$y_{11} = \mu - \alpha - \beta + (\alpha\beta) + e_{11} \tag{13}$$

$$y_{12} = \mu - \alpha + \beta - (\alpha\beta) + e_{12} \tag{14}$$

$$y_{21} = \mu + \alpha - \beta - (\alpha\beta) + e_{21} \tag{15}$$

$$y_{22} = \mu + \alpha + \beta + (\alpha\beta) + e_{21} \tag{16}$$

We should guess the values of μ, α, β, and $(\alpha\beta)$ from the obtained data y_{ij}. The predicted value can be determined so that the error is the minimum. The sum of error Q_e^2 can be evaluated as

$$Q_e^2 = e_{11}^2 + e_{12}^2 + e_{21}^2 + e_{22}^2$$
$$= \{y_{11} - [\mu - \alpha - \beta + (\alpha\beta)]\}^2$$
$$+ \{y_{12} - [\mu - \alpha + \beta - (\alpha\beta)]\}^2$$
$$+ \{y_{21} - [\mu + \alpha - \beta - (\alpha\beta)]\}^2$$
$$+ \{y_{22} - [\mu + \alpha + \beta + (\alpha\beta)]\}^2 \tag{17}$$

Differentiating Eq. (17) with respect to μ and set it to 0, we obtain

$$\frac{\partial Q_e^2}{\partial \mu} = -2\{y_{11} - [\mu - \alpha - \beta + (\alpha\beta)]\}$$
$$-2\{y_{12} - [\mu - \alpha + \beta - (\alpha\beta)]\}$$
$$-2\{y_{21} - [\mu + \alpha - \beta - (\alpha\beta)]\}$$
$$-2\{y_{22} - [\mu + \alpha + \beta + (\alpha\beta)]\}$$
$$= -2[(y_{11} + y_{12} + y_{21} + y_{22}) - 4\hat{\mu}]$$
$$= 0 \tag{18}$$

We then obtain

$$\hat{\mu} = \frac{1}{n}(y_{11} + y_{12} + y_{21} + y_{22}) \tag{19}$$

where we use hat $\hat{}$ to express that it is the predicted value. We utilize this notation in the followings.

Differentiating Eq. (17) with respect to α and set it to 0, we obtain

$$\frac{\partial Q_e^2}{\partial \alpha} = 2\{y_{11} - [\mu - \alpha - \beta + (\alpha\beta)]\}$$
$$+2\{y_{12} - [\mu - \alpha + \beta - (\alpha\beta)]\}$$
$$-2\{y_{21} - [\mu + \alpha - \beta - (\alpha\beta)]\}$$
$$-2\{y_{22} - [\mu + \alpha + \beta + (\alpha\beta)]\}$$
$$= 2[(y_{11} + y_{12} - y_{21} - y_{22}) + 4\hat{\alpha}]$$
$$= 0 \tag{20}$$

We then obtain

$$\hat{\alpha} = \frac{1}{n}\left(-y_{11} - y_{12} + y_{21} + y_{22}\right) \tag{21}$$

Differentiating Eq. (17) with respect to β and set it to 0, we obtain

$$\begin{aligned}
\frac{\partial Q_e^2}{\partial \beta} &= 2\left\{y_{11} - \left[\mu - \alpha - \beta + (\alpha\beta)\right]\right\} \\
&\quad -2\left\{y_{12} - \left[\mu - \alpha + \beta - (\alpha\beta)\right]\right\} \\
&\quad +2\left\{y_{21} - \left[\mu + \alpha - \beta - (\alpha\beta)\right]\right\} \\
&\quad -2\left\{y_{22} - \left[\mu + \alpha + \beta + (\alpha\beta)\right]\right\} \\
&= 2\left[\left(y_{11} - y_{12} + y_{21} - y_{22}\right) + 4\hat{\beta}\right] \\
&= 0
\end{aligned} \tag{22}$$

We then obtain

$$\hat{\beta} = \frac{1}{n}\left(-y_{11} + y_{12} - y_{21} + y_{22}\right) \tag{23}$$

Differentiating Eq. (17) with respect to $(\alpha\beta)$ and set it to 0, we obtain

$$\begin{aligned}
\frac{\partial Q_e^2}{\partial (\alpha\beta)} &= -2\left\{y_{11} - \left[\mu - \alpha - \beta + (\alpha\beta)\right]\right\} \\
&\quad +2\left\{y_{12} - \left[\mu - \alpha + \beta - (\alpha\beta)\right]\right\} \\
&\quad +2\left\{y_{21} - \left[\mu + \alpha - \beta - (\alpha\beta)\right]\right\} \\
&\quad -2\left\{y_{22} - \left[\mu + \alpha + \beta + (\alpha\beta)\right]\right\} \\
&= -2\left[\left(y_{11} - y_{12} - y_{21} + y_{22}\right) - 4(\alpha\beta)\right] \\
&= 0
\end{aligned} \tag{24}$$

We then obtain

$$(\alpha\beta) = \frac{1}{n}(y_{11} - y_{12} - y_{21} + y_{22}) \tag{25}$$

We define the variables below

$$Q_{A+} = y_{11} + y_{12} \tag{26}$$

$$Q_{A-} = y_{21} + y_{22} \tag{27}$$

$$Q_{B+} = y_{12} + y_{22} \tag{28}$$

$$Q_{B-} = y_{11} + y_{21} \tag{29}$$

$$Q_{\alpha\beta+} = y_{11} + y_{22} \tag{30}$$

$$Q_{\alpha\beta-} = y_{12} + y_{21} \tag{31}$$

The total average μ is given

$$\mu = \hat{\mu} = \frac{1}{n}(y_{11} + y_{12} + y_{21} + y_{22}) \tag{32}$$

This is also expressed by

$$\mu = \frac{1}{n}(Q_{\xi+} + Q_{\xi-}) \tag{33}$$

where ξ is a dummy variable and $\xi = \alpha, \beta, \alpha\beta$.

The average associated with each factor of levels are expressed by

$$\mu_{\xi-} = \frac{1}{\left(\frac{n}{2}\right)} Q_{\xi-} \tag{34}$$

$$\mu_{\xi+} = \frac{1}{\left(\frac{n}{2}\right)} Q_{\xi+} \tag{35}$$

Let us consider the variance associated with a factor A denoted by S_A^2, which is evaluated as

$$S_A^2 = \frac{\frac{n}{2}\left[\left(\mu_{A+} - \mu\right)^2 + \left(\mu_{A+} - \mu\right)^2\right]}{n} \tag{36}$$

This can be further reduced to

$$\begin{aligned} S_A^2 &= \frac{\frac{n}{2}\left[\left(\mu_{A-} - \mu\right)^2 + \left(\mu_{A+} - \mu\right)^2\right]}{n} \\ &= \frac{1}{2}\left\{\left[\frac{Q_{A-}}{\left(\frac{n}{2}\right)} - \frac{1}{n}\left(Q_{A-} + Q_{A+}\right)\right]^2 + \left[\frac{Q_{A+}}{\left(\frac{n}{2}\right)} - \frac{1}{n}\left(Q_{A-} + Q_{A+}\right)\right]^2\right\} \\ &= \frac{1}{2}\left\{\left[\frac{1}{n}\left(Q_{A-} - Q_{A+}\right)\right]^2 + \left[\frac{1}{n}\left(Q_{A+} - Q_{A-}\right)\right]^2\right\} \\ &= \frac{1}{n^2}\left(Q_{A+} - Q_{A-}\right)^2 \end{aligned} \tag{37}$$

The other variances are evaluated similar way as

$$S_B^2 = \frac{1}{n^2}\left(Q_{B+} - Q_{B-}\right)^2 \tag{38}$$

$$S_C^2 = \frac{1}{n^2}\left(Q_{B+} - Q_{B-}\right)^2 \tag{39}$$

$$S_{AB}^2 = \frac{1}{n^2}\left(Q_{AB+} - Q_{AB-}\right)^2 \tag{40}$$

We can modify the table related to the above equations.

First, we can add column associated with AB as in Table 5. The final table can be formed using this data as shown Table 6.

Table 5. Two column expression for factors adding interaction, and related data

A	B	AB	Data
-1	-1	1	y_{11}
-1	1	-1	y_{12}
1	-1	-1	y_{21}
1	1	1	y_{22}

Table 6. Two column expression for factors adding interaction, and related data

Factor	A		B		(AB)	
Level	A-	A+	B-	B+	(AB)-	(AB)+
Data	y_{11}	y_{21}	y_{11}	y_{12}	y_{12}	y_{11}
	y_{12}	y_{22}	y_{21}	y_{22}	y_{21}	y_{22}
Sum	Q_{A-}	Q_{A+}	Q_{B-}	Q_{B+}	$Q_{(AB)-}$	$Q_{(AB)+}$
Variance	$S^{(2)}{}_A$		$S^{(2)}{}_B$		$S^{(2)}{}_{(AB)}$	

The freedom is all 1, that is,

$$\phi_A = \phi_B = \phi_{AB} = 1 \tag{41}$$

and the unbiased variance is the same as the variance. The unbiased variances are denoted as small s character and are given by

$$s_A^2 = \frac{n}{\phi_A} S_A^2, s_B^2 = \frac{n}{\phi_B} S_B^2, s_{AB}^2 = \frac{n}{\phi_{AB}} S_{AB}^2 \tag{42}$$

We cannot separate the variance associated with error. However, if the interaction can be ignored, that is $(\alpha\beta) = 0$, the related variance can be regarded as the one for error. Therefore,

$$s_e^2 \leftarrow s_{AB}^2 \tag{43}$$

We can then test the effectiveness the factor using a F distribution as

$$F_A = \frac{s_A^2}{s_e^2} \tag{44}$$

$$F_B = \frac{s_B^2}{s_e^2} \tag{45}$$

The critical value for the F distribution F_{crit} is given by

$$F_{crit} = F\left(1,1,0.95\right) \tag{46}$$

3. Three Factor Analysis without Repetition

We extend the analysis in the previous section to three factors of A, B, and C.

We assume that the data y_{ijk} is expressed by the contribution of the factor A_i denoted as α_i, the factor B_j denoted as β_j, the factor C_k denoted as γ_k, the interaction between each factor denoted as $\left(\alpha\beta\right)_{ij}$, $\left(\alpha\gamma\right)_{ik}$, $\left(\beta\gamma\right)_{jk}$, $\left(\alpha\beta\gamma\right)_{ijk}$, the independent constant number denoted as μ, and the error denoted as e_{ijk}.

The number of y_{ijk} is denoted as n, and

$$n = 2^3 = 8 \tag{47}$$

and it is expressed as

$$y_{ijk} = \mu + \alpha_i + \beta_j + \gamma_k + \left(\alpha\beta\right)_{ij} + \left(\alpha\gamma\right)_{ik} + \left(\beta\gamma\right)_{jk} + \left(\alpha\beta\gamma\right)_{ijk} + e_{ijk} \tag{48}$$

We impose restrictions to the variables as

$$\alpha_1 + \alpha_2 = 0 \tag{49}$$

$$\beta_1 + \beta_2 = 0 \tag{50}$$

$$\gamma_1 + \gamma_2 = 0 \quad (51)$$

$$(\alpha\beta)_{11} + (\alpha\beta)_{12} = 0 \quad (52)$$

$$(\alpha\beta)_{21} + (\alpha\beta)_{22} = 0 \quad (53)$$

$$(\alpha\beta)_{11} + (\alpha\beta)_{21} = 0 \quad (54)$$

$$(\alpha\beta)_{12} + (\alpha\beta)_{22} = 0 \quad (55)$$

$$(\alpha\gamma)_{11} + (\alpha\gamma)_{12} = 0 \quad (56)$$

$$(\alpha\gamma)_{21} + (\alpha\gamma)_{22} = 0 \quad (57)$$

$$(\alpha\gamma)_{11} + (\alpha\gamma)_{21} = 0 \quad (58)$$

$$(\alpha\gamma)_{12} + (\alpha\gamma)_{22} = 0 \quad (59)$$

$$(\beta\gamma)_{11} + (\beta\gamma)_{12} = 0 \quad (60)$$

$$(\beta\gamma)_{21} + (\beta\gamma)_{22} = 0 \quad (61)$$

$$(\beta\gamma)_{11} + (\beta\gamma)_{21} = 0 \quad (62)$$

$$(\beta\gamma)_{12} + (\beta\gamma)_{22} = 0 \quad (63)$$

These restrictions do not vanish the generality. We assume that

$$\begin{aligned} &s_1 : i + j(even) \\ &s_2 : i + j(odd) \end{aligned} \quad (64)$$

and impose

$$\left(\alpha\beta\gamma\right)_{s_1 1}+\left(\alpha\beta\gamma\right)_{s_1 2}=0 \tag{65}$$

$$\left(\alpha\beta\gamma\right)_{s_2 1}+\left(\alpha\beta\gamma\right)_{s_2 2}=0 \tag{66}$$

$$\left(\alpha\beta\gamma\right)_{s_1 1}+\left(\alpha\beta\gamma\right)_{s_2 1}=0 \tag{67}$$

$$\left(\alpha\beta\gamma\right)_{s_1 2}+\left(\alpha\beta\gamma\right)_{s_2 2}=0 \tag{68}$$

Eqs. (65)-(68) can be summarized as

$$\left(\alpha\beta\gamma\right)_{u_1}+\left(\alpha\beta\gamma\right)_{u_2}=0 \tag{69}$$

where

$$\begin{aligned} &u_1 : i+j+k\left(even\right) \\ &u_2 : i+j+k\left(odd\right) \end{aligned} \tag{70}$$

Therefore, we decrease the variables number as

$$-\alpha_1=\alpha_2\equiv\alpha \tag{71}$$

$$-\beta_1=\beta_2\equiv\beta \tag{72}$$

$$-\gamma_1=\gamma_2\equiv\gamma \tag{73}$$

$$\left(\alpha\beta\right)_{11}=-\left(\alpha\beta\right)_{21}=-\left(\alpha\beta\right)_{21}=\left(\alpha\beta\right)_{22}=\left(\alpha\beta\right) \tag{74}$$

$$\left(\alpha\gamma\right)_{11}=-\left(\alpha\gamma\right)_{21}=-\left(\alpha\gamma\right)_{21}=\left(\alpha\gamma\right)_{22}=\left(\alpha\gamma\right) \tag{75}$$

$$\left(\beta\gamma\right)_{11}=-\left(\beta\gamma\right)_{21}=-\left(\beta\gamma\right)_{21}=\left(\beta\gamma\right)_{22}=\left(\beta\gamma\right) \tag{76}$$

$$(\alpha\beta\gamma)_{u_1} = -(\alpha\beta\gamma)_{u_2} = (\alpha\beta\gamma) \tag{77}$$

Each data y_{ijk} is then expressed by

$$y_{111} = \mu - \alpha - \beta + (\alpha\beta) - \gamma + (\alpha\gamma) + (\beta\gamma) - (\alpha\beta\gamma) + e_{111} \tag{78}$$

$$y_{112} = \mu - \alpha - \beta + (\alpha\beta) + \gamma - (\alpha\gamma) - (\beta\gamma) + (\alpha\beta\gamma) + e_{112} \tag{79}$$

$$y_{121} = \mu - \alpha + \beta - (\alpha\beta) - \gamma + (\alpha\gamma) - (\beta\gamma) + (\alpha\beta\gamma) + e_{121} \tag{80}$$

$$y_{122} = \mu - \alpha + \beta - (\alpha\beta) + \gamma - (\alpha\gamma) + (\beta\gamma) - (\alpha\beta\gamma) + e_{122} \tag{81}$$

$$y_{211} = \mu + \alpha - \beta - (\alpha\beta) - \gamma - (\alpha\gamma) + (\beta\gamma) + (\alpha\beta\gamma) + e_{211} \tag{82}$$

$$y_{212} = \mu + \alpha - \beta - (\alpha\beta) + \gamma + (\alpha\gamma) - (\beta\gamma) - (\alpha\beta\gamma) + e_{212} \tag{83}$$

$$y_{221} = \mu + \alpha + \beta + (\alpha\beta) - \gamma - (\alpha\gamma) - (\beta\gamma) - (\alpha\beta\gamma) + e_{221} \tag{84}$$

$$y_{222} = \mu + \alpha + \beta + (\alpha\beta) + \gamma + (\alpha\gamma) + (\beta\gamma) + (\alpha\beta\gamma) + e_{222} \tag{85}$$

Based on the above analysis, we can construct an orthogonal table which is called as $L_8(2^7)$. This can accommodate the case with factors up to 6. 8 in the notation of $L_8(2^7)$ comes from 2^3 and 3 comes from the number of character A, B, C. 2 in the notation of $L_8(2^7)$ comes from the level number. 7 in the notation of $L_8(2^7)$ comes from the column number.

The column number is assigned to the character as

$A \rightarrow 2^0 = 1$: first column
$B \rightarrow 2^1 = 2$: second column
$C \rightarrow 2^2 = 4$: fourth column

This is related to the direct effects of each factor.

The other column is related to AB, AC, BC, ABC as below.

$$AB \to 1+2=3$$
$$AC \to 1+4=5$$
$$BC \to 2+4=6$$
$$ABC \to 1+2+4=7$$

We can generate the y_{ijk} in order as shown in the table. The level of column 1(A), 2(B), and 4(C) are directly related to the subscript of y_{ijk}, where it is -1 when the subscript is 1 and 1 when the subscript is 2. This can also be appreciated that it is -1 when the subscript is odd number and 1 when the subscript is even number.

The level for clum3 (AB) is determined from the sum of $i+j$ is an odd or even number. The other column levels are also determined with the same way.

Table 7. Orthogonal table $L_8\left(2^7\right)$

Column No.		1	2	3	4	5	6	7	Data	Value
Combination No	1	-1	-1	1	-1	1	1	-1	y_{111}	2.3
	2	-1	-1	1	1	-1	-1	1	y_{112}	3.4
	3	-1	1	-1	-1	1	-1	1	y_{121}	4.5
	4	-1	1	-1	1	-1	1	-1	y_{122}	5.6
	5	1	-1	-1	-1	-1	1	1	y_{211}	7.5
	6	1	-1	-1	1	1	-1	-1	y_{212}	8.9
	7	1	1	1	-1	-1	-1	-1	y_{221}	9.7
	8	1	1	1	1	1	1	1	y_{222}	8.9
Factor		A	B	AB	C	AC	BC	ABC		

Let us perform dot product of any combination of columns. For example levels of column number 1 and 2. We perform dot product of columns 1 and 2 are

$$\begin{pmatrix} 1 & 1 & 1 & 1 & -1 & -1 & -1 & -1 \end{pmatrix} \begin{pmatrix} 1 \\ 1 \\ -1 \\ -1 \\ 1 \\ 1 \\ -1 \\ -1 \end{pmatrix} = 1+1-1-1-1-1+1+1$$
$$= 0 \tag{86}$$

We can verify this result for any combination of columns. This expresses the orthogonality of the table.

We should guess the value of μ, α, β, γ, and $(\alpha\beta)$, $(\alpha\gamma)$, $(\beta\gamma)$, and $(\alpha\beta\gamma)$ from the obtained data y_{ijk}. The predicted value can be determined so that the error is the minimum. The error Q_e can be evaluated as

$$\begin{aligned}
Q_e^2 &= \sum_{i=1}^{2}\sum_{j=1}^{2}\sum_{k=1}^{2} e_{ijk}^2 \\
&= \left\{y_{111} - \left[\mu - \alpha - \beta + (\alpha\beta) - \gamma + (\alpha\gamma) + (\beta\gamma) - (\alpha\beta\gamma)\right]\right\}^2 \\
&+ \left\{y_{112} - \left[\mu - \alpha - \beta + (\alpha\beta) + \gamma - (\alpha\gamma) - (\beta\gamma) + (\alpha\beta\gamma)\right]\right\}^2 \\
&+ \left\{y_{121} - \left[\mu - \alpha + \beta - (\alpha\beta) - \gamma + (\alpha\gamma) - (\beta\gamma) + (\alpha\beta\gamma)\right]\right\}^2 \\
&+ \left\{y_{122} - \left[\mu - \alpha + \beta - (\alpha\beta) + \gamma - (\alpha\gamma) + (\beta\gamma) - (\alpha\beta\gamma)\right]\right\}^2 \\
&+ \left\{y_{211} - \left[\mu + \alpha - \beta - (\alpha\beta) - \gamma - (\alpha\gamma) + (\beta\gamma) + (\alpha\beta\gamma)\right]\right\}^2 \\
&+ \left\{y_{212} - \left[\mu + \alpha - \beta - (\alpha\beta) + \gamma + (\alpha\gamma) - (\beta\gamma) - (\alpha\beta\gamma)\right]\right\}^2 \\
&+ \left\{y_{221} - \left[\mu + \alpha + \beta + (\alpha\beta) - \gamma - (\alpha\gamma) - (\beta\gamma) - (\alpha\beta\gamma)\right]\right\}^2 \\
&+ \left\{y_{222} - \left[\mu + \alpha + \beta + (\alpha\beta) + \gamma + (\alpha\gamma) + (\beta\gamma) + (\alpha\beta\gamma)\right]\right\}^2
\end{aligned} \tag{87}$$

Differentiating Eq. (87) with respect to μ and set it to 0, we obtain

$$\begin{aligned}
\frac{\partial Q_e^2}{\partial \mu} &= -2\left\{y_{111} - \left[\mu - \alpha - \beta + (\alpha\beta) - \gamma + (\alpha\gamma) + (\beta\gamma) - (\alpha\beta\gamma)\right]\right\} \\
&-2\left\{y_{112} - \left[\mu - \alpha - \beta + (\alpha\beta) + \gamma - (\alpha\gamma) - (\beta\gamma) + (\alpha\beta\gamma)\right]\right\} \\
&-2\left\{y_{121} - \left[\mu - \alpha + \beta - (\alpha\beta) - \gamma + (\alpha\gamma) - (\beta\gamma) + (\alpha\beta\gamma)\right]\right\} \\
&-2\left\{y_{122} - \left[\mu - \alpha + \beta - (\alpha\beta) + \gamma - (\alpha\gamma) + (\beta\gamma) - (\alpha\beta\gamma)\right]\right\} \\
&-2\left\{y_{211} - \left[\mu + \alpha - \beta - (\alpha\beta) - \gamma - (\alpha\gamma) + (\beta\gamma) + (\alpha\beta\gamma)\right]\right\} \\
&-2\left\{y_{212} - \left[\mu + \alpha - \beta - (\alpha\beta) + \gamma + (\alpha\gamma) - (\beta\gamma) - (\alpha\beta\gamma)\right]\right\} \\
&-2\left\{y_{221} - \left[\mu + \alpha + \beta + (\alpha\beta) - \gamma - (\alpha\gamma) - (\beta\gamma) - (\alpha\beta\gamma)\right]\right\} \\
&-2\left\{y_{222} - \left[\mu + \alpha + \beta + (\alpha\beta) + \gamma + (\alpha\gamma) + (\beta\gamma) + (\alpha\beta\gamma)\right]\right\} \\
&= -2\left\{(y_{111} + y_{112} + y_{121} + y_{122} + y_{211} + y_{212} + y_{221} + y_{222}) - 8\mu\right\} \\
&= 0
\end{aligned} \tag{88}$$

We then obtain

$$\hat{\mu} = \frac{1}{n}\left(y_{111} + y_{112} + y_{121} + y_{122} + y_{211} + y_{212} + y_{221} + y_{222}\right) \tag{89}$$

Differentiating Eq. (87) with respect to α and set it to 0, we obtain

$$\begin{aligned}
\frac{\partial Q_e^2}{\partial \alpha} &= 2\left\{y_{111} - \left[\mu - \alpha - \beta + (\alpha\beta) - \gamma + (\alpha\gamma) + (\beta\gamma) - (\alpha\beta\gamma)\right]\right\} \\
&+2\left\{y_{112} - \left[\mu - \alpha - \beta + (\alpha\beta) + \gamma - (\alpha\gamma) - (\beta\gamma) + (\alpha\beta\gamma)\right]\right\} \\
&+2\left\{y_{121} - \left[\mu - \alpha + \beta - (\alpha\beta) - \gamma + (\alpha\gamma) - (\beta\gamma) + (\alpha\beta\gamma)\right]\right\} \\
&+2\left\{y_{122} - \left[\mu - \alpha + \beta - (\alpha\beta) + \gamma - (\alpha\gamma) + (\beta\gamma) - (\alpha\beta\gamma)\right]\right\} \\
&-2\left\{y_{211} - \left[\mu + \alpha - \beta - (\alpha\beta) - \gamma - (\alpha\gamma) + (\beta\gamma) + (\alpha\beta\gamma)\right]\right\} \\
&-2\left\{y_{212} - \left[\mu + \alpha - \beta - (\alpha\beta) + \gamma + (\alpha\gamma) - (\beta\gamma) - (\alpha\beta\gamma)\right]\right\} \\
&-2\left\{y_{221} - \left[\mu + \alpha + \beta + (\alpha\beta) - \gamma - (\alpha\gamma) - (\beta\gamma) - (\alpha\beta\gamma)\right]\right\} \\
&-2\left\{y_{222} - \left[\mu + \alpha + \beta + (\alpha\beta) + \gamma + (\alpha\gamma) + (\beta\gamma) + (\alpha\beta\gamma)\right]\right\} \\
&= -2\left\{\left(y_{111} + y_{112} + y_{121} + y_{122} - y_{211} - y_{212} - y_{221} - y_{222}\right) + 8\alpha\right\} \\
&= 0
\end{aligned} \tag{90}$$

We then obtain

$$\hat{\alpha} = \frac{1}{n}\left(-y_{111} - y_{112} - y_{121} - y_{122} + y_{211} + y_{212} + y_{221} + y_{222}\right) \tag{91}$$

Similarly, we obtain

$$\hat{\beta} = \frac{1}{n}\left(-y_{111} - y_{112} + y_{121} + y_{122} - y_{211} - y_{212} + y_{221} + y_{222}\right) \tag{92}$$

$$\hat{\gamma} = \frac{1}{n}\left(-y_{111} + y_{112} - y_{121} + y_{122} - y_{211} + y_{212} - y_{221} + y_{222}\right) \tag{93}$$

$$\alpha\beta = \frac{1}{n}\left(y_{111} + y_{112} - y_{121} - y_{122} - y_{211} - y_{212} + y_{221} + y_{222}\right) \tag{94}$$

$$\alpha\gamma = \frac{1}{n}\left(y_{111} - y_{112} + y_{121} - y_{122} - y_{211} + y_{212} - y_{221} + y_{222}\right) \tag{95}$$

$$\beta\gamma = \frac{1}{n}\left(y_{111} - y_{112} - y_{121} + y_{122} + y_{211} - y_{212} - y_{221} + y_{222}\right) \tag{96}$$

$$\alpha\beta\gamma = \frac{1}{n}\left(-y_{111} + y_{112} + y_{121} - y_{122} + y_{211} - y_{212} - y_{221} + y_{222}\right) \tag{97}$$

We define the variables below

$$Q_{A+} = y_{211} + y_{212} + y_{221} + y_{222} \tag{98}$$

$$Q_{A-} = y_{111} + y_{112} + y_{121} + y_{122} \tag{99}$$

$$Q_{B+} = y_{121} + y_{122} + y_{221} + y_{222} \tag{100}$$

$$Q_{B-} = y_{111} + y_{112} + y_{211} + y_{212} \tag{101}$$

$$Q_{C+} = y_{112} + y_{122} + y_{212} + y_{222} \tag{102}$$

$$Q_{C-} = y_{111} + y_{121} + y_{211} + y_{221} \tag{103}$$

$$Q_{\alpha\beta+} = y_{111} + y_{112} + y_{221} + y_{222} \tag{104}$$

$$Q_{\alpha\beta-} = y_{121} + y_{122} + y_{211} + y_{212} \tag{105}$$

$$Q_{\alpha\gamma+} = y_{111} + y_{121} + y_{212} + y_{222} \tag{106}$$

$$Q_{\alpha\gamma-} = y_{112} + y_{122} + y_{211} + y_{221} \tag{107}$$

$$Q_{\beta\gamma+} = y_{111} + y_{122} + y_{211} + y_{222} \tag{108}$$

$$Q_{\beta\gamma-} = y_{112} + y_{121} + y_{212} + y_{221} \tag{109}$$

$$Q_{\alpha\beta\gamma+} = y_{112} + y_{121} + y_{211} + y_{222} \tag{110}$$

$$Q_{\alpha\beta\gamma-} = y_{111} + y_{122} + y_{212} + y_{221} \tag{111}$$

The total average μ is given

$$\mu = \frac{1}{n}\left(y_{111} + y_{112} + y_{121} + y_{122} + y_{211} + y_{212} + y_{221} + y_{222} \right) \tag{112}$$

This is also expressed by

$$\mu = \frac{1}{n}\left(Q_{\xi+} + Q_{\xi-} \right) \tag{113}$$

where ξ is a dummy variable and $\xi = \alpha, \beta, \cdots, \alpha\beta\gamma$.

The average associated with each factor of levels are expressed by

$$\mu_{\xi-} = \frac{1}{\left(\frac{n}{2}\right)} Q_{\xi-} \tag{114}$$

$$\mu_{\xi+} = \frac{1}{\left(\frac{n}{2}\right)} Q_{\xi+} \tag{115}$$

Let us consider the variance associated with a factor A denoted by S_A^2, which is evaluated as

$$S_A^2 = \frac{\frac{n}{2}\left[\left(\mu_{A+} - \mu \right)^2 + \left(\mu_{A+} - \mu \right)^2 \right]}{n} \tag{116}$$

This can be further reduced to

$$\begin{aligned} S_A^2 &= \frac{\frac{n}{2}\left[\left(\mu_{A-} - \mu \right)^2 + \left(\mu_{A+} - \mu \right)^2 \right]}{n} \\ &= \frac{1}{2}\left\{ \left[\frac{Q_{A-}}{\left(\frac{n}{2}\right)} - \frac{1}{n}\left(Q_{A-} + Q_{A+} \right) \right]^2 + \left[\frac{Q_{A+}}{\left(\frac{n}{2}\right)} - \frac{1}{n}\left(Q_{A-} + Q_{A+} \right) \right]^2 \right\} \\ &= \frac{1}{2}\left\{ \left[\frac{1}{n}\left(Q_{A-} - Q_{A+} \right) \right]^2 + \left[\frac{1}{n}\left(Q_{A+} - Q_{A-} \right) \right]^2 \right\} \\ &= \frac{1}{n^2}\left(Q_{A+} - Q_{A-} \right)^2 \end{aligned} \tag{117}$$

The other variances are evaluated with the similar way as

$$S_B^2 = \frac{1}{n^2}\left(Q_{B+} - Q_{B-}\right)^2 \tag{118}$$

$$S_C^2 = \frac{1}{n^2}\left(Q_{B+} - Q_{B-}\right)^2 \tag{119}$$

$$S_{AB}^2 = \frac{1}{n^2}\left(Q_{AB+} - Q_{AB-}\right)^2 \tag{120}$$

$$S_{AC}^2 = \frac{1}{n^2}\left(Q_{AC+} - Q_{AC-}\right)^2 \tag{121}$$

$$S_{BC}^2 = \frac{1}{n^2}\left(Q_{BC+} - Q_{BC-}\right)^2 \tag{122}$$

$$S_{ABC}^2 = \frac{1}{n^2}\left(Q_{ABC+} - Q_{ABC-}\right)^2 \tag{123}$$

The freedom of each variance is all 1 and their variances are the same as the variances given by

$$s_A^2 = \frac{n}{\phi_A} S_A^2, s_B^2 = \frac{n}{\phi_B} S_B^2, s_c^2 = \frac{n}{\phi_C} S_C^2 \tag{124}$$

$$s_{AB}^2 = \frac{n}{\phi_{AB}} S_{AB}^2, s_{AC}^2 = \frac{n}{\phi_{AC}} S_{AC}^2, s_{BC}^2 = \frac{n}{\phi_{BC}} S_{BC}^2 \tag{125}$$

$$s_{ABC}^2 = \frac{n}{\phi_{ABC}} S_{ABC}^2 \tag{126}$$

Based on the above process, we can form the Table 8, and evaluated value is shown in Table 9.

Table 8. Orthogonal table $L_8(2^7)$

Column No.	1		2		3		4		5		6		7	
Factor	A		B		AB		C		AC		BC		ABC	
Level	-1	1	-1	1	-1	1	-1	1	-1	1	-1	1	-1	1
Data	y_{111}	y_{211}	y_{111}	y_{121}	y_{121}	y_{111}	y_{111}	y_{112}	y_{112}	y_{111}	y_{112}	y_{111}	y_{111}	y_{112}
	y_{112}	y_{212}	y_{112}	y_{122}	y_{122}	y_{112}	y_{121}	y_{122}	y_{122}	y_{121}	y_{121}	y_{122}	y_{122}	y_{121}
	y_{121}	y_{221}	y_{211}	y_{221}	y_{211}	y_{221}	y_{211}	y_{212}	y_{211}	y_{212}	y_{212}	y_{211}	y_{212}	y_{211}
	y_{122}	y_{222}	y_{212}	y_{222}	y_{212}	y_{222}	y_{221}	y_{222}	y_{221}	y_{222}	y_{221}	y_{222}	y_{221}	y_{222}
Sum	Q_{A-}	Q_{A+}	Q_{B-}	Q_{B+}	$Q_{(AB)-}$	$Q_{(AB)+}$	Q_{C-}	Q_{C+}	$Q_{(AC)-}$	$Q_{(AC)+}$	$Q_{(BC)-}$	$Q_{(BC)+}$	$Q_{(ABC)-}$	$Q_{(ABC)+}$
Variance	$S^{(2)}{}_{A}$		$S^{(2)}{}_{B}$		$S^{(2)}{}_{AB}$		$S^{(2)}{}_{C}$		$S^{(2)}{}_{AC}$		$S^{(2)}{}_{BC}$		$S^{(2)}{}_{ABC}$	

Table 9. Values for Orthogonal table $L_8(2^7)$

Column No.	1		2		3		4		5		6		7	
Factor	A		B		AB		C		AC		BC		ABC	
Level	-1	1	-1	1	-1	1	-1	1	-1	1	-1	1	-1	1
Data	2.3	7.5	2.3	4.5	4.5	2.3	2.3	3.4	3.4	2.3	3.4	2.3	2.3	3.4
	3.4	8.9	3.4	5.6	5.6	3.4	4.5	5.6	5.6	4.5	4.5	5.6	5.6	4.5
	4.5	9.7	7.5	9.7	7.5	9.7	7.5	8.9	7.5	8.9	8.9	7.5	8.9	7.5
	5.6	8.9	8.9	8.9	8.9	8.9	9.7	8.9	9.7	8.9	9.7	8.9	9.7	8.9
Sum	15.8	35	22.1	28.7	26.5	24.3	24	26.8	26.2	24.6	26.5	24.3	26.5	24.3
Variance	5.76		0.68		0.08		0.12		0.04		0.08		0.08	
Unbiased variance	46.08		5.445		0.605		0.98		0.32		0.605		0.605	

We can utilize this table more than its original meaning.

The table is for three factors with full interaction. Therefore, we have no data associated with error.

In the practical usage, we do not consider the whole interaction.

3.1. No Interaction

If we do not consider any interaction, the direct effect is expressed with columns 1, 2, and 4.

We regard the other column as the error one, and the corresponding variance S_e^2 is given by

$$S_e^2 = S_{AB}^2 + S_{AC}^2 + S_{BC}^2 + S_{ABC}^2 = 0.27 \tag{127}$$

The corresponding freedom is the sum of the each variance and is $\phi_e = 4$ in this case, and the unbiased error variance is given by

$$s_e^2 = \frac{n}{\phi_e} S_e^2 = 0.58 \tag{128}$$

The each factor effect validity is evaluated as

$$F_A = \frac{s_A^2}{s_e^2} = \frac{46.08}{0.53} = 86.33 \tag{129}$$

$$F_B = \frac{s_B^2}{s_e^2} = \frac{5.45}{0.53} = 10.20 \tag{130}$$

$$F_C = \frac{s_C^2}{s_e^2} = \frac{0.98}{0.53} = 1.84 \tag{131}$$

The corresponding critical F value is the same for all factors F_{crit} as

$$F_{crit} = F(1,4,0.95) = 7.71 \tag{132}$$

Therefore, factors A and B are effective, and factor C is not effective.

3.2. The Other Factors

Although the table is for three factors, we can utilize it for more factors if we neglect some interaction factors.

Let us consider that we do not consider no interaction.

We want to consider 4 factors, that is, factors A, B, C, and D. We can select any columns for factor D among the columns 3, 5, 6, and 7.

We assign column 5 (AC) for the factor D, and set the experimental condition based on the Table 7.

We assume that we obtain the same value for simplicity.

$$s_D^2 = s_{AC}^2 \tag{133}$$

The variance associated with error is then given by

$$S_e^2 = S_{AB}^2 + S_{BC}^2 + S_{ABC}^2 = 0.23 \tag{134}$$

The corresponding freedom is the sum of the each variance and is $\phi_e = 3$ in this case, and the unbiased error variance is given by

$$s_e^2 = \frac{n}{\phi_e} S_e^2 = 0.61 \tag{135}$$

The each factor effect validity is evaluated as

$$F_A = \frac{s_A^2}{s_e^2} = \frac{46.08}{0.61} = 76.2 \tag{136}$$

$$F_B = \frac{s_B^2}{s_e^2} = \frac{5.45}{0.61} = 9.00 \tag{137}$$

$$F_C = \frac{s_C^2}{s_e^2} = \frac{0.98}{0.61} = 1.62 \tag{138}$$

$$F_D = \frac{s_C^2}{s_e^2} = \frac{0.32}{0.61} = 0.53 \tag{139}$$

The corresponding critical F value is the same for all factors F_{crit} as

$$F_{crit} = F\left(1,3,0.95\right) = 10.13 \tag{140}$$

Therefore, factors A is effective, and factors B, C, and D are not effective.
We need one column for error at least. Therefore, we can add four factors as maximum.

3.3. Interaction

We can consider 3 interactions as maximum, for factors A, B, and C. If we consider the interaction, we need to use the corresponding column. We consider the interaction of AC, and the other factor D. We use the column 5 for the interaction of AC, and assign the column 3 for the factor D. The other columns of 6 and 7 are the ones for error.

$$S_e^2 = S_{BC}^2 + S_{ABC}^2 = 0.15 \tag{141}$$

The corresponding freedom $\phi_e = 2$, and the unbiased variance is

$$s_e^2 = \frac{n}{\phi_e} S_e^2 = 0.61 \tag{142}$$

$$F_A = \frac{s_A^2}{s_e^2} = \frac{46.08}{0.61} = 76.2 \tag{143}$$

$$F_B = \frac{s_B^2}{s_e^2} = \frac{5.45}{0.61} = 9.00 \tag{144}$$

$$F_C = \frac{s_C^2}{s_e^2} = \frac{0.98}{0.61} = 1.62 \tag{145}$$

$$F_{AC} = \frac{s_{AC}^2}{s_e^2} = \frac{0.32}{0.61} = 0.07 \tag{146}$$

$$F_D = \frac{s_C^2}{s_e^2} = \frac{0.32}{0.61} = 0.53 \tag{147}$$

The corresponding critical F value is the same for all factors F_{crit} as

$$F_{crit} = F(1,2,0.95) = 18.51 \tag{148}$$

Therefore, only factor A is effective.

4. Three Factor Analysis with Repetition

We extend the analysis in the previous section to the data with repetition, where we have r data for one given condition.

The data is denoted as y_{ijk_r}. The number of y_{ijk} is denoted as n, and is given by

$$n = 2^3 \times r \tag{149}$$

We use $r = 3$ here, and hence $n = 24$. However, the analysis procedure is applicable for any number of r.

The orthogonal table is shown in Table 10. There are three data for the one same condition.

Table 10. Orthogonal table $L_8(2^7)$ with repeated data

Column No.		1	2	3	4	5	6	7	Data1	Data2	Data3
Combination No	1	-1	-1	1	-1	1	1	-1	y_{111_1}	y_{111_2}	y_{111_3}
	2	-1	-1	1	1	-1	-1	1	y_{112_1}	y_{112_2}	y_{112_3}
	3	-1	1	-1	-1	1	-1	1	y_{121_1}	y_{121_2}	y_{121_3}
	4	-1	1	-1	1	-1	1	-1	y_{122_1}	y_{122_2}	y_{122_3}
	5	1	-1	-1	-1	-1	1	1	y_{211_1}	y_{211_2}	y_{211_3}
	6	1	-1	-1	1	1	-1	-1	y_{212_1}	y_{212_2}	y_{212_3}
	7	1	1	1	-1	-1	-1	-1	y_{221_1}	y_{221_2}	y_{221_3}
	8	1	1	1	1	1	1	1	y_{222_1}	y_{222_2}	y_{222_3}
Factor		A	B	AB	C	AC	BC	ABC			

Therefore, we have additional variation associated error. The average associated with one condition is given by

$$\mu_{ijk} = \frac{1}{r}\sum_{p=1}^{r} y_{ijk\_p} \tag{150}$$

The corresponding 8 averages are shown in Table 11.

Table 11. Averages of repeated data

Data1	Data2	Data3	Mean
y_{111_1}	y_{111_2}	y_{111_3}	$(y_{111\_1}+y_{111\_2}+y_{111\_3})/3$
y_{112_1}	y_{112_2}	y_{112_3}	$(y_{112\_1}+y_{112\_2}+y_{112\_3})/3$
y_{121_1}	y_{121_2}	y_{121_3}	$(y_{121\_1}+y_{121\_2}+y_{121\_3})/3$
y_{122_1}	y_{122_2}	y_{122_3}	$(y_{122\_1}+y_{122\_2}+y_{122\_3})/3$
y_{211_1}	y_{211_2}	y_{211_3}	$(y_{211\_1}+y_{211\_2}+y_{211\_3})/3$
y_{212_1}	y_{212_2}	y_{212_3}	$(y_{212\_1}+y_{212\_2}+y_{212\_3})/3$
y_{221_1}	y_{221_2}	y_{221_3}	$(y_{221\_1}+y_{221\_2}+y_{221\_3})/3$
y_{222_1}	y_{222_2}	y_{222_3}	$(y_{222\_1}+y_{222\_2}+y_{222\_3})/3$

We can obtain variance associated with this S_{ein}^2 as

$$S_{ein}^2 = \frac{1}{n}\sum_{i=1}^{2}\sum_{j=1}^{2}\sum_{k=1}^{2}\sum_{r=1}^{3}\left(y_{ijk\_r} - \mu_{ijk}\right)^2 \tag{151}$$

The freedom of the variance ϕ_{ein} is given by

$$\begin{aligned}\phi_{ein} &= 2\times 2\times 2\times\left(r-1\right)\\ &= 2\times 2\times 2\times\left(3-1\right)\\ &= 16\end{aligned} \tag{152}$$

The unbiased variance is given by

$$s_{ein}^2 = \frac{n}{\phi_{ein}} S_{ein}^2 \tag{153}$$

After we obtain the data shown in the Table 10, we construct the table. We simply add the data to the corresponding columns.

Table 12. Orthogonal table $L_8\left(2^7\right)$

Column No.	1		2		3		4		5		6		7	
Factor	A		B		AB		C		AC		BC		ABC	
Level	-1	1	-1	1	-1	1	-1	1	-1	1	-1	1	-1	1
Data1	y_{111_1}	y_{211_1}	y_{111_1}	y_{121_1}	y_{121_1}	y_{111_1}	y_{111_1}	y_{112_1}	y_{112_1}	y_{111_1}	y_{112_1}	y_{111_1}	y_{111_1}	y_{112_1}
	y_{112_1}	y_{212_1}	y_{112_1}	y_{122_1}	y_{122_1}	y_{112_1}	y_{121_1}	y_{122_1}	y_{122_1}	y_{121_1}	y_{121_1}	y_{122_1}	y_{122_1}	y_{121_1}
	y_{121_1}	y_{221_1}	y_{211_1}	y_{221_1}	y_{211_1}	y_{221_1}	y_{211_1}	y_{212_1}	y_{211_1}	y_{212_1}	y_{212_1}	y_{211_1}	y_{212_1}	y_{211_1}
	y_{122_1}	y_{222_1}	y_{212_1}	y_{222_1}	y_{212_1}	y_{222_1}	y_{221_1}	y_{222_1}	y_{221_1}	y_{222_1}	y_{221_1}	y_{222_1}	y_{221_1}	y_{222_1}
Data2	y_{111_2}	y_{211_2}	y_{111_2}	y_{121_2}	y_{121_2}	y_{111_2}	y_{111_2}	y_{112_2}	y_{112_2}	y_{111_2}	y_{112_2}	y_{111_2}	y_{111_2}	y_{112_2}
	y_{112_2}	y_{212_2}	y_{112_2}	y_{122_2}	y_{122_2}	y_{112_2}	y_{121_2}	y_{122_2}	y_{122_2}	y_{121_2}	y_{121_2}	y_{122_2}	y_{122_2}	y_{121_2}
	y_{121_2}	y_{221_2}	y_{211_2}	y_{221_2}	y_{211_2}	y_{221_2}	y_{211_2}	y_{212_2}	y_{211_2}	y_{212_2}	y_{212_2}	y_{211_2}	y_{212_2}	y_{211_2}
	y_{122_2}	y_{222_2}	y_{212_2}	y_{222_2}	y_{212_2}	y_{222_2}	y_{221_2}	y_{222_2}	y_{221_2}	y_{222_2}	y_{221_2}	y_{222_2}	y_{221_2}	y_{222_2}
Data3	y_{111_3}	y_{211_3}	y_{111_3}	y_{121_3}	y_{121_3}	y_{111_3}	y_{111_3}	y_{112_3}	y_{112_3}	y_{111_3}	y_{112_3}	y_{111_3}	y_{111_3}	y_{112_3}
	y_{112_3}	y_{212_3}	y_{112_3}	y_{122_3}	y_{122_3}	y_{112_3}	y_{121_3}	y_{122_3}	y_{122_3}	y_{121_3}	y_{121_3}	y_{122_3}	y_{122_3}	y_{121_3}
	y_{121_3}	y_{221_3}	y_{211_3}	y_{221_3}	y_{211_3}	y_{221_3}	y_{211_3}	y_{212_3}	y_{211_3}	y_{212_3}	y_{212_3}	y_{211_3}	y_{212_3}	y_{211_3}
	y_{122_3}	y_{222_3}	y_{212_3}	y_{222_3}	y_{212_3}	y_{222_3}	y_{221_3}	y_{222_3}	y_{221_3}	y_{222_3}	y_{221_3}	y_{222_3}	y_{221_3}	y_{222_3}
Sum	Q_{A-}	Q_{A+}	Q_{B-}	Q_{B+}	$Q_{(AB)-}$	$Q_{(AB)+}$	Q_{C-}	Q_{C+}	$Q_{(AC)-}$	$Q_{(AC)+}$	$Q_{(BC)-}$	$Q_{(BC)+}$	$Q_{(ABC)-}$	$Q_{(ABC)+}$
Variance	$S^{(2)}_{A}$		$S^{(2)}_{B}$		$S^{(2)}_{AB}$		$S^{(2)}_{C}$		$S^{(2)}_{AC}$		$S^{(2)}_{BC}$		$S^{(2)}_{ABC}$	
Freedom	1		1		1		1		1		1		1	

We define the variables below

$$Q_{A+} = \sum_{r=1}^{3} \left(y_{211\_r} + y_{212\_r} + y_{221\_r} + y_{222\_r} \right) \tag{154}$$

$$Q_{A-} = \sum_{r=1}^{3} \left(y_{111\_r} + y_{112\_r} + y_{121\_r} + y_{122\_r} \right) \tag{155}$$

$$Q_{B+} = \sum_{r=1}^{3} \left(y_{121\_r} + y_{122\_r} + y_{221\_r} + y_{222\_r} \right) \tag{156}$$

$$Q_{B-} = \sum_{r=1}^{3} \left(y_{111\_r} + y_{112\_r} + y_{211\_r} + y_{212\_r} \right) \tag{157}$$

$$Q_{C+} = \sum_{r=1}^{3} \left(y_{112\_r} + y_{122\_r} + y_{212\_r} + y_{222\_r} \right) \tag{158}$$

$$Q_{C-} = \sum_{r=1}^{3} \left(y_{111\_r} + y_{121\_r} + y_{211\_r} + y_{221\_r} \right) \tag{159}$$

$$Q_{\alpha\beta+} = \sum_{r=1}^{3} \left(y_{111\_r} + y_{112\_r} + y_{221\_r} + y_{222\_r} \right) \tag{160}$$

$$Q_{\alpha\beta-} = \sum_{r=1}^{3} \left(y_{121\_r} + y_{122\_r} + y_{211\_r} + y_{212\_r} \right) \tag{161}$$

$$Q_{\alpha\gamma+} = \sum_{r=1}^{3} \left(y_{111\_r} + y_{121\_r} + y_{212\_r} + y_{222\_r} \right) \tag{162}$$

$$Q_{\alpha\gamma-} = \sum_{r=1}^{3} \left(y_{112\_r} + y_{122\_r} + y_{211\_r} + y_{221\_r} \right) \tag{163}$$

$$Q_{\beta\gamma+} = \sum_{r=1}^{3} \left(y_{111\_r} + y_{122\_r} + y_{211\_r} + y_{222\_r} \right) \tag{164}$$

$$Q_{\beta\gamma-} = \sum_{r=1}^{3} \left(y_{112\_r} + y_{121\_r} + y_{212\_r} + y_{221\_r} \right) \tag{165}$$

$$Q_{\alpha\beta\gamma+} = \sum_{r=1}^{3}\left(y_{112\_r} + y_{121\_r} + y_{211\_r} + y_{222\_r}\right) \tag{166}$$

$$Q_{\alpha\beta\gamma-} = \sum_{r=1}^{3}\left(y_{111\_r} + y_{122\_r} + y_{212\_r} + y_{221\_r}\right) \tag{167}$$

The total average μ is given

$$\mu = \frac{1}{n}\sum_{r=1}^{3}\left(y_{111\_r} + y_{112\_r} + y_{121\_r} + y_{122\_r} + y_{211\_r} + y_{212\_r} + y_{221\_r} + y_{222\_r}\right) \tag{168}$$

This is also expressed by

$$\mu = \frac{1}{n}\left(Q_{\xi+} + Q_{\xi-}\right) \tag{169}$$

where ξ is a dummy variable and $\xi = \alpha, \beta, \cdots, \alpha\beta\gamma$.

The average associated with each factor of levels are expressed by

$$\mu_{\xi-} = \frac{1}{\left(\frac{n}{2}\right)} Q_{\xi-} \tag{170}$$

$$\mu_{\xi+} = \frac{1}{\left(\frac{n}{2}\right)} Q_{\xi+} \tag{171}$$

Let us consider the variance associated with a factor A denoted by S_A^2, which is evaluated as

$$S_A^2 = \frac{\frac{n}{2}\left[\left(\mu_{A+} - \mu\right)^2 + \left(\mu_{A+} - \mu\right)^2\right]}{n} \tag{172}$$

This can be further reduced to

$$S_A^2 = \frac{\frac{n}{2}\left[\left(\mu_{A-} - \mu\right)^2 + \left(\mu_{A+} - \mu\right)^2\right]}{n}$$
$$= \frac{1}{2}\left\{\left[\frac{Q_{A-}}{\left(\frac{n}{2}\right)} - \frac{1}{n}\left(Q_{A-} + Q_{A+}\right)\right]^2 + \left[\frac{Q_{A+}}{\left(\frac{n}{2}\right)} - \frac{1}{n}\left(Q_{A-} + Q_{A+}\right)\right]^2\right\}$$
$$= \frac{1}{2}\left\{\left[\frac{1}{n}\left(Q_{A-} - Q_{A+}\right)\right]^2 + \left[\frac{1}{n}\left(Q_{A+} - Q_{A-}\right)\right]^2\right\}$$
$$= \frac{1}{n^2}\left(Q_{A+} - Q_{A-}\right)^2 \tag{173}$$

The other variances are evaluated with a similar way as

$$S_B^2 = \frac{1}{n^2}\left(Q_{B+} - Q_{B-}\right)^2 \tag{174}$$

$$S_C^2 = \frac{1}{n^2}\left(Q_{B+} - Q_{B-}\right)^2 \tag{175}$$

$$S_{AB}^2 = \frac{1}{n^2}\left(Q_{AB+} - Q_{AB-}\right)^2 \tag{176}$$

$$S_{AC}^2 = \frac{1}{n^2}\left(Q_{AC+} - Q_{AC-}\right)^2 \tag{177}$$

$$S_{BC}^2 = \frac{1}{n^2}\left(Q_{BC+} - Q_{BC-}\right)^2 \tag{178}$$

$$S_{ABC}^2 = \frac{1}{n^2}\left(Q_{ABC+} - Q_{ABC-}\right)^2 \tag{179}$$

The freedom of each variance is all 1 and their variances are the same as the variances given by

$$s_A^2 = \frac{n}{\phi_A} S_A^2, s_B^2 = \frac{n}{\phi_B} S_B^2, s_c^2 = \frac{n}{\phi_C} S_C^2 \tag{180}$$

$$s_{AB}^2 = \frac{n}{\phi_{AB}} S_{AB}^2, s_{AC}^2 = \frac{n}{\phi_{AC}} S_{AC}^2, s_{BC}^2 = \frac{n}{\phi_{BC}} S_{BC}^2 \tag{181}$$

$$s_{ABC}^{(2)} = \frac{n}{\phi_{ABC}} S_{ABC}^{(2)} \tag{182}$$

If we neglect the some interactions, we should regard them as the extrinsic error. For example, we neglect the interaction BC and ABC, the corresponding variance $S_{e'}^2$ is given by

$$S_{e'}^2 = S_{BC}^2 + S_{ABC}^2 \tag{183}$$

The corresponding freedom $\phi_{e'}$ is given by

$$\phi_{e'} = \phi_{BC} + \phi_{ABC} \tag{184}$$

The total variance associated with the error is given by

$$S_e^2 = S_{ein}^2 + S_{e'}^2 \tag{185}$$

The corresponding freedom ϕ_e is given by

$$\phi_e = \phi_{ein} + \phi_{e'} \tag{186}$$

The unbiased variance is then given by

$$s_e^2 = \frac{n}{\phi_e} S_e^2 \tag{187}$$

The F values are given by

$$F_A = \frac{s_A^2}{s_e^2} \tag{188}$$

$$F_B = \frac{s_B^2}{s_e^2} \tag{189}$$

$$F_C = \frac{s_C^2}{s_e^2} \tag{190}$$

$$F_{AB} = \frac{s_{AB}^2}{s_e^2} \tag{191}$$

$$F_{AC} = \frac{s_{AC}^2}{s_e^2} \tag{192}$$

The corresponding critical F value is the same for all factors F_{crit} as

$$F_{crit} = F\left(1, \phi_e, 0.95\right) \tag{193}$$

SUMMARY

To summarize the results in this chapter:

In the $L_8\left(2^7\right)$ table, we have 7 columns and 8 rows, and basically we consider three factors of A, B, C without repeated data.

The column number is assigned to the characters as

$A \rightarrow 2^0 = 1$: first column
$B \rightarrow 2^1 = 2$: second column
$C \rightarrow 2^2 = 4$: fourth column

This is related to the direct effects of each factor.

The other column is related to AB, AC, BC, ABC as below.

$$AB \rightarrow 1+2=3$$
$$AC \rightarrow 1+4=5$$
$$BC \rightarrow 2+4=6$$
$$ABC \rightarrow 1+2+4=7$$

We have 8 data denoted as y_{ijk} where i corresponds to A , j corresponds to B , and k corresponds to C, and we have $y_{111}, y_{112}, y_{121}, \cdots, y_{222}$. We focus on the corresponding number related to the column. For example, the column is assigned to AB , we form $i+j$, and assign -1 or 1 for the row in each column depending on that $i+j$ is odd or even.

$$y_{111} \rightarrow 1+1=2 \rightarrow -1$$
$$y_{112} \rightarrow 1+1=2 \rightarrow -1$$
$$y_{121} \rightarrow 1+2=3 \rightarrow 1$$
$$\cdots$$
$$y_{222} \rightarrow 2+2=4 \rightarrow -1$$

We denote each column as ξ , and denote each column with -1 as ξ_- and 1 as ξ_+.

We denote the sum of the signed column as Q_{ξ_-} or Q_{ξ_-} . The corresponding variance associated with a factor ξ is given by

$$S_\xi^{(2)} = \frac{1}{n^2}\left(Q_{\xi_+} - Q_{\xi_-}\right)^2$$

The freedom of each factor ϕ_ξ is all 1. Therefore, the corresponding unbiased variance is given by

$$s_\xi^2 = \frac{nS_\xi^2}{\phi_\xi} = nS_\xi^2$$

The variance associated with the error corresponds to the one we do not focus on. We denote it as ζ . Therefore, the variance associated with the error is given by

$$S_e^2 = \sum_\zeta S_\zeta^2$$

The unbiased variance is given by

$$s_e^2 = \frac{nS_e^2}{n_\zeta}$$

where n_ζ is the number of factors we do not focus on.

We evaluate

$$F_\xi = \frac{s_\xi^2}{s_e^2}$$

We compare this with $F\left(1, n_\zeta, P\right)$.

We then treat repeated data. The repeated number is denoted as r. The data number n then becomes $n = 2^3 \times r$.

Each variance associated with the factor has the same expression as

$$S_\xi^2 = \frac{1}{n^2}\left(Q_{\xi_+} - Q_{\xi_-}\right)^2$$

The unbiased variance has also the same expression given by

$$s_\xi^2 = \frac{nS_\xi^2}{\phi_\xi} = nS_\xi^2$$

In this case, we have r data for given i, j, k, that is, each data are expressed with y_{ijk_r} and we can define averages for each level as

$$\mu_{ijk} = \frac{1}{r}\sum_{p=1}^{r} y_{ijk\_p}$$

We can then obtain the corresponding variance as

$$S_{ein}^2 = \frac{1}{n}\sum_{i=1}^{2}\sum_{j=1}^{2}\sum_{k=1}^{2}\sum_{p=1}^{r}\left(y_{ijk\_p} - \mu_{ijk}\right)^2$$

The corresponding freedom ϕ_{ein} is given by

$$\phi_{ein} = 2^3 (r-1)$$

We regard that the total error variance is given by

$$S_e^2 = S_{ein}^2 + \sum_{\zeta} S_{\zeta}^2$$

The corresponding freedom is given by

$$\phi_e = \phi_{ein} + n_{\zeta}$$

Therefore, the corresponding unbiased variance is given by

$$s_e^2 = \frac{nS_e^2}{\phi_e}$$

We evaluate

$$F_{\xi} = \frac{s_{\xi}^2}{s_e^2}$$

We compare this with $F(1, \phi_e, P)$.

Chapter 10

PRINCIPAL COMPONENT ANALYSIS

ABSTRACT

We evaluate one subject from various points of view, and want to evaluate each subject totally. Principal component analysis enables us to evaluate it. We compose variables by linearly combining the original variables. In the principal component analysis, the variables are composed so that the variance has the maximum value. We can judge that the value of the first component expresses the total score. We show the procedure to compose the variables. The method is applied to the regression where the explanatory variables have strong interaction. The variables are composed by the principal component analysis. We can then perform the regression using the composed variable, which gives us stable results.

Keywords: principal analysis, principal component, eigenvalue, eigenvector; regression

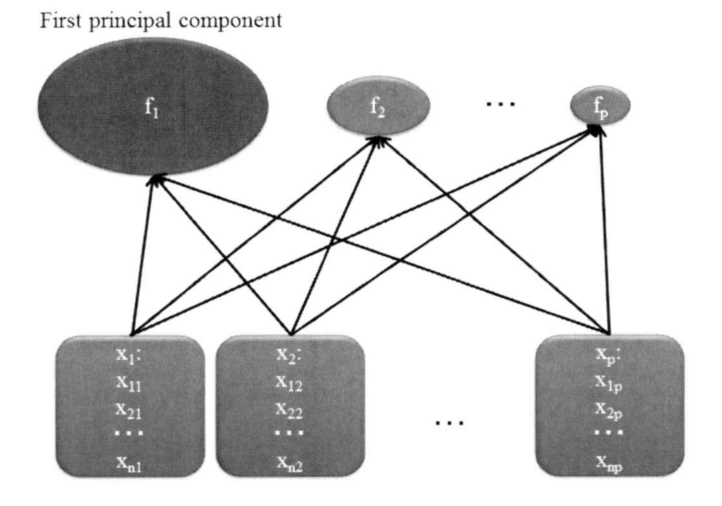

Figure 1. Path diagram for principal analysis.

1. INTRODUCTION

We want to evaluate a subject from some different point of views in many cases. It is a rarely case that one subject is superior to the other subjects in all points. Therefore, we cannot judge which is better totally considering all points of views. In the principal component analysis, we construct composite variables, and decide which is better by simply evaluating the first composite variable. The schematic path diagram is shown in Figure 1. In the figure, we use p kinds of variables, and we theoretically have the same number kinds of principal variables.

There is the other usage of principal component analysis. As we mentioned in Chapter 4 of volume 2 associated with multiple regression, the multiple regression factor coefficient becomes very complicated, unstable, and inaccurate when the interaction between explanatory variables are significant. In that case, we can construct a composite variable which consists of the explanatory variables with strong interaction. Using the composite variable, we can apply the multiple regression analysis, which makes the results simple, stable, and clear.

We treat two variables first and extend the analysis to the multiple ones.

We perform matrix operation in this chapter. The basics of the matrix operation are described in Chapter 15 of volume 3.

2. PRINCIPAL COMPONENT ANALYSIS WITH TWO VARIABLES

We consider principal analysis with two variables $x_j \left(j = 1, 2 \right)$. The sample number is n, and each data are denoted as x_{i1}, x_{i2} $\left(i = 1, 2, \cdots, n \right)$ which is shown in Table 1.

Table 1. Data form for two variable principal analysis

ID	x_1	x_1
1	x_{11}	x_{11}
2	x_{21}	x_{12}
...	...	...
i	x_{i1}	x_{i2}
...	...	...
n	x_{n1}	x_{n2}

The average and unbiased variance of x_1 are denoted as $\bar{x}_1$ and $s_1^{(2)}$, respectively. Those of x_2 are denoted as $\bar{x}_2$ and $s_2^{(2)}$, respectively. These parameters are given by

$$\bar{x}_1 = \frac{1}{n}\sum_{i=1}^{n} x_{i1} \tag{1}$$

$$s_1^{(2)} = \frac{1}{n-1}\sum_{i=1}^{n}\left(x_{i1} - \bar{x}_1\right)^2 \tag{2}$$

$$\bar{x}_2 = \frac{1}{n}\sum_{i=1}^{n} x_{i2} \tag{3}$$

$$s_2^{(2)} = \frac{1}{n-1}\sum_{i=1}^{n}\left(x_{i2} - \bar{x}_2\right)^2 \tag{4}$$

The correlation factor between x_1 and x_2 is denoted as r_{12} and is given by

$$r_{12} = \frac{1}{n-1}\sum_{i=1}^{n}\frac{\left(x_1 - \bar{x}_1\right)\left(x_2 - \bar{x}_2\right)}{\sqrt{s_1^{(2)} s_2^{(2)}}} \tag{5}$$

First of all, we normalize the variables as

$$u_1 = \frac{x_1 - \bar{x}_1}{\sqrt{s_1^{(2)}}} \tag{6}$$

$$u_2 = \frac{x_2 - \bar{x}_2}{\sqrt{s_2^{(2)}}} \tag{7}$$

The normalized variables hold below.

$$\bar{u}_1 = \bar{u}_2 = 0 \tag{8}$$

$$\sum_{i=1}^{n} u_{i1}^{\ 2} = n-1 \tag{9}$$

$$\sum_{i=1}^{n} u_{i2}^{2} = n-1 \tag{10}$$

$$\sum_{i=1}^{n} u_{i1}u_{i2} = (n-1) r_{x_1 x_2} \tag{11}$$

We construct a composite variable z below.

$$z = a_1 u_1 + a_2 u_2 \tag{12}$$

We impose that the unbiased variance of z has the maximum value to determine the factors a_1 and a_2. The unbiased variance s_z^2 is evaluated as

$$\begin{aligned} s_z^{(2)} &= \frac{1}{n-1}\sum_{i=1}^{n}\left(z_i - \bar{z}\right)^2 \\ &= \frac{1}{n-1}\sum_{i=1}^{n} z_i^{2} \\ &= \frac{1}{n-1}\sum_{i=1}^{n}\left(a_1 u_{i1} + a_2 u_{i2}\right)^2 \\ &= \frac{1}{n-1}\left\{ a_1^2 \sum_{i=1}^{n} u_{i1}^2 + 2a_1 a_2 \sum_{i=1}^{n} u_{i1}u_{i2} + a_2^2 \sum_{i=1}^{n} u_{i2}^2 \right\} \\ &= a_1^2 + a_2^2 + 2a_1 a_2 r_{x_1 x_2} \end{aligned} \tag{13}$$

This value increases with increasing a_1 and a_2. Therefore, we need a certain restriction to determine the value of a_1 and a_2. It is given by

$$a_1^2 + a_2^2 = 1 \tag{14}$$

Consequently, the problem is defined as follows.

We determine the factors of a_1 and a_2 so that the unbiased variance has the maximum value under the restriction of Eq. (14).

We apply Lagrange's method of undetermined multipliers to solve this problem. We introduce a Lagrange function given by

$$f\left(a_1, a_2, \lambda\right) = a_1^2 + a_2^2 + 2a_1 a_2 r_{12} - \lambda\left(a_1^2 + a_2^2 - 1\right) \tag{15}$$

Partial differentiating Eq. (15) with respect to a_1 and set it to 0, we obtain

$$\frac{\partial f}{\partial a_1}(a_1, a_2, \lambda) = 2a_1 + 2a_2 r_{12} - 2\lambda a_1 = 0 \tag{16}$$

Arranging this, we obtain

$$a_1 + r_{12} a_2 = \lambda a_1 \tag{17}$$

Partial differentiating Eq. (15) with respect to a_1 and set it to 0, we obtain

$$\frac{\partial f}{\partial a_2}(a_1, a_2, \lambda) = 2a_2 + 2a_2 r_{12} - 2\lambda a_2 = 0 \tag{18}$$

Arranging this, we obtain

$$a_2 + r_{12} a_1 = \lambda a_2 \tag{19}$$

Eqs. (17) and (19) are expressed by a matrix form as

$$\begin{pmatrix} 1 & r_{12} \\ r_{12} & 1 \end{pmatrix} \begin{pmatrix} a_1 \\ a_2 \end{pmatrix} = \lambda \begin{pmatrix} a_1 \\ a_2 \end{pmatrix} \tag{20}$$

This can also be expressed by

$$\mathbf{R}\boldsymbol{a} = \lambda \boldsymbol{a} \tag{21}$$

where

$$\mathbf{R} = \begin{pmatrix} 1 & r_{12} \\ r_{12} & 1 \end{pmatrix} \tag{22}$$

$$\boldsymbol{a} = \begin{pmatrix} a_1 \\ a_2 \end{pmatrix} \tag{23}$$

$$\boldsymbol{a}^T = (a_1, a_2) \tag{24}$$

Let us consider the meaning of λ. Multiplying Eq. (24) to Eq. (20), we obtain

$$(a_1, a_2)\begin{pmatrix} 1 & r_{12} \\ r_{12} & 1 \end{pmatrix}\begin{pmatrix} a_1 \\ a_2 \end{pmatrix} = (a_1, a_2)\lambda\begin{pmatrix} a_1 \\ a_2 \end{pmatrix} \quad (25)$$

Expanding this, we obtain

$$\begin{aligned} left\ side &= (a_1, a_2)\begin{pmatrix} 1 & r_{12} \\ r_{12} & 1 \end{pmatrix}\begin{pmatrix} a_1 \\ a_2 \end{pmatrix} \\ &= (a_1, a_2)\begin{pmatrix} a_1 + r_{12}a_2 \\ r_{12}a_1 + a_2 \end{pmatrix} \\ &= a_1^2 + r_{12}a_1a_2 + r_{12}a_1a_2 + a_2^2 \\ &= a_1^2 + +a_2^2 + 2r_{12}a_1a_2 \\ &= s_z^{(2)} \end{aligned} \quad (26)$$

$$\begin{aligned} right\ side &= (a_1, a_2)\lambda\begin{pmatrix} a_1 \\ a_2 \end{pmatrix} \\ &= \lambda\left(a_1^2 + a_2^2\right) \\ &= \lambda \end{aligned} \quad (27)$$

Therefore, we obtain

$$\lambda = s_z^{(2)} \quad (28)$$

λ is equal to the variance of z. Large λ means a large variance. This λ is called as an eigenvalue. Since our target for the principal analysis is to construct a variable that makes clear the difference of each member, the large variance is the expected one. Therefore, the large λ is our requested one, and we denote the maximum one as first eigenvalue, and second, and so on.

The eigenvalue λ can be evaluated as follows.

Modifying Eq. (20), we obtain

$$\begin{pmatrix} 1-\lambda & r_{12} \\ r_{12} & 1-\lambda \end{pmatrix}\begin{pmatrix} a_1 \\ a_2 \end{pmatrix} = 0 \quad (29)$$

This has roots only when it holds below.

$$\begin{vmatrix} 1-\lambda & r_{12} \\ r_{12} & 1-\lambda \end{vmatrix} = 0 \tag{30}$$

This is reduced to

$$(1-\lambda)^2 - r_{12}^2 = 0 \tag{31}$$

Therefore, we have two roots given by

$$\lambda_1 = 1 + r_{12} \tag{32}$$

$$\lambda_1 = 1 - r_{x1x2} \tag{33}$$

We can obtain a set of (a_1, a_2) for each λ.
Substituting $\lambda = 1 + r_{12}$ into (20), we obtain

$$a_1 + r_{12} a_2 = (1 + r_{12}) a_1 \tag{34}$$

$$a_2 + r_{12} a_1 = (1 + r_{12}) a_2 \tag{35}$$

These two equations are reduced to the same result of

$$a_1 = a_2 \tag{36}$$

Imposing the restriction of Eq. (14), we obtain

$$2a_1^2 = 1 \tag{37}$$

Therefore, we obtain

$$a_1 = \pm \frac{1}{\sqrt{2}} \tag{38}$$

We adopt a positive factor, and the composite number z_1 is given by

$$z_1 = \frac{1}{\sqrt{2}} u_1 + \frac{1}{\sqrt{2}} u_2 \tag{39}$$

Similarly, we obtain for $\lambda = 1 - r_{12}$ as

$$a_1 + r_{12} a_2 = (1 - r_{12}) a_1 \tag{40}$$

$$a_2 + r_{x_1 x_2} a_1 = (1 - r_{12}) a_2 \tag{41}$$

This is reduced to

$$a_1 = -a_2 \tag{42}$$

Considering the restriction of Eq. (14), we obtain

$$2a_1^2 = 1 \tag{43}$$

Therefore, we obtain

$$a_1 = \pm \frac{1}{\sqrt{2}} \tag{44}$$

We adopt the positive one, and the composite number z_2 is given by

$$z_2 = \frac{1}{\sqrt{2}} u_1 - \frac{1}{\sqrt{2}} u_2 \tag{45}$$

In the case of $r_{12} > 0$, z_1 is called as the first principal component, and z_2 is the second principal component. In the case of $r_{12} < 0$, z_2 is called as the first principal component, and z_1 is the second principal component.

Assuming $r_{12} > 0$, we summarize the results.

The first principal component eigenvalue and eigenvector are given by

$$\lambda_1 = 1 + r_{12} \tag{46}$$

$$a^{(1)} = \begin{pmatrix} \frac{1}{\sqrt{2}} \\ \frac{1}{\sqrt{2}} \end{pmatrix} \tag{47}$$

The second principal component eigenvalue and eigenvector are given by

$$\lambda_2 = 1 - r_{12} \tag{48}$$

$$\boldsymbol{a}^{(2)} = \begin{pmatrix} \frac{1}{\sqrt{2}} \\ -\frac{1}{\sqrt{2}} \end{pmatrix} \tag{49}$$

The normalized composite values are given as bellows.
The first principal component is given by

$$z_i^{(1)} = \frac{1}{\sqrt{2}} u_{i1} + \frac{1}{\sqrt{2}} u_{i2} \tag{50}$$

The second principal component is given by

$$z_i^{(2)} = \frac{1}{\sqrt{2}} u_{i1} - \frac{1}{\sqrt{2}} u_{i2} \tag{51}$$

The contribution of i component tot the total is evaluated as

$$\frac{\lambda_i}{s_1^{(2)} + s_2^{(2)}} = \frac{\lambda_i}{1 + r_{12} + 1 - r_{12}} = \frac{\lambda_i}{2} \tag{52}$$

The total sum of the eigenvalue is equal to the variable number in general.

3. Principal Component Analysis with Multi Variables

We extend the analysis of two variables in the previous section to m variables $x_1, x_2, \cdots, x_m$, where m is bigger than two. The sample data number is assumed to be n. Therefore, the data are denoted as $x_{i1}, x_{i2}, \cdots, x_{im} \ (i = 1, 2, \cdots, n)$.

Normalizing these variables, we obtain

$$u_1 = \frac{x_1 - \overline{x}_1}{\sqrt{s_1^{(2)}}}, u_2 = \frac{x_2 - \overline{x}_2}{\sqrt{s_2^{(2)}}}, \cdots, u_p = \frac{x_p - \overline{x}_p}{\sqrt{s_p^{(2)}}} \tag{53}$$

The averages and variances of the variables are 0 and 1 for all.
We construct a composited variable given by

$$z = a_1 u_1 + a_2 u_2 + \cdots + a_m u_m \tag{54}$$

The corresponding unbiased variance is given by

$$\begin{aligned}
s_z^{(2)} &= \frac{1}{n-1}\sum_{i=1}^{n}\left(z_i - \overline{z}\right)^2 \\
&= \frac{1}{n-1}\sum_{i=1}^{n} z_i^2 \\
&= \frac{1}{n-1}\sum_{i=1}^{n}\left(a_1 u_{i1} + a_2 u_{i2} + \cdots + a_m u_{im}\right)^2 \\
&= a_1^2 + a_2^2 + \cdots + a_m^2 \\
&\quad 2a_1 a_2 r_{12} + 2a_1 a_3 r_{13} + \cdots + 2a_1 a_m r_{1m} \\
&\quad +2a_2 a_3 r_{23} + 2a_2 a_4 r_{24} + \cdots + 2a_2 a_m r_{2m} \\
&\quad +2a_3 a_4 r_{34} + 2a_3 a_5 r_{35} + \cdots + 2a_3 a_m r_{2m} \\
&\quad \cdots \\
&\quad +2a_{m-1} a_m r_{m-1,m}
\end{aligned} \tag{55}$$

The constrain is expressed by

$$a_1^2 + a_2^2 + \cdots + a_m^2 = 1 \tag{56}$$

Lagrange function given by

$$\begin{aligned}
f &= a_1^2 + a_2^2 + \cdots + a_m^2 \\
&\quad 2a_1 a_2 r_{12} + 2a_1 a_3 r_{13} + \cdots + 2a_1 a_m r_{1m} \\
&\quad +2a_2 a_3 r_{23} + 2a_2 a_4 r_{24} + \cdots + 2a_2 a_m r_{2m} \\
&\quad +2a_3 a_4 r_{34} + 2a_3 a_5 r_{35} + \cdots + 2a_3 a_m r_{2m} \\
&\quad \cdots \\
&\quad +2a_{m-1} a_m r_{m-1,m} \\
&\quad -\lambda\left(a_1^2 + a_2^2 + \cdots + a_m^2 - 1\right)
\end{aligned} \tag{57}$$

Partial differentiating Eq. (57) with respect to a_1 and set it 0, we obtain

$$\frac{\partial f}{\partial a_1} = 2a_1 + 2a_2 r_{12} + 2a_3 r_{13} + \cdots + 2a_m r_{1m} - 2\lambda a_1 = 0 \tag{58}$$

Partial differentiating Eq. (57) with respect to a_2 and set it 0, we obtain

$$\frac{\partial f}{\partial a_2} = 2a_1 r_{12} + 2a_2 + 2a_3 r_{23} + 2a_4 r_{24} + \cdots + 2a_m r_{2m} - 2\lambda a_2 = 0 \tag{59}$$

Repeating the same process to the a_m, and we finally obtain the results with a matrix form, which is given by

$$\begin{pmatrix} 1 & r_{12} & \cdots & r_{1m} \\ r_{21} & 1 & \cdots & r_{2m} \\ \vdots & \vdots & \ddots & \vdots \\ r_{m1} & r_{m2} & \cdots & 1 \end{pmatrix} \begin{pmatrix} a_1 \\ a_2 \\ \vdots \\ a_m \end{pmatrix} = \lambda \begin{pmatrix} a_1 \\ a_2 \\ \vdots \\ a_m \end{pmatrix} \tag{60}$$

We obtain m kinds of eigenvalues from

$$\begin{vmatrix} 1-\lambda & r_{12} & \cdots & r_{1m} \\ r_{21} & 1-\lambda & \cdots & r_{2m} \\ \vdots & \vdots & \ddots & \vdots \\ r_{m1} & r_{m2} & \cdots & 1-\lambda \end{vmatrix} = 0 \tag{61}$$

We show briefly how we can derive eigenvalues and eigenvectors. We assume that

$$\lambda_1 > \lambda_2 > \cdots > \lambda_m \tag{62}$$

λ_i corresponds to the $i-th$ principal component. The corresponding eigenvector can be evaluated from

$$\begin{pmatrix} 1 & r_{12} & \cdots & r_{1m} \\ r_{21} & 1 & \cdots & r_{2m} \\ \vdots & \vdots & \ddots & \vdots \\ r_{m1} & r_{m2} & \cdots & 1 \end{pmatrix} \begin{pmatrix} a_1 \\ a_2 \\ \vdots \\ a_m \end{pmatrix} = \lambda_i \begin{pmatrix} a_1 \\ a_2 \\ \vdots \\ a_m \end{pmatrix} \tag{63}$$

We cannot obtain the value of the elements of the eigenvector but the ratio of each element. We consider the ratio with respect to a_1, and set the other factor as

$$a_i = \rho_i a_1 \tag{64}$$

We can evaluate ρ_i from Eq. (63), where a_1 is not determined. From the constraint of Eq. (54), we can decide a_1 as

$$\begin{aligned} a_1^2 + \rho_2^2 a_1^2 + \cdots + \rho_m^2 a_m^2 &= \left(1 + \rho_2^2 + \cdots + \rho_m^2\right) a_1^2 \\ &= 1 \end{aligned} \tag{65}$$

We select positive one and obtain

$$a_1 = \frac{1}{\sqrt{1 + \rho_2^2 + \cdots + \rho_m^2}} \tag{66}$$

Therefore, we obtain

$$z = \frac{1}{\sqrt{1 + \rho_2^2 + \cdots + \rho_m^2}} \left(u_1 + \rho_2 u_2 + \cdots + \rho_m u_m\right) \tag{67}$$

The contribution ratio is given by

$$\frac{\lambda_i}{m} \tag{68}$$

In the principal component analysis, we usually consider up to the second component. It is due to the fact that we can utilize a two-dimensional plot using them. The validity of only using two principal components can be evaluated by the sum of the component given by

$$\frac{\lambda_1 + \lambda_2}{m} \tag{69}$$

The critical value for the accuracy is not clear. However, we can roughly estimate if the value is bigger than 0.8 or not for judging the validity of the analysis.

The data are plotted on the two dimensional plane with the first and second principal component axes.

Each component variable for $j = 1, 2, \cdots, m$ is plotted as

$$\begin{aligned} &\left(a_1^{(1)}, a_1^{(2)}\right) \\ &\left(a_2^{(1)}, a_2^{(2)}\right) \\ &\cdots \\ &\left(a_m^{(1)}, a_m^{(2)}\right) \end{aligned} \tag{70}$$

The each member is plotted as

$$\left(z_i^{(1)}, z_i^{(2)}\right) \tag{71}$$

where

$$z_i^{(1)} = a_1^{(1)} u_{i1} + a_2^{(1)} u_{i2} \tag{72}$$

$$z_i^{(2)} = a_1^{(2)} u_{i1} + a_2^{(2)} u_{i2} \tag{73}$$

We treat the first and second components identically up to here. However, it should be weighted depending on the eigenvalue. We should remind that the eigenvalue has the meaning of the variance, and hence it is plausible to weight the square root of the eigenvalue which corresponds to the standard deviation. That is, we multiply $\sqrt{\lambda_1}$ to the first component, and $\sqrt{\lambda_2}$ to the second component.

We can then define the distance between two members i and j denoted as d_{ij}, and it is given by

$$d_{ij} = \sqrt{\left(\sqrt{\lambda_1} z_i^{(1)} - \sqrt{\lambda_1} z_j^{(1)}\right)^2 + \left(\sqrt{\lambda_2} z_i^{(2)} - \sqrt{\lambda_2} z_j^{(2)}\right)^2} \tag{74}$$

4. EXAMPLE FOR PRINCIPAL COMPONENT ANALYSIS (SCORE EVALUATION)

We apply principal component analysis to the score evaluation. We obtain the score data as shown in Table 2.

Table 2. Score of subjects

Member ID	Japanese x1	English x2	Mathematics x3	Science x4
1	86	79	67	68
2	71	75	78	84
3	42	43	39	44
4	62	58	98	95
5	96	97	61	63
6	39	33	45	50
7	50	53	64	72
8	78	66	52	47
9	51	44	76	72
10	89	92	93	91
Mean	66.4	64.0	67.3	68.6
Stdev.	20.5	21.6	19.4	18.0

Table 3. Normalized scores and corresponding first and second principal components

Member ID	Japanese u1	English u2	Mathematics u3	Science u4	First component	Second component
1	0.954	0.696	-0.015	-0.033	0.796	0.857
2	0.224	0.510	0.552	0.857	1.073	-0.348
3	-1.188	-0.974	-1.461	-1.368	-2.493	0.321
4	-0.214	-0.278	1.585	1.469	1.283	-1.765
5	1.441	1.531	-0.325	-0.312	1.165	1.802
6	-1.334	-1.438	-1.151	-1.035	-2.479	-0.297
7	-0.798	-0.510	-0.170	0.189	-0.643	-0.678
8	0.565	0.093	-0.790	-1.202	-0.671	1.342
9	-0.750	-0.928	0.449	0.189	-0.518	-1.148
10	1.100	1.299	1.327	1.246	2.488	-0.086

The normalized values are shown in Table 3. The correlation matrix is then evaluated as

$$\mathbf{R} = \begin{pmatrix} 1 & 0.967 & 0.376 & 0.311 \\ 0.967 & 1 & 0.415 & 0.398 \\ 0.376 & 0.415 & 1 & 0.972 \\ 0.311 & 0.398 & 0.972 & 1 \end{pmatrix} \tag{75}$$

The corresponding eigenvalues can be evaluated with a standard matrix operation shown in Chapter 15 of volume 3, and are given by

$$\lambda_1 = 2.721, \lambda_2 = 1.222, \lambda_3 = 0.052, \lambda_4 = 0.005 \tag{76}$$

The eigenvectors are given by

$$z_1 = 0.487u_1 + 0.511u_2 + 0.508u_3 + 0.493u_4 \tag{77}$$

$$z_2 = 0.527u_1 + 0.474u_2 - 0.481u_3 - 0.516u_4 \tag{78}$$

$$z_3 = -0.499u_1 + 0.539u_2 - 0.504u_3 + 0.455u_4 \tag{79}$$

$$z_4 = 0.485u_1 - 0.474u_2 - 0.506u_3 + 0.533u_4 \tag{80}$$

The contribution of each principal component is given by

$$\frac{\lambda_1}{4} = 0.680, \frac{\lambda_2}{4} = 0.306, \frac{\lambda_3}{4} = 0.013, \frac{\lambda_4}{4} = 0.001 \tag{81}$$

The contribution summing up to the second component is 98% and is more than 80%, and we can focus on the two components neglecting the other ones. The first and second components are shown in Table 3.

The first and second component for each subject is given by

$$\begin{aligned} \text{Japanese:}&(0.487, 0.527) \\ \text{English:}&(0.511, 0.474) \\ \text{Mathematics:}&(0.508, -0.481) \\ \text{Science:}&(0.493, -0.516) \end{aligned} \tag{82}$$

We treat the first and second component identically above. The weighted ones depending on the eigenvalue can also be evaluated. That is, we multiply $\sqrt{\lambda_1} = \sqrt{2.721}$ to the first component, and $\sqrt{\lambda_2} = \sqrt{1.222}$ to the second component. The resultant data are given by

$$\begin{aligned} \text{Japanese:}&(0.803, 0.583) \\ \text{English:}&(0.843, 0.524) \\ \text{Mathematics:}&(0.838, -0.532) \\ \text{Science:}&(0.813, -0.570) \end{aligned} \tag{83}$$

Let us inspect the first component z_1, where the each factor is almost the same. Therefore, it means simple summation.

Let us inspect the second component z_2. u_1 and u_2 are the almost the same, and u_3 and u_4 are the almost the same with the negative sign. The group u_1 and u_2 corresponds to Japanese and English, and the group u_3 and u_4 corresponds to mathematics and science. Therefore, this expresses the quantitative course / qualitative course. The naming is not straight forward. We should do it by inspecting the value of factors.

We further weight each score as shown in Table 4, and the corresponding two-dimensional plot is shown in Figure 2.

Table 4. Principal components and weighted principal components

Member ID	First component	Second component	Weighted first component	Weighted second component
1	0.796	0.857	1.313	0.948
2	1.073	-0.348	1.770	-0.385
3	-2.493	0.321	-4.113	0.355
4	1.283	-1.765	2.116	-1.951
5	1.165	1.802	1.922	1.992
6	-2.479	-0.297	-4.090	-0.328
7	-0.643	-0.678	-1.060	-0.750
8	-0.671	1.342	-1.107	1.483
9	-0.518	-1.148	-0.854	-1.270
10	2.488	-0.086	4.104	-0.095

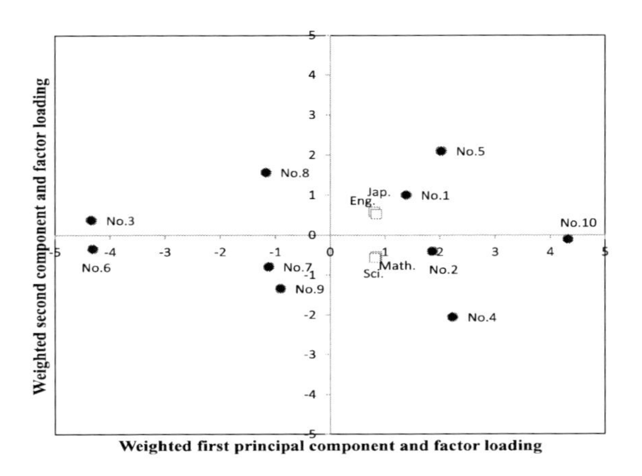

Figure 2. Weighted principal components and factor loadings.

We first simply inspect the horizontal direction which corresponds to the first principal component. No. 10 has high score, and the second group is No. 4 and No. 5, and the last group is No. 3 and No. 6.

We then focus on the vertical direction which corresponds to the second principal component. If it is positive it is qualitative characteristics and negative it is quantitative characteristics. Let us compare the second group No. 4 and 5 based on the second principal component, where both have almost the same first principal component. No. 4 has a quantitative character, and No. 5 has a quality character. We can also see that No. 10 is an all-rounder.

The distance in the plane of Figure 2 corresponds to the resemblance of each member. The distances are summarized in Table 5. No. 1 resembles to No. 5, and No.2 to No.1, and so on.

Table 5. Distance between each member

	No.1	No.2	No.3	No.4	No.5	No.6	No.7	No.8	No.9	No.10
No.1		1.5	5.8	3.2	1.3	5.9	3.1	2.6	3.3	3.1
No.2	1.5		6.3	1.7	2.5	6.2	3.0	3.6	2.9	2.5
No.3	5.8	6.3		7.0	6.6	0.7	3.4	3.4	3.8	8.7
No.4	3.2	1.7	7.0		4.2	6.8	3.6	5.0	3.2	2.9
No.5	1.3	2.5	6.6	4.2		6.8	4.3	3.2	4.5	3.2
No.6	5.9	6.2	0.7	6.8	6.8		3.2	3.7	3.6	8.6
No.7	3.1	3.0	3.4	3.6	4.3	3.2		2.4	0.6	5.5
No.8	2.6	3.6	3.4	5.0	3.2	3.7	2.4		2.9	5.7
No.9	3.3	2.9	3.8	3.2	4.5	3.6	0.6	2.9		5.4
No.10	3.1	2.5	8.7	2.9	3.2	8.6	5.5	5.7	5.4	

5. MULTIPLE REGRESSION USING PRINCIPAL COMPONENT ANALYSIS

In the multiple regression analysis, the analysis becomes unstable and inaccurate when the interaction between explanatory variables is significant as shown in Chapter 4 of volume 2. In that case, we can utilize the principal analysis to overcome the problem.

Let us consider the case where the interaction between two explanatory variables x_1 and x_2 is significant. We construct a composite variable given by

$$z = a_1 u_1 + a_2 u_2 \tag{84}$$

where u_1 and u_2 are normalized variables. We then obtain the first principal component as

$$z_i^{(1)} = \frac{1}{\sqrt{2}} u_{i1} + \frac{1}{\sqrt{2}} u_{i2} \tag{85}$$

We should evaluate the contribution of the component given by

$$\frac{\lambda_1}{2} \tag{86}$$

This should be more than 0.8.

We assume that the contribution is more than 0.8 and continue the analysis below.

The average of the first principal component is zero.

The variance $\sigma_{z1}^{(2)}$ can be evaluated below.

$$\begin{aligned}
\sigma_{z1}^{(2)} &= \frac{1}{n-1}\sum_{i=1}^{n} z_{i1}{}^2 \\
&= \frac{1}{n-1}\sum_{i=1}^{n}\left(\frac{1}{\sqrt{2}}u_{i1} + \frac{1}{\sqrt{2}}u_{i2}\right)^2 \\
&= \frac{1}{2}\frac{1}{n-1}\sum_{i=1}^{n}\left(u_{i1}{}^2 + u_{i2}{}^2 + 2u_{i1}u_{i2}\right) \\
&= 1 + r_{12}
\end{aligned} \tag{87}$$

The correlation factor between objective variables and the first principal component is given by

$$\begin{aligned}
r_{yz1} &= \frac{1}{n-1}\sum_{i=1}^{n}\frac{v_i z_i^{(1)}}{\sqrt{\sigma_{z1}^{(2)}}} \\
&= \frac{1}{\sqrt{1+r_{12}}}\frac{1}{n-1}\sum_{i=1}^{n} v_i\left(\frac{1}{\sqrt{2}}u_{i1} + \frac{1}{\sqrt{2}}u_{i2}\right) \\
&= \frac{1}{\sqrt{2(1+r_{12})}}\left(r_{y1} + r_{y2}\right)
\end{aligned} \tag{88}$$

When r_{12} is zero, it becomes average of each correlation factor, that is, the two variables becomes one variable.

The regression is given by

$$\begin{aligned}
v &= r_{yz1} z^{(1)} \\
&= \frac{1}{\sqrt{2(1+r_{12})}}\left(r_{y1} + r_{y2}\right)\left(\frac{1}{\sqrt{2}}u_1 + \frac{1}{\sqrt{2}}u_2\right) \\
&= \frac{r_{y1} + r_{y2}}{2\sqrt{1+r_{12}}}\left(u_1 + u_2\right)
\end{aligned} \tag{89}$$

Therefore, the regression can be expressed by the average correlation factor, which is decreased by the interaction between the two explanatory variables.

We can easily extend this procedure to multi variables. We can obtain the first component variable that consists of m variables where the interaction is significant. That is,

$$z = \sum_{i=1}^{m} a_i^{(1)} u_i \tag{90}$$

We then analyze the multiple regression analysis using this lumped variable as well as the other variables.

SUMMARY

To summarize the results in this chapter—

We normalize the data as

$$u_{ij}\left(j = 1, 2, \cdots, p; i = 1, 2, \cdots, n\right)$$

We then obtain the matrix relationship given by

$$\begin{pmatrix} 1 & r_{12} & \cdots & r_{1m} \\ r_{21} & 1 & \cdots & r_{2m} \\ \vdots & \vdots & \ddots & \vdots \\ r_{m1} & r_{m2} & \cdots & 1 \end{pmatrix} \begin{pmatrix} a_1 \\ a_2 \\ \vdots \\ a_m \end{pmatrix} = \lambda \begin{pmatrix} a_1 \\ a_2 \\ \vdots \\ a_m \end{pmatrix}$$

where r_{ij} is the correlation factor between factor i and j. We can obtain eigenvalues and eigenvectors given by

$$\lambda_1; a^{(1)} = \left(k_1^{(1)}, k_2^{(1)}, \cdots, k_p^{(1)}\right)$$

$$\lambda_2; a^{(2)} = \left(k_1^{(2)}, k_2^{(2)}, \cdots, k_p^{(2)}\right)$$

• • •

$$\lambda_p; a^{(p)} = \left(k_1^{(p)}, k_2^{(p)}, \cdots, k_p^{(p)}\right)$$

We then express the data as

$$z_i^{(1)} = k_1^{(1)} u_{i1} + k_2^{(1)} u_{i2} + \cdots + k_p^{(1)} u_{ip}$$

$$z_i^{(2)} = k_1^{(2)} u_{i1} + k_2^{(2)} u_{i2} + \cdots + k_p^{(2)} u_{ip}$$

$$\cdot \cdot \cdot$$

$$z_i^{(p)} = k_1^{(p)} u_{i1} + k_2^{(p)} u_{i2} + \cdots + k_p^{(p)} u_{ip}$$

This is only the different expression of the same data.

However, the difference exits where we focus on only the first two components. The effectiveness of the two component selection can be evaluated as

$$\frac{\lambda_1 + \lambda_2}{\lambda_1 + \lambda_2 + \cdots + \lambda_p} = \frac{\lambda_1 + \lambda_2}{p}$$

$$\left(z_i^{(1)}, z_i^{(2)} \right)$$

Each item can be plotted in the two dimensional plane as

$$\begin{aligned} &Item\ 1: \left(k_1^{(1)}, k_1^{(2)} \right) \\ &Item\ 2: \left(k_2^{(1)}, k_2^{(2)} \right) \\ &\cdots \\ &Item\ p: \left(k_p^{(1)}, k_p^{(2)} \right) \end{aligned}$$

The score of each member is plotted in the two-dimensional plane as

$$\left(\sqrt{\lambda_1} z_i^{(1)}, \sqrt{\lambda_2} z_i^{(2)} \right)$$

We can judge the goodness of the member by evaluating the first term of $\sqrt{\lambda_1} z_i^{(1)}$, and characteristics from the second term of $\sqrt{\lambda_1} z_i^{(2)}$.

The closeness of each member can be evaluated from the distance given by

$$d_{zij} = \sqrt{\left(\sqrt{\lambda_1} z_i^{(1)} - \sqrt{\lambda_1} z_j^{(1)}\right)^2 + \left(\sqrt{\lambda_2} z_i^{(2)} - \sqrt{\lambda_2} z_j^{(2)}\right)^2}$$

The each item is plotted in the two-dimensional plane as

$$\left(\sqrt{\lambda_1} k_j^{(1)}, \sqrt{\lambda_2} k_j^{(2)}\right)$$

The closeness of each item can be evaluated from the distance given by

$$d_{kij} = \sqrt{\left(\sqrt{\lambda_1} k_i^{(1)} - \sqrt{\lambda_1} k_j^{(1)}\right)^2 + \left(\sqrt{\lambda_2} k_i^{(2)} - \sqrt{\lambda_2} k_j^{(2)}\right)^2}$$

We can also evaluate the closeness of the member and the item from the distance given by

$$d_{zkij} = \sqrt{\left(\sqrt{\lambda_1} z_i^{(1)} - \sqrt{\lambda_1} k_j^{(1)}\right)^2 + \left(\sqrt{\lambda_2} z_i^{(2)} - \sqrt{\lambda_2} k_j^{(2)}\right)^2}$$

When the multiple regression suffers high interaction between explanatory variables, we perform the principal component analysis and for a composite variable given by

$$z = \sum_{i=1}^{m} a_i^{(1)} u_i$$

We then perform one variable regression for each pair.

Chapter 11

FACTOR ANALYSIS

ABSTRACT

We obtain various data for many factors. We assume the output data comes from some few fundamental factors. Factor analysis gives us the procedure to decide the factors, where the factor number is less than the parameter numbers in general. Whole data are then expressed by the fewer factors, which are regarded as causes of the original data.

Keywords: factor analysis, factor point, factor loading, independent load, Valimax method, rotation matrix

1. INTRODUCTION

We obtain various kinds of data. We assume that there are some common causes for all data, and each data can be reproduced by the cause factors. The procedure to evaluate the common cause is called a factor analysis.

There are some confusion between principal component analysis and factor analysis, and we utilize the principal component analysis in the factor analysis. However, both analyses are quite different as explained below.

In the principal component analysis, we extract the same kind number of the variable as that of the original variables, but focus on only the first principal component and treat it as though it is the target variable of the original variables, which is shown in Figure **1** (a).

In the factor analysis, we want to extract variables which are regarded as the cause for the all original variables. We set the number of the variable for the cause from the beginning, and determine the variables. There is no order between the extracted variables. The schematic path diagram is shown in Figure 1 (b).

Consequently, in principal component analysis, and factor analysis, the focused variable number is fewer than the original variable number, and the analysis is quite similar. However, the role of the extracted variables is quite different. The ones for the principal analysis are objective one for the obtained data, and the ones for the factor analysis are the explanatory variables.

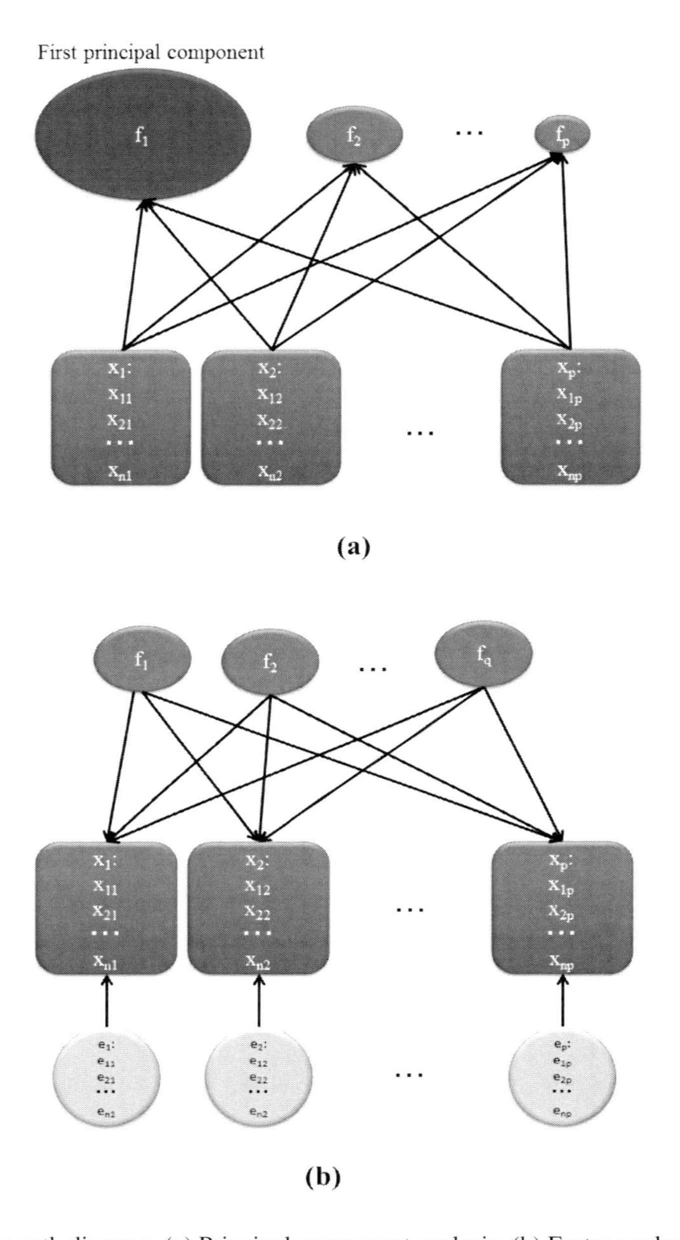

Figure 1. Schematic path diagram. (a) Principal component analysis, (b) Factor analysis.

In this chapter, we start with one factor, and then treat two factors, and finally treat arbitrarily factor numbers. We repeat almost the same step three times, where some steps are added associated with the number. Performing the steps, we may be able to obtain a clear image of the factor analysis.

We perform matrix operation in this chapter. The basic matrix operation is described in Chapter 15 of volume 3.

2. ONE FACTOR ANALYSIS

We focus on one variable, that is, we assume that only one variable contributes to the whole original data.

2.1. Basic Equation for One Variable Factor Analysis

We assume that one member has three kinds of scores of x, y, z. We also assume that the three variables are all normalized.

We denote f as the factor point, and factor loadings as (a_x, a_y, a_z), and independent loads as (e_x, e_y, e_z). Then the original variables are expressed by

$$\begin{pmatrix} x \\ y \\ z \end{pmatrix} = f \begin{pmatrix} a_x \\ a_y \\ a_z \end{pmatrix} + \begin{pmatrix} e_x \\ e_y \\ e_z \end{pmatrix} \tag{1}$$

We denote each data using a subscript i, and they are expressed by

$$\begin{pmatrix} x_i \\ y_i \\ z_i \end{pmatrix} = f_i \begin{pmatrix} a_x \\ a_y \\ a_z \end{pmatrix} + \begin{pmatrix} e_{ix} \\ e_{iy} \\ e_{iz} \end{pmatrix} \tag{2}$$

The factor loadings (a_x, a_y, a_z) are independent of the member, but depend on x, y, z. The factor point f depends on the member, but independent of x, y, z. The independent variables e_x, e_y, e_z depend on both x, y, z and the member.

The terms on right side of Eq. (2) are all unknown variables. We need some restrictions on them to decide the terms.

We assume that a factor point f is normalized, that is

$$E[f]=0 \tag{3}$$

$$V[f]=1 \tag{4}$$

The average of the independent load is zero, that is,

$$E[e_\alpha]=0 \tag{5}$$

where $\alpha = x, y, z$.

The covariance between f, e_α are zero, that is,

$$Cov[f, e_\alpha]=0 \tag{6}$$

$$Cov[e_\alpha, e_\beta]=0 \quad for\ \alpha \neq \beta \tag{7}$$

We define the vectors below.

$$\boldsymbol{f} = \begin{pmatrix} f_1 \\ f_2 \\ \vdots \\ f_n \end{pmatrix} \tag{8}$$

$$\boldsymbol{e}_x = \begin{pmatrix} e_{1x} \\ e_{2x} \\ \vdots \\ e_{nx} \end{pmatrix} \tag{9}$$

$$\boldsymbol{e}_y = \begin{pmatrix} e_{1y} \\ e_{2y} \\ \vdots \\ e_{ny} \end{pmatrix} \tag{10}$$

$$\boldsymbol{e}_z = \begin{pmatrix} e_{1z} \\ e_{2z} \\ \vdots \\ e_{nz} \end{pmatrix} \tag{11}$$

$$X = \begin{pmatrix} x_1 & x_2 & \cdots & x_n \\ y_1 & y_2 & \cdots & y_n \\ z_1 & z_2 & \cdots & z_n \end{pmatrix} \tag{12}$$

$$A = \begin{pmatrix} a_x \\ a_y \\ a_z \end{pmatrix} \tag{13}$$

$$F = \begin{pmatrix} f_1 & f_2 & \cdots & f_n \end{pmatrix} \tag{14}$$

$$\Gamma = \begin{pmatrix} e_{1x} & e_{2x} & \cdots & e_{nx} \\ e_{1y} & e_{2y} & \cdots & e_{ny} \\ e_{1z} & e_{2z} & \cdots & e_{nz} \end{pmatrix} \tag{15}$$

Eq. (1) can then be expressed by

$$X = AF + \Gamma \tag{16}$$

Multiplying X^T to the Eq. (16) from the right side, and expanding it, we obtain

$$\begin{aligned} XX^T &= \begin{pmatrix} x_1 & x_2 & \cdots & x_n \\ y_1 & y_2 & \cdots & y_n \\ z_1 & z_2 & \cdots & z_n \end{pmatrix} \begin{pmatrix} x_1 & y_1 & z_1 \\ x_2 & y_2 & z_2 \\ \vdots & \vdots & \vdots \\ x_n & y_n & z_n \end{pmatrix} \\ &= \begin{pmatrix} nS_{xx} & nS_{xy} & nS_{xz} \\ nS_{xy} & nS_{yy} & nS_{yz} \\ nS_{xz} & nS_{yz} & nS_{zz} \end{pmatrix} \\ &= n \begin{pmatrix} 1 & r_{xy} & r_{xz} \\ r_{xy} & 1 & r_{yz} \\ r_{xz} & r_{yz} & 1 \end{pmatrix} \end{aligned} \tag{17}$$

where X^T is the transverse matrix of X.

Similarly, we multiply transverse matrix to Eq. (16) from the right side, we obtain

$$\begin{aligned}
&(AF+\Gamma)(AF+\Gamma)^T \\
&=(AF+\Gamma)\left(F^T A^T+\Gamma^T\right) \\
&=AFF^T A^T + AF\Gamma^T + \Gamma F^T A^T + \Gamma\Gamma^T \\
&=A\left(FF^T\right)A^T + A\left(F\Gamma^T\right)+\left(F\Gamma^T\right)^T A^T + \Gamma\Gamma^T
\end{aligned} \tag{18}$$

We evaluate the terms in Eq. (18) as below.

$$\begin{aligned}
FF^T &= \sum_{i=1}^{n} f_i^2 \\
&= n
\end{aligned} \tag{19}$$

$$\begin{aligned}
F\Gamma^T &= \begin{pmatrix} f_1 & f_2 & \cdots & f_n \end{pmatrix} \begin{pmatrix} e_{1x} & e_{1y} & e_{1z} \\ e_{2x} & e_{2y} & e_{2z} \\ \vdots & \vdots & \vdots \\ e_{nx} & e_{ny} & e_{nz} \end{pmatrix} \\
&= \begin{pmatrix} \sum_{i=1}^{n} f_i e_{ix} & \sum_{i=1}^{n} f_i e_{iy} & \sum_{i=1}^{n} f_i e_{iz} \end{pmatrix} \\
&= \begin{pmatrix} 0 & 0 & 0 \end{pmatrix}
\end{aligned} \tag{20}$$

Substituting Eqs. (17), (19), and (20) into Eq. (16), we obtain

$$\frac{1}{n} XX^T = \begin{pmatrix} 1 & r_{xy} & r_{xz} \\ r_{xy} & 1 & r_{yz} \\ r_{xz} & r_{yz} & 1 \end{pmatrix} = AA^T + \frac{1}{n}\Gamma\Gamma^T \tag{21}$$

Evaluating the terms in the right side of Eq.(21), we obtain

$$AA^T = \begin{pmatrix} a_x \\ a_y \\ a_z \end{pmatrix} \begin{pmatrix} a_x & a_y & a_z \end{pmatrix} \tag{22}$$

$$\frac{1}{n}\Gamma\Gamma^T = \frac{1}{n}\begin{pmatrix} e_{1x} & e_{2x} & \cdots & e_{nx} \\ e_{1y} & e_{2y} & \cdots & e_{ny} \\ e_{1z} & e_{2z} & \cdots & e_{nz} \end{pmatrix}\begin{pmatrix} e_{1x} & e_{1y} & e_{1z} \\ e_{2x} & e_{2y} & e_{2z} \\ \vdots & \vdots & \vdots \\ e_{nx} & e_{ny} & e_{nz} \end{pmatrix}$$

$$= \frac{1}{n}\begin{pmatrix} \sum_{i=1}^{n} e_{ix}{}^2 & \sum_{i=1}^{n} e_{ix}e_{iy} & \sum_{i=1}^{n} e_{ix}e_{iz} \\ \sum_{i=1}^{n} e_{iy}e_{ix} & \sum_{i=1}^{n} e_{iy}{}^2 & \sum_{i=1}^{n} e_{iy}e_{iz} \\ \sum_{i=1}^{n} e_{iz}e_{ix} & \sum_{i=1}^{n} e_{iz}e_{iy} & \sum_{i=1}^{n} e_{iz}{}^2 \end{pmatrix}$$

$$= \begin{pmatrix} \frac{1}{n}\sum_{i=1}^{n} e_{ix}{}^2 & 0 & 0 \\ 0 & \frac{1}{n}\sum_{i=1}^{n} e_{iy}{}^2 & 0 \\ 0 & 0 & \frac{1}{n}\sum_{i=1}^{n} e_{iz}{}^2 \end{pmatrix}$$

$$= \begin{pmatrix} S_{e_x}^{(2)} & 0 & 0 \\ 0 & S_{e_y}^{(2)} & 0 \\ 0 & 0 & S_{e_z}^{(2)} \end{pmatrix} \tag{23}$$

Substituting Eqs. (22) and (23) into Eq. (21), we obtain

$$\begin{pmatrix} 1 & r_{xy} & r_{xz} \\ r_{xy} & 1 & r_{yz} \\ r_{xz} & r_{yz} & 1 \end{pmatrix} = \begin{pmatrix} a_x \\ a_y \\ a_z \end{pmatrix}\begin{pmatrix} a_x & a_y & a_z \end{pmatrix} + \begin{pmatrix} S_{e_x}^{(2)} & 0 & 0 \\ 0 & S_{e_y}^{(2)} & 0 \\ 0 & 0 & S_{e_z}^{(2)} \end{pmatrix} \tag{24}$$

This is the basic equation for one factor analysis.

Let us compare the case for principal component analysis, where the covariance matrix is expanded as follows.

$$\begin{pmatrix} 1 & r_{xy} & r_{xz} \\ r_{xy} & 1 & r_{yz} \\ r_{xz} & r_{yz} & 1 \end{pmatrix} = \lambda_1 \boldsymbol{v}_1 \boldsymbol{v}_1{}^T + \lambda_2 \boldsymbol{v}_2 \boldsymbol{v}_2{}^T + \lambda_3 \boldsymbol{v}_3 \boldsymbol{v}_3{}^T \tag{25}$$

The right side terms are all determined clearly.

If we can assume

$$\boldsymbol{a} = \begin{pmatrix} a_x \\ a_y \\ a_z \end{pmatrix} = \sqrt{\lambda_1}\boldsymbol{v}_1\boldsymbol{v}_1^T \tag{26}$$

$$\begin{pmatrix} S_{e_x}^{(2)} & 0 & 0 \\ 0 & S_{e_y}^{(2)} & 0 \\ 0 & 0 & S_{e_z}^{(2)} \end{pmatrix} = \lambda_2\boldsymbol{v}_2\boldsymbol{v}_2^T + \lambda_3\boldsymbol{v}_3\boldsymbol{v}_3^T \tag{27}$$

The first principal analysis and factor analysis become identical. It is valid when the first principal component dominates all. However, it is not true in general.

As we mentioned, we can determine whole terms of Eq. (25) in the principal component analysis. On the other hand, we cannot solve the terms in Eq. (24) shown below in general.

$$\begin{pmatrix} a_x \\ a_y \\ a_z \end{pmatrix}, \begin{pmatrix} S_{e_x}^{(2)} & 0 & 0 \\ 0 & S_{e_y}^{(2)} & 0 \\ 0 & 0 & S_{e_z}^{(2)} \end{pmatrix} \tag{28}$$

We solve Eq. (24) with the aid of the principal component analysis, which is shown in the next section.

2.2. Factor Load

We can expand the covariance matrix as below as shown in Chapter 10 for the principal component analysis.

$$\begin{aligned} \begin{pmatrix} 1 & r_{xy} & r_{xz} \\ r_{xy} & 1 & r_{yz} \\ r_{xz} & r_{yz} & 1 \end{pmatrix} &= \lambda_1\boldsymbol{v}_1\boldsymbol{v}_1^T + \lambda_2\boldsymbol{v}_2\boldsymbol{v}_2^T + \lambda_3\boldsymbol{v}_3\boldsymbol{v}_3^T \\ &= \left(\sqrt{\lambda_1}\boldsymbol{v}_1\right)\left(\sqrt{\lambda_1}\boldsymbol{v}_1\right)^T + \left(\sqrt{\lambda_2}\boldsymbol{v}_2\right)\left(\sqrt{\lambda_2}\boldsymbol{v}_2\right)^T + \left(\sqrt{\lambda_3}\boldsymbol{v}_3\right)\left(\sqrt{\lambda_3}\boldsymbol{v}_3\right)^T \end{aligned} \tag{29}$$

We regards the first term $\left(\sqrt{\lambda_1}\boldsymbol{v}_1\right)\left(\sqrt{\lambda_1}\boldsymbol{v}_1\right)^T$ this as factor load in the factor analysis, that is,

$$\begin{pmatrix} 1 & r_{xy} & r_{xz} \\ r_{xy} & 1 & r_{yz} \\ r_{xz} & r_{yz} & 1 \end{pmatrix} \approx \left(\sqrt{\lambda_{1\_1}}\boldsymbol{v}_{1\_1}\right)\left(\sqrt{\lambda_{1\_1}}\boldsymbol{v}_{1\_1}\right)^T + \begin{pmatrix} S_{e_x}^{(2)} & 0 & 0 \\ 0 & S_{e_y}^{(2)} & 0 \\ 0 & 0 & S_{e_z}^{(2)} \end{pmatrix} \tag{30}$$

Therefore, we approximate the factor load as

$$\begin{pmatrix} a_x \\ a_y \\ a_z \end{pmatrix} = \sqrt{\lambda_{1\_1}}\boldsymbol{v}_{1\_1} \tag{31}$$

This is not the final form factor loadings, but is the first approximated one. The first approximated independent load can be evaluated as

$$\begin{pmatrix} S_{e_x\_1}^{(2)} & 0 & 0 \\ 0 & S_{e_y\_1}^{(2)} & 0 \\ 0 & 0 & S_{e_z\_1}^{(2)} \end{pmatrix} = \begin{pmatrix} 1 & r_{xy} & r_{xz} \\ r_{xy} & 1 & r_{yz} \\ r_{xz} & r_{yz} & 1 \end{pmatrix} - \left(\sqrt{\lambda_{1\_1}}\boldsymbol{v}_{1\_1}\right)\left(\sqrt{\lambda_{1\_1}}\boldsymbol{v}_{1\_1}\right)^T \tag{32}$$

The non-diagonal elements on the right side of Eq. (32) are not 0 in general. Therefore, Eq. (32) is invalid from the rigorous point of view. However, we ignore the non-diagonal elements and regard the diagonal elements on both sides are the same. We then form the new covariance matrix given by

$$\boldsymbol{R}_1 = \begin{pmatrix} 1 - S_{e_x\_1}^{(2)} & r_{xy} & r_{xz} \\ r_{xy} & 1 - S_{e_y\_1}^{(2)} & r_{yz} \\ r_{xz} & r_{yz} & 1 - S_{e_z\_1}^{(2)} \end{pmatrix} \tag{33}$$

We evaluate the eigenvalue and eigenvectors of the matrix, and evaluate below.

$$\begin{pmatrix} S^{(2)}_{e_x\_2} & 0 & 0 \\ 0 & S^{(2)}_{e_y\_2} & 0 \\ 0 & 0 & S^{(2)}_{e_z\_2} \end{pmatrix} = \begin{pmatrix} 1 & r_{xy} & r_{xz} \\ r_{xy} & 1 & r_{yz} \\ r_{xz} & r_{yz} & 1 \end{pmatrix} - \left(\sqrt{\lambda_{1\_2}}\boldsymbol{v}_{1\_2}\right)\left(\sqrt{\lambda_{1\_2}}\boldsymbol{v}_{1\_2}\right)^T \tag{34}$$

We evaluate the difference between the approximated independent factors given by

$$\Delta = \begin{pmatrix} S^{(2)}_{e_x\_2} - S^{(2)}_{e_x\_1} \\ S^{(2)}_{e_y\_2} - S^{(2)}_{e_y\_1} \\ S^{(2)}_{e_z\_2} - S^{(2)}_{e_z\_1} \end{pmatrix} \tag{35}$$

If the value below is less than a certain value

$$\delta = \frac{1}{3}\left\{\left[S^{(2)}_{e_x\_2} - S^{(2)}_{e_x\_1}\right]^2 + \left[S^{(2)}_{e_y\_2} - S^{(2)}_{e_y\_1}\right]^2 + \left[S^{(2)}_{e_z\_2} - S^{(2)}_{e_z\_1}\right]^2\right\} \tag{36}$$

we regard the component is the load factor given by

$$\boldsymbol{a} = \sqrt{\lambda_{1\_m}}\boldsymbol{v}_{1\_m} \tag{37}$$

If the value is more than a certain value, we construct the updated covariance matrix given by

$$\boldsymbol{R}_2 = \begin{pmatrix} 1 - S^{(2)}_{e_x\_2} & r_{xy} & r_{xz} \\ r_{xy} & 1 - S^{(2)}_{e_y\_2} & r_{yz} \\ r_{xz} & r_{yz} & 1 - S^{(2)}_{e_z\_2} \end{pmatrix} \tag{38}$$

We evaluate the eigenvalue and eigenvector of it and the updated one denote as $\lambda_{1\_3}, \boldsymbol{v}_{1\_3}$.

We can then evaluate the updated independent matrix given by

$$\begin{pmatrix} S^{(2)}_{e_x\_3} & 0 & 0 \\ 0 & S^{(2)}_{e_y\_3} & 0 \\ 0 & 0 & S^{(2)}_{e_z\_3} \end{pmatrix} = \begin{pmatrix} 1 & r_{xy} & r_{xz} \\ r_{xy} & 1 & r_{yz} \\ r_{xz} & r_{yz} & 1 \end{pmatrix} - \left(\sqrt{\lambda_{1\_3}}\boldsymbol{v}_{1\_3}\right)\left(\sqrt{\lambda_{1\_3}}\boldsymbol{v}_{1\_3}\right)^T \tag{39}$$

We further evaluate the updated difference of the independent element matrix below.

$$\Delta = \begin{pmatrix} S^{(2)}_{e_x\_3} - S^{(2)}_{e_x\_2} \\ S^{(2)}_{e_y\_3} - S^{(2)}_{e_y\_2} \\ S^{(2)}_{e_z\_3} - S^{(2)}_{e_z\_2} \end{pmatrix} \tag{40}$$

and evaluate below.

$$\delta = \frac{1}{3}\left\{\left[S^{(2)}_{e_x\_3} - S^{(2)}_{e_x\_2}\right]^2 + \left[S^{(2)}_{e_y\_3} - S^{(2)}_{e_y\_2}\right]^2 + \left[S^{(2)}_{e_z\_3} - S^{(2)}_{e_z\_2}\right]^2\right\} \tag{41}$$

We repeat this process until is less than the certain critical value. When we obtain this value after the m times process, we obtain the load factor as

$$\boldsymbol{a} = \sqrt{\lambda_{1\_m}}\boldsymbol{v}_{1\_m} \tag{42}$$

The corresponding diagonal elements are given by

$$d_x^2 = 1 - S^{(2)}_{e_x\_m}, d_y^2 = 1 - S^{(2)}_{e_y\_m}, d_z^2 = 1 - S^{(2)}_{e_z\_m} \tag{43}$$

which we utilize later.

2.3. Factor Point and Independent Load

We review the process up to here. The basic equation for one variable factor analysis is given by

$$\begin{pmatrix} x_i \\ y_i \\ z_i \end{pmatrix} = f_i \begin{pmatrix} a_x \\ a_y \\ a_z \end{pmatrix} + \begin{pmatrix} e_{ix} \\ e_{iy} \\ e_{iz} \end{pmatrix} \tag{44}$$

We decide a factor load, and the variances of e_x, e_y, e_z. We further need to evaluate f_i and e_x, e_y, e_z themselves.

We assume that x, y, z are commonly expressed with f. This means that the predictive value of f is denoted as $\hat{f}_i$ and it can be expressed with the linear sum of x, y, z given by

$$\hat{f}_i = h_{fx} x_i + h_{fy} y_i + h_{fz} z_i \tag{45}$$

The deviation from the true value f_i is expressed by

$$\begin{aligned} Q_f &= \sum_{i=1}^{n} \left(f_i - \hat{f}_i \right)^2 \\ &= \sum_{i=1}^{n} \left[f_i - \left(h_{fx} x_i + h_{fy} y_i + h_{fz} z_i \right) \right]^2 \end{aligned} \tag{46}$$

We decide the factors h_{fx}, h_{fy}, h_{fz} so that Q_f has the minimum value.

Partial differentiate Eq. (46) with respect to h_{fx}, and set it to 0, we obtain

$$\begin{aligned} \frac{\partial Q_f}{\partial h_{fx}} &= \frac{\partial}{\partial h_{fx}} \sum_{i=1}^{n} \left[f_i - \left(h_{fx} x_i + h_{fy} y_i + h_{fz} z_i \right) \right]^2 \\ &= -2 \sum_{i=1}^{n} \left[f_i - \left(h_{fx} x_i + h_{fy} y_i + h_{fz} z_i \right) \right] x_i = 0 \end{aligned} \tag{47}$$

We then obtain

$$\sum_{i=1}^{n} \left(h_{fx} x_i x_i + h_{fy} y_i x_i + h_{fz} z_i x_i \right) = \sum_{i=1}^{n} f_i x_i \tag{48}$$

The left hand side of Eq. (48) is further modified as

$$\sum_{i=1}^{n}\left(h_{fx}x_ix_i + h_{fy}y_ix_i + h_{fz}z_ix_i\right) = h_{fx}\sum_{i=1}^{n}x_ix_i + h_{fy}\sum_{i=1}^{n}y_ix_i + h_{fz}\sum_{i=1}^{n}z_ix_i$$
$$= n\left(h_{fx}r_{xx} + h_{fy}r_{xy} + h_{fz}r_{xz}\right) \tag{49}$$

The right side of Eq. (48) is further modified as

$$\sum_{i=1}^{n}f_ix_i = \sum_{i=1}^{n}f_i\left(a_xf_i + b_xg_i + e_{xi}\right)$$
$$= na_x \tag{50}$$

Therefore, we obtain

$$h_{fx}r_{xx} + h_{fy}r_{xy} + h_{fz}r_{xz} = a_x \tag{51}$$

Similarly, we obtain with respect to h_{fy} as

$$\frac{\partial Q_f}{\partial h_{fy}} = \frac{\partial}{\partial h_{fx}}\sum_{i=1}^{n}\left[f_i - \left(h_{fx}x_i + h_{fy}y_i + h_{fz}z_i\right)\right]^2$$
$$= -2\sum_{i=1}^{n}\left[f_i - \left(h_{fx}x_i + h_{fy}y_i + h_{fz}z_i\right)\right]y_i = 0 \tag{52}$$

We then obtain

$$h_{fx}r_{xy} + h_{fy}r_{yy} + h_{fz}r_{yz} = a_y \tag{53}$$

Similarly, we obtain with respect to h_{fz} as

$$h_{fx}r_{zx} + h_{fy}r_{yz} + h_{fz}r_{zz} = a_z \tag{54}$$

Summarizing these, we obtain

$$\begin{pmatrix} r_{xx} & r_{xy} & r_{zx} \\ r_{xy} & r_{yy} & r_{yz} \\ r_{zx} & r_{yz} & r_{zz} \end{pmatrix}\begin{pmatrix} h_{fx} \\ h_{fy} \\ h_{fz} \end{pmatrix} = \begin{pmatrix} a_x \\ a_y \\ a_z \end{pmatrix} \tag{55}$$

We then obtain

$$\begin{pmatrix} h_{fx} \\ h_{fy} \\ h_{fz} \end{pmatrix} = \begin{pmatrix} r_{xx} & r_{xy} & r_{zx} \\ r_{xy} & r_{yy} & r_{yz} \\ r_{zx} & r_{yz} & r_{zz} \end{pmatrix}^{-1} \begin{pmatrix} a_x \\ a_y \\ a_z \end{pmatrix} \tag{56}$$

Using the h_{fx}, h_{fy}, h_{fz}, the factor point is given by

$$\hat{f}_i = h_{fx} x_i + h_{fy} y_i + h_{fz} z_i \tag{57}$$

Independent load can be evaluated as

$$\begin{pmatrix} e_{xi} \\ e_{yi} \\ e_{zi} \end{pmatrix} = \begin{pmatrix} x_i \\ y_i \\ z_i \end{pmatrix} - f_i \begin{pmatrix} a_x \\ a_y \\ a_z \end{pmatrix} \tag{58}$$

3. Two-Factor Analysis

3.1. Basic Equation

We treat a two-factor analysis with three normalized variables x, y, z in this section. The one factor point is denoted as f, and the other as g. The factor load associated with f is denoted as (a_x, a_y, a_z), and the factor load associated with g is denoted as (b_x, b_y, b_z). The independent load is denoted as (e_x, e_y, e_z). The variable is then expressed as

$$\begin{pmatrix} x \\ y \\ z \end{pmatrix} = f \begin{pmatrix} a_x \\ a_y \\ a_z \end{pmatrix} + g \begin{pmatrix} b_x \\ b_y \\ b_z \end{pmatrix} + \begin{pmatrix} e_x \\ e_y \\ e_z \end{pmatrix} \tag{59}$$

Expressing each data as i, we have

$$\begin{pmatrix} x_i \\ y_i \\ z_i \end{pmatrix} = f_i \begin{pmatrix} a_x \\ a_y \\ a_z \end{pmatrix} + g_i \begin{pmatrix} b_x \\ b_y \\ b_z \end{pmatrix} + \begin{pmatrix} e_{ix} \\ e_{iy} \\ e_{iz} \end{pmatrix} \tag{60}$$

The factor loadings (a_x, a_y, a_z) and (b_x, b_y, b_z) are independent of the member, but depend on x, y, z. The factor point f and g depends on the member, but independent on x, y, z. The independent variables e_x, e_y, e_z depend on both x, y, z and member.

The terms on right side of Eq. (60) are all unknown variables. We need some restrictions on them to decide the terms

We assume that factor points f and g are normalized, that is

$$E[f] = 0 \tag{61}$$

$$V[f] = 1 \tag{62}$$

$$E[g] = 0 \tag{63}$$

$$V[g] = 1 \tag{64}$$

The average of the independent load is zero, that is,

$$E[e_\alpha] = 0 \tag{65}$$

where $\alpha = x, y, z$.

The covariance between f, g, e_α are zero, that is,

$$Cov[f, g] = 0 \tag{66}$$

$$Cov[f, e_\alpha] = 0 \tag{67}$$

$$Cov[g, e_\alpha] = 0 \tag{68}$$

$$Cov[e_\alpha, e_\beta] = 0 \quad for\ \alpha \neq \beta \tag{69}$$

We define the vectors below.

$$\boldsymbol{f} = \begin{pmatrix} f_1 \\ f_2 \\ \vdots \\ f_n \end{pmatrix} \tag{70}$$

$$\boldsymbol{g} = \begin{pmatrix} g_1 \\ g_2 \\ \vdots \\ g_n \end{pmatrix} \tag{71}$$

$$\boldsymbol{e}_x = \begin{pmatrix} e_{1x} \\ e_{2x} \\ \vdots \\ e_{nx} \end{pmatrix} \tag{72}$$

$$\boldsymbol{e}_y = \begin{pmatrix} e_{1y} \\ e_{2y} \\ \vdots \\ e_{ny} \end{pmatrix} \tag{73}$$

$$\boldsymbol{e}_z = \begin{pmatrix} e_{1z} \\ e_{2z} \\ \vdots \\ e_{nz} \end{pmatrix} \tag{74}$$

$$X = \begin{pmatrix} x_1 & x_2 & \cdots & x_n \\ y_1 & y_2 & \cdots & y_n \\ z_1 & z_2 & \cdots & z_n \end{pmatrix} \tag{75}$$

$$A = \begin{pmatrix} a_x & b_x \\ a_y & b_y \\ a_z & b_z \end{pmatrix} \tag{76}$$

$$F = \begin{pmatrix} f_1 & f_2 & \cdots & f_n \\ g_1 & g_2 & \cdots & g_n \end{pmatrix} \tag{77}$$

$$\Gamma = \begin{pmatrix} e_{1x} & e_{2x} & \cdots & e_{nx} \\ e_{1y} & e_{2y} & \cdots & e_{ny} \\ e_{1z} & e_{2z} & \cdots & e_{nz} \end{pmatrix} \tag{78}$$

Eq. (59) can then be expressed by

$$X = AF + \Gamma \tag{79}$$

The final form is exactly the same as the one for one factor analysis.

Multiplying X^T to the Eq. (90) from the right side, and expanding it, we obtain

$$XX^T = \begin{pmatrix} x_1 & x_2 & \cdots & x_n \\ y_1 & y_2 & \cdots & y_n \\ z_1 & z_2 & \cdots & z_n \end{pmatrix} \begin{pmatrix} x_1 & y_1 & z_1 \\ x_2 & y_2 & z_2 \\ \vdots & \vdots & \vdots \\ x_n & y_n & z_n \end{pmatrix}$$
$$= \begin{pmatrix} nS_{xx} & nS_{xy} & nS_{xz} \\ nS_{xy} & nS_{yy} & nS_{yz} \\ nS_{xz} & nS_{yz} & nS_{zz} \end{pmatrix}$$
$$= n \begin{pmatrix} 1 & r_{xy} & r_{xz} \\ r_{xy} & 1 & r_{yz} \\ r_{xz} & r_{yz} & 1 \end{pmatrix} \tag{80}$$

where X^T is the transverse matrix of X.

Similarly, we multiply transverse matrix to Eq. (16) from the right side, we obtain

$$\begin{aligned} &(AF + \Gamma)(AF + \Gamma)^T \\ &= (AF + \Gamma)(F^T A^T + \Gamma^T) \\ &= AFF^T A^T + AF\Gamma^T + \Gamma F^T A^T + \Gamma\Gamma^T \\ &= A(FF^T)A^T + A(F\Gamma^T) + (F\Gamma^T)^T A^T + \Gamma\Gamma^T \end{aligned} \tag{81}$$

We evaluate the terms in Eq. (81) as below.

$$FF^T = \begin{pmatrix} f_1 & f_2 & \cdots & f_n \\ g_1 & g_2 & \cdots & g_n \end{pmatrix} \begin{pmatrix} f_1 & g_1 \\ f_2 & g_2 \\ \vdots & \vdots \\ f_n & g_n \end{pmatrix}$$
$$= \begin{pmatrix} \sum_{i=1}^{n} f_i^2 & \sum_{i=1}^{n} f_i g_i \\ \sum_{i=1}^{n} g_i f_i & \sum_{i=1}^{n} g_i^{\ 2} \end{pmatrix}$$
$$= n \begin{pmatrix} 1 & 0 \\ 0 & 1 \end{pmatrix} \tag{82}$$

$$F\Gamma^T = \begin{pmatrix} f_1 & f_2 & \cdots & f_n \\ g_1 & g_2 & \cdots & g_n \end{pmatrix} \begin{pmatrix} e_{1x} & e_{1y} & e_{1z} \\ e_{2x} & e_{2y} & e_{2z} \\ \vdots & \vdots & \vdots \\ e_{nx} & e_{ny} & e_{nz} \end{pmatrix}$$
$$= \begin{pmatrix} \sum_{i=1}^{n} f_i e_{ix} & \sum_{i=1}^{n} f_i e_{iy} & \sum_{i=1}^{n} f_i e_{iz} \\ \sum_{i=1}^{n} g_i e_{ix} & \sum_{i=1}^{n} g_i e_{iy} & \sum_{i=1}^{n} g_i e_{iz} \end{pmatrix}$$
$$= \begin{pmatrix} 0 & 0 & 0 \\ 0 & 0 & 0 \end{pmatrix} \tag{83}$$

Therefore, we obtain

$$\frac{1}{n} XX^T = \begin{pmatrix} 1 & r_{xy} & r_{xz} \\ r_{xy} & 1 & r_{yz} \\ r_{xz} & r_{yz} & 1 \end{pmatrix} = AA^T + \frac{1}{n}\Gamma\Gamma^T \tag{84}$$

We further expand the term AA^T and $(1/n)\Gamma\Gamma^T$ as

$$AA^T = \begin{pmatrix} a_x & b_x \\ a_y & b_y \\ a_z & b_z \end{pmatrix} \begin{pmatrix} a_x & a_y & a_z \\ b_x & b_y & b_z \end{pmatrix}$$
$$= \begin{pmatrix} a_x a_x + b_x b_x & a_x a_y + b_x b_y & a_x a_z + b_x b_z \\ a_y a_x + b_y b_x & a_y a_y + b_y b_y & a_y a_z + b_y b_z \\ a_z a_x + b_z b_x & a_z a_y + b_z b_y & a_z a_z + b_z b_z \end{pmatrix}$$
$$= \begin{pmatrix} a_x a_x & a_x a_y & a_x a_z \\ a_y a_x & a_y a_y & a_y a_z \\ a_z a_x & a_z a_y & a_z a_z \end{pmatrix} + \begin{pmatrix} b_x b_x & b_x b_y & b_x b_z \\ b_y b_x & b_y b_y & b_y b_z \\ b_z b_x & b_z b_y & b_z b_z \end{pmatrix}$$
$$= \begin{pmatrix} a_x \\ a_y \\ a_z \end{pmatrix} \begin{pmatrix} a_x & a_y & a_z \end{pmatrix} + \begin{pmatrix} b_x \\ b_y \\ b_z \end{pmatrix} \begin{pmatrix} b_x & b_y & b_z \end{pmatrix} \tag{85}$$

$$\frac{1}{n}\Gamma\Gamma^T = \frac{1}{n}\begin{pmatrix} e_{1x} & e_{2x} & \cdots & e_{nx} \\ e_{1y} & e_{2y} & \cdots & e_{ny} \\ e_{1z} & e_{2z} & \cdots & e_{nz} \end{pmatrix}\begin{pmatrix} e_{1x} & e_{1y} & e_{1z} \\ e_{2x} & e_{2y} & e_{2z} \\ \vdots & \vdots & \vdots \\ e_{nx} & e_{ny} & e_{nz} \end{pmatrix}$$

$$= \frac{1}{n}\begin{pmatrix} \sum_{i=1}^{n} e_{ix}^2 & \sum_{i=1}^{n} e_{ix}e_{iy} & \sum_{i=1}^{n} e_{ix}e_{iz} \\ \sum_{i=1}^{n} e_{iy}e_{ix} & \sum_{i=1}^{n} e_{iy}^2 & \sum_{i=1}^{n} e_{iy}e_{iz} \\ \sum_{i=1}^{n} e_{iz}e_{ix} & \sum_{i=1}^{n} e_{iz}e_{iy} & \sum_{i=1}^{n} e_{iz}^2 \end{pmatrix}$$

$$= \begin{pmatrix} \frac{1}{n}\sum_{i=1}^{n} e_{ix}^2 & 0 & 0 \\ 0 & \frac{1}{n}\sum_{i=1}^{n} e_{iy}^2 & 0 \\ 0 & 0 & \frac{1}{n}\sum_{i=1}^{n} e_{iz}^2 \end{pmatrix}$$

$$= \begin{pmatrix} S_{e_x}^{(2)} & 0 & 0 \\ 0 & S_{e_y}^{(2)} & 0 \\ 0 & 0 & S_{e_z}^{(2)} \end{pmatrix} \tag{86}$$

Finally, we obtain

$$\begin{pmatrix} 1 & r_{xy} & r_{xz} \\ r_{xy} & 1 & r_{yz} \\ r_{xz} & r_{yz} & 1 \end{pmatrix} = \begin{pmatrix} a_x \\ a_y \\ a_z \end{pmatrix}\begin{pmatrix} a_x & a_y & a_z \end{pmatrix} + \begin{pmatrix} b_x \\ b_y \\ b_z \end{pmatrix}\begin{pmatrix} b_x & b_y & b_z \end{pmatrix} + \begin{pmatrix} S_{e_x}^{(2)} & 0 & 0 \\ 0 & S_{e_y}^{(2)} & 0 \\ 0 & 0 & S_{e_z}^{(2)} \end{pmatrix} \tag{87}$$

This is the basic equation for two factor analysis.

Let us compare the case for principal component analysis, where the covariance matrix is expanded as follows.

$$\begin{pmatrix} 1 & r_{xy} & r_{xz} \\ r_{xy} & 1 & r_{yz} \\ r_{xz} & r_{yz} & 1 \end{pmatrix} = \lambda_1 \boldsymbol{v}_1 \boldsymbol{v}_1^T + \lambda_2 \boldsymbol{v}_2 \boldsymbol{v}_2^T + \lambda_3 \boldsymbol{v}_3 \boldsymbol{v}_3^T \tag{88}$$

The right side terms are all determined clearly. We utilize this equation for principal component analysis as shown in the one factor analysis.

One subject is added for two factor analysis.

We will finally obtain X, A, F, Γ in the end, which satisfy

$$X = AF + \Gamma \tag{89}$$

However, in the mathematical stand point of view, Eq. (89) is identical to

$$X = ATT^{-1}F + \Gamma \tag{90}$$

We assume that T is the rotation matrix, Eq. (90) can be modified as

$$X = AT(\theta)T(-\theta)F + \Gamma \tag{91}$$

where

$$\begin{cases} A' = AT(\theta) \\ F' = T(-\theta)F \end{cases} \tag{92}$$

Therefore, if A snd F are the root of Eq. (89), A', F' are also the root of it. We will discuss this uncertainty later.

3.2. Factor Load for Two-Factor Variables

We can expand the covariance matrix as below as shown in the one factor analysis given by

$$\begin{pmatrix} 1 & r_{xy} & r_{xz} \\ r_{xy} & 1 & r_{yz} \\ r_{xz} & r_{yz} & 1 \end{pmatrix} = \lambda_1 \boldsymbol{v}_1 \boldsymbol{v}_1^T + \lambda_2 \boldsymbol{v}_2 \boldsymbol{v}_2^T + \lambda_3 \boldsymbol{v}_3 \boldsymbol{v}_3^T$$
$$= \left(\sqrt{\lambda_1}\boldsymbol{v}_1\right)\left(\sqrt{\lambda_1}\boldsymbol{v}_1\right)^T + \left(\sqrt{\lambda_2}\boldsymbol{v}_2\right)\left(\sqrt{\lambda_2}\boldsymbol{v}_2\right)^T + \left(\sqrt{\lambda_3}\boldsymbol{v}_3\right)\left(\sqrt{\lambda_3}\boldsymbol{v}_3\right)^T \tag{93}$$

We regards the first term $\left(\sqrt{\lambda_1}\boldsymbol{v}_1\right)\left(\sqrt{\lambda_1}\boldsymbol{v}_1\right)^T + \left(\sqrt{\lambda_2}\boldsymbol{v}_2\right)\left(\sqrt{\lambda_2}\boldsymbol{v}_2\right)^T$ as a factor load in the factor analysis, that is,

$$\begin{pmatrix} 1 & r_{xy} & r_{xz} \\ r_{xy} & 1 & r_{yz} \\ r_{xz} & r_{yz} & 1 \end{pmatrix} \approx \left(\sqrt{\lambda_{1\_1}}\boldsymbol{v}_{1\_1}\right)\left(\sqrt{\lambda_{1\_1}}\boldsymbol{v}_{1\_1}\right)^T + \left(\sqrt{\lambda_{2\_1}}\boldsymbol{v}_{2\_1}\right)\left(\sqrt{\lambda_{2\_1}}\boldsymbol{v}_{2\_1}\right)^T + \begin{pmatrix} S_{e_x}^{(2)} & 0 & 0 \\ 0 & S_{e_y}^{(2)} & 0 \\ 0 & 0 & S_{e_z}^{(2)} \end{pmatrix} \tag{94}$$

Therefore, we approximate the factor load as

$$\begin{pmatrix} a_x \\ a_y \\ a_z \end{pmatrix} = \sqrt{\lambda_{1\_1}}\boldsymbol{v}_{1\_1}, \begin{pmatrix} b_x \\ b_y \\ b_z \end{pmatrix} = \sqrt{\lambda_{2\_1}}\boldsymbol{v}_{2\_1} \tag{95}$$

This is not the final form factor loadings, but is the first approximated one. The first approximated independent load can be evaluated as

$$\begin{pmatrix} S_{e_x\_1}^{(2)} & 0 & 0 \\ 0 & S_{e_y\_1}^{(2)} & 0 \\ 0 & 0 & S_{e_z\_1}^{(2)} \end{pmatrix} = \begin{pmatrix} 1 & r_{xy} & r_{xz} \\ r_{xy} & 1 & r_{yz} \\ r_{xz} & r_{yz} & 1 \end{pmatrix} - \left[\left(\sqrt{\lambda_{1\_1}}\boldsymbol{v}_{1\_1}\right)\left(\sqrt{\lambda_{1\_1}}\boldsymbol{v}_{1\_1}\right)^T + \left(\sqrt{\lambda_{2\_1}}\boldsymbol{v}_{2\_1}\right)\left(\sqrt{\lambda_{2\_1}}\boldsymbol{v}_{2\_1}\right)^T\right] \tag{96}$$

The non-diagonal elements on the right side of Eq. (96) are not 0 in general. Therefore, Eq. (96)is invalid from the rigorous point of view. However, we ignore the non-diagonal elements and regard the diagonal element on both sides are the same.

We then form the new covariance matrix given by

$$\boldsymbol{R}_1 = \begin{pmatrix} 1 - S_{e_x\_1}^{(2)} & r_{xy} & r_{xz} \\ r_{xy} & 1 - S_{e_y\_1}^{(2)} & r_{yz} \\ r_{xz} & r_{yz} & 1 - S_{e_z\_1}^{(2)} \end{pmatrix} \tag{97}$$

We evaluate the eigenvalue and eigenvectors of the matrix, and evaluate below.

$$\begin{pmatrix} S_{e_x\_2}^{(2)} & 0 & 0 \\ 0 & S_{e_y\_2}^{(2)} & 0 \\ 0 & 0 & S_{e_z\_2}^{(2)} \end{pmatrix}$$

$$= \begin{pmatrix} 1 & r_{xy} & r_{xz} \\ r_{xy} & 1 & r_{yz} \\ r_{xz} & r_{yz} & 1 \end{pmatrix} - \left(\sqrt{\lambda_{1\_2}}\boldsymbol{v}_{1\_2}\right)\left(\sqrt{\lambda_{1\_2}}\boldsymbol{v}_{1\_2}\right)^T \tag{98}$$

We evaluate the difference between the approximated independent factors given by

$$\Delta = \begin{pmatrix} S^{(2)}_{e_x\_2} - S^{(2)}_{e_x\_1} \\ S^{(2)}_{e_y\_2} - S^{(2)}_{e_y\_1} \\ S^{(2)}_{e_z\_2} - S^{(2)}_{e_z\_1} \end{pmatrix} \tag{99}$$

If the value below is less than a certain value

$$\delta = \frac{1}{3}\left\{\left[S^{(2)}_{e_x\_2} - S^{(2)}_{e_x\_1}\right]^2 + \left[S^{(2)}_{e_y\_2} - S^{(2)}_{e_y\_1}\right]^2 + \left[S^{(2)}_{e_z\_2} - S^{(2)}_{e_z\_1}\right]^2\right\} \tag{100}$$

We regard the component is the load factor given by

$$\boldsymbol{a} = \sqrt{\lambda_{1\_2}}\boldsymbol{v}_{1\_2}, \boldsymbol{b} = \sqrt{\lambda_{2\_2}}\boldsymbol{v}_{2\_2} \tag{101}$$

If the value is more than a certain value, we construct the updated covariance matrix given by

$$\boldsymbol{R}_2 = \begin{pmatrix} 1 - S^{(2)}_{e_x\_2} & r_{xy} & r_{xz} \\ r_{xy} & 1 - S^{(2)}_{e_y\_2} & r_{yz} \\ r_{xz} & r_{yz} & 1 - S^{(2)}_{e_z\_2} \end{pmatrix} \tag{102}$$

We evaluate the eigenvalue and eigenvector of it and the updated one denote as $\lambda_{1\_3}, \boldsymbol{v}_{1\_3}, \lambda_{2\_3}, \boldsymbol{v}_{2\_3}$.

We can then evaluate the updated independent matrix given by

$$\begin{pmatrix} S^{(2)}_{e_x\_3} & 0 & 0 \\ 0 & S^{(2)}_{e_y\_3} & 0 \\ 0 & 0 & S^{(2)}_{e_z\_3} \end{pmatrix} = \begin{pmatrix} 1 & r_{xy} & r_{xz} \\ r_{xy} & 1 & r_{yz} \\ r_{xz} & r_{yz} & 1 \end{pmatrix} - \left[\left(\sqrt{\lambda_{1\_3}}\boldsymbol{v}_{1\_3}\right)\left(\sqrt{\lambda_{1\_3}}\boldsymbol{v}_{1\_3}\right)^T + \left(\sqrt{\lambda_{2\_3}}\boldsymbol{v}_{2\_3}\right)\left(\sqrt{\lambda_{2\_3}}\boldsymbol{v}_{2\_3}\right)^T\right] \tag{103}$$

We further evaluate the updated difference of the independent element matrix below.

$$\Delta = \begin{pmatrix} S^{(2)}_{e_x\_3} - S^{(2)}_{e_x\_2} \\ S^{(2)}_{e_y\_3} - S^{(2)}_{e_y\_2} \\ S^{(2)}_{e_z\_3} - S^{(2)}_{e_z\_2} \end{pmatrix} \tag{104}$$

and evaluate below.

$$\delta = \frac{1}{3}\left\{\left[S^{(2)}_{e_x\_3} - S^{(2)}_{e_x\_2}\right]^2 + \left[S^{(2)}_{e_y\_3} - S^{(2)}_{e_y\_2}\right]^2 + \left[S^{(2)}_{e_z\_3} - S^{(2)}_{e_z\_2}\right]^2\right\} \tag{105}$$

We repeat this process until it is less than the certain value. When we obtain this value after the m times process, we obtain the load factor as

$$\boldsymbol{a} = \sqrt{\lambda_{1\_m}}\boldsymbol{v}_{1\_m}, \boldsymbol{b} = \sqrt{\lambda_{2\_m}}\boldsymbol{v}_{2\_m} \tag{106}$$

The corresponding diagonal elements are given by

$$d_x^2 = 1 - S^{(2)}_{e_x\_m}, d_y^2 = 1 - S^{(2)}_{e_y\_m}, d_z^2 = 1 - S^{(2)}_{e_z\_m} \tag{107}$$

which we utilize later.

3.3. Uncertainty of Load Factor and Rotation

There is uncertainty in load factors as we mentioned before. We solve the uncertainty in this section.

We assume that we have factor loads a_x, a_y, a_z and b_x, b_y, b_z using the showed procedure, and obtain factor points $f_i, g_i \left(i = 1, 2, \cdots, n\right)$. The procedure to obtain the factor points will be shown later. We then obtain

$$\begin{pmatrix} x_i \\ y_i \\ z_i \end{pmatrix} = \begin{pmatrix} a_x & b_x \\ a_y & b_y \\ a_z & b_z \end{pmatrix} \begin{pmatrix} f_i \\ g_i \end{pmatrix} + \begin{pmatrix} e_{ix} \\ e_{iy} \\ e_{iz} \end{pmatrix} \tag{108}$$

We rotate f, g with an angle of θ, and the factor point (f_i, g_i) is mapped to $(\hat{f}_i, \hat{g}_i)$, and both are related to

$$\begin{pmatrix} \tilde{f}_i \\ \tilde{g}_i \end{pmatrix} = \begin{pmatrix} \cos\theta & \sin\theta \\ -\sin\theta & \cos\theta \end{pmatrix} \begin{pmatrix} f_i \\ g_i \end{pmatrix} \tag{109}$$

$$\begin{pmatrix} f_i \\ g_i \end{pmatrix} = \begin{pmatrix} \cos\theta & -\sin\theta \\ \sin\theta & \cos\theta \end{pmatrix} \begin{pmatrix} \tilde{f}_i \\ \tilde{g}_i \end{pmatrix} \tag{110}$$

Therefore, we obtain

$$\begin{aligned} \begin{pmatrix} a_x & b_x \\ a_y & b_y \\ a_z & b_z \end{pmatrix} \begin{pmatrix} f_i \\ g_i \end{pmatrix} &= \begin{pmatrix} a_x & b_x \\ a_y & b_y \\ a_z & b_z \end{pmatrix} \begin{pmatrix} \cos\theta & -\sin\theta \\ \sin\theta & \cos\theta \end{pmatrix} \begin{pmatrix} \tilde{f}_i \\ \tilde{g}_i \end{pmatrix} \\ &= \begin{pmatrix} a_x \cos\theta + b_x \sin\theta & -a_x \sin\theta + b_x \cos\theta \\ a_y \cos\theta + b_y \sin\theta & -a_y \sin\theta + b_y \cos\theta \\ a_z \cos\theta + b_z \sin\theta & -a_z \sin\theta + b_z \cos\theta \end{pmatrix} \begin{pmatrix} \tilde{f}_i \\ \tilde{g}_i \end{pmatrix} \\ &= \begin{pmatrix} \tilde{a}_x & \tilde{b}_x \\ \tilde{a}_y & \tilde{b}_y \\ \tilde{a}_z & \tilde{b}_z \end{pmatrix} \begin{pmatrix} \tilde{f}_i \\ \tilde{g}_i \end{pmatrix} \end{aligned} \tag{111}$$

where

$$\begin{cases} \tilde{a}_x = a_x \cos\theta + b_x \sin\theta \\ \tilde{a}_y = a_y \cos\theta + b_y \sin\theta \\ \tilde{a}_z = a_z \cos\theta + b_z \sin\theta \end{cases} \tag{112}$$

$$\begin{cases} \tilde{b}_x = -a_x \sin\theta + b_x \cos\theta \\ \tilde{b}_y = -a_y \sin\theta + b_y \cos\theta \\ \tilde{b}_z = -a_z \sin\theta + b_z \cos\theta \end{cases} \tag{113}$$

The above operation is the product of load factor matrix $\boldsymbol{A}$ and rotation matrix denoted as $T(\theta)$ and we obtain

$$AT(\theta) = \tilde{A} \tag{114}$$

We can select any rotation as the stand point of root. How, we can select the angle? We select the angle so that factor points have strong relationship to the factor load.

We introduce a Valimax method, where the elements in $\tilde{A}$ has the maximum and minimum values. We introduce a variable

$$V = \frac{1}{3}\sum_{\alpha=x,y,z}\left[(\tilde{a}_\alpha)^2\right]^2 - \frac{1}{3^2}\left[\sum_{\alpha=x,y,z}(\tilde{a}_\alpha)^2\right]^2 + \frac{1}{3}\sum_{\alpha=x,y,z}\left[(\tilde{b}_\alpha)^2\right]^2 - \frac{1}{3^2}\left[\sum_{\alpha=x,y,z}(\tilde{b}_\alpha)^2\right]^2 \tag{115}$$

and evaluate the angle that givens the maximum V, which is called as a crude Valimax method.

We can evaluate the angle that maximize V numerically. However, we can obtain its analytical one as shown below.

We consider the first and third terms on the right side of Eq. (115), and modify them below.

$$\begin{aligned}
&\sum_{\alpha=x,y,z}\left[(\tilde{a}_\alpha)^2\right]^2 + \sum_{\alpha=x,y,z}\left[(\tilde{b}_\alpha)^2\right]^2 \\
&= \sum_{\alpha=x,y,z}\left[(a_\alpha\cos\theta + b_\alpha\sin\theta)^2\right]^2 + \sum_{\alpha=x,y,z}\left[(-a_\alpha\sin\theta + b_\alpha\cos\theta)^2\right]^2 \\
&= \sum_{\alpha=x,y,z}\left(a_\alpha{}^4\cos^4\theta + 4a_\alpha{}^3b_\alpha\cos^3\theta\sin\theta + 6a_\alpha{}^2b_\alpha{}^2\cos^2\theta\sin^2\theta + 4a_\alpha b_\alpha{}^3\cos\theta\sin^3\theta + b_\alpha{}^4\sin^4\theta\right) \\
&+ \sum_{\alpha=x,y,z}\left(a_\alpha{}^4\sin^4\theta - 4a_\alpha{}^3b_\alpha\sin^3\theta\cos\theta + 6a_\alpha{}^2b_\alpha{}^2\sin^2\theta\cos^2\theta - 4a_\alpha b_\alpha{}^3\sin\theta\cos^3\theta + b_\alpha{}^4\cos^4\theta\right) \\
&= \sum_{\alpha=x,y,z}\left(a_\alpha{}^4 + b_\alpha{}^4\right)\cos^4\theta \\
&+4\sum_{\alpha=x,y,z}\left(a_\alpha{}^3b_\alpha - a_\alpha b_\alpha{}^3\right)\cos^3\theta\sin\theta \\
&+12\sum_{\alpha=x,y,z}a_\alpha{}^2b_\alpha{}^2\cos^2\theta\sin^2\theta \\
&-4\sum_{\alpha=x,y,z}\left(a_\alpha{}^3b_\alpha - a_\alpha b_\alpha{}^3\right)\cos\theta\sin^3\theta \\
&+\sum_{\alpha=x,y,z}\left(a_\alpha{}^4 + b_\alpha{}^4\right)\sin^4\theta
\end{aligned} \tag{116}$$

We then consider the second term of the right side of Eq. (115), and modify it as below.

$$
\begin{aligned}
&\left[\sum_{\alpha=x,y,z}(\tilde{a}_\alpha)^2\right]^2 \\
&=\left[\sum_{\alpha=x,y,z}\left(a_\alpha\cos\theta+b_\alpha\sin\theta\right)^2\right]^2 \\
&=\left[\sum_{\alpha=x,y,z}\left(a_\alpha{}^2\cos^2\theta+2a_\alpha b_\alpha\cos\theta\sin\theta+b_\alpha{}^2\sin^2\theta\right)\right]^2 \\
&=\sum_{\alpha=x,y,z}\left(a_\alpha{}^2\cos^2\theta+2a_\alpha b_\alpha\cos\theta\sin\theta+b_\alpha{}^2\sin^2\theta\right) \\
&\times\sum_{\beta=x,y,z}\left(a_\beta{}^2\cos^2\theta+2a_\beta b_\beta\cos\theta\sin\theta+b_\beta{}^2\sin^2\theta\right) \\
&=\left(\sum_{\alpha=x,y,z}a_\alpha{}^2\sum_{\beta=x,y,z}a_\beta{}^2\right)\cos^4\theta \\
&+2\left(\sum_{\alpha=x,y,z}a_\alpha b_\alpha\sum_{\beta=x,y,z}a_\beta{}^2\right)\cos^3\theta\sin\theta \\
&+\left(\sum_{\alpha=x,y,z}b_\alpha{}^2\sum_{\beta=x,y,z}a_\beta{}^2\right)\cos^2\theta\sin^2\theta \\
&+2\left(\sum_{\alpha=x,y,z}a_\alpha{}^2\sum_{\beta=x,y,z}a_\beta b_\beta\right)\cos^3\theta\sin\theta \\
&+4\left(\sum_{\alpha=x,y,z}a_\alpha b_\alpha\sum_{\beta=x,y,z}a_\beta b_\beta\right)\cos^2\theta\sin^2\theta \\
&+2\left(\sum_{\alpha=x,y,z}b_\alpha{}^2\sum_{\beta=x,y,z}a_\beta b_\beta\right)\cos\theta\sin^3\theta \\
&+\left(\sum_{\alpha=x,y,z}a_\alpha{}^2\sum_{\beta=x,y,z}b_\beta{}^2\right)\cos^2\theta\sin^2\theta \\
&+2\left(\sum_{\alpha=x,y,z}a_\alpha b_\alpha\sum_{\beta=x,y,z}b_\beta{}^2\right)\cos\theta\sin^3\theta \\
&+\left(\sum_{\alpha=x,y,z}b_\alpha{}^2\sum_{\beta=x,y,z}b_\beta{}^2\right)\sin^4\theta
\end{aligned}
\tag{117}
$$

Rearranging them with respect to the same term associated with θ, we obtain

$$\left[\sum_{\alpha=x,y,z}(\tilde{a}_\alpha)^2\right]^2$$
$$=\left(\sum_{\alpha=x,y,z}a_\alpha{}^2\right)^2\cos^4\theta$$
$$+4\left(\sum_{\alpha=x,y,z}a_\alpha{}^2\sum_{\alpha=x,y,z}a_\alpha b_\alpha\right)\cos^3\theta\sin\theta$$
$$+\left[2\left(\sum_{\alpha=x,y,z}a_\alpha{}^2\sum_{\alpha=x,y,z}b_\alpha{}^2\right)+4\left(\sum_{\alpha=x,y,z}a_\alpha b_\alpha\right)^2\right]\cos^2\theta\sin^2\theta$$
$$+4\left(\sum_{\alpha=x,y,z}b_\alpha{}^2\sum_{\alpha=x,y,z}a_\alpha b_\alpha\right)\cos\theta\sin^3\theta$$
$$+\left(\sum_{\alpha=x,y,z}b_\alpha{}^2\right)^2\sin^4\theta \tag{118}$$

$$\left[\sum_{\alpha=x,y,z}(\tilde{b}_\alpha)^2\right]^2$$
$$=\left[\sum_{\alpha=x,y,z}\left(-a_\alpha\sin\theta+b_\alpha\cos\theta\right)^2\right]^2$$
$$=\left[\sum_{\alpha=x,y,z}\left(a_\alpha{}^2\sin^2\theta-2a_\alpha b_\alpha\sin\theta\cos\theta+b_\alpha{}^2\cos^2\theta\right)\right]^2$$
$$=\sum_{\alpha=x,y,z}\left(a_\alpha{}^2\sin^2\theta-2a_\alpha b_\alpha\sin\theta\cos\theta+b_\alpha{}^2\cos^2\theta\right)$$
$$\times\sum_{\beta=x,y,z}\left(a_\beta{}^2\sin^2\theta-2a_\beta b_\beta\sin\theta\cos\theta+b_\beta{}^2\cos^2\theta\right)$$
$$=\left(\sum_{\alpha=x,y,z}a_\alpha{}^2\sum_{\beta=x,y,z}a_\beta{}^2\right)\sin^4\theta$$
$$-2\left(\sum_{\alpha=x,y,z}a_\alpha b_\alpha\sum_{\beta=x,y,z}a_\beta{}^2\right)\sin^3\theta\cos\theta$$
$$+\left(\sum_{\alpha=x,y,z}b_\alpha{}^2\sum_{\beta=x,y,z}a_\beta{}^2\right)\sin^2\theta\cos^2\theta$$
$$-2\left(\sum_{\alpha=x,y,z}a_\alpha{}^2\sum_{\beta=x,y,z}a_\beta b_\beta\right)\sin^3\theta\cos\theta$$
$$+4\left(\sum_{\alpha=x,y,z}a_\alpha b_\alpha\sum_{\beta=x,y,z}a_\beta b_\beta\right)\sin^2\theta\cos^2\theta$$
$$-2\left(\sum_{\alpha=x,y,z}b_\alpha{}^2\sum_{\beta=x,y,z}a_\beta b_\beta\right)\sin\theta\cos^3\theta$$
$$+\left(\sum_{\alpha=x,y,z}a_\alpha{}^2\sum_{\beta=x,y,z}b_\beta{}^2\right)\sin^2\theta\cos^2\theta$$
$$-2\left(\sum_{\alpha=x,y,z}a_\alpha b_\alpha\sum_{\beta=x,y,z}b_\beta{}^2\right)\sin\theta\cos^3\theta$$
$$+\left(\sum_{\alpha=x,y,z}b_\alpha{}^2\sum_{\beta=x,y,z}b_\beta{}^2\right)\cos^4\theta \tag{119}$$

We then consider the fourth term of the right side of Eq. (115), and modify it as below. Rearranging them with respect to the same term associated with θ, we obtain

$$\begin{aligned}
&\left[\sum_{\alpha=x,y,z}\left(\tilde{b}_\alpha\right)^2\right]^2 \\
&=\left(\sum_{\alpha=x,y,z} a_\alpha^{\ 2}\right)^2 \sin^4\theta \\
&-4\left(\sum_{\alpha=x,y,z} a_\alpha b_\alpha \sum_{\alpha=x,y,z} a_\alpha^{\ 2}\right)\sin^3\theta\cos\theta \\
&+2\left(\sum_{\alpha=x,y,z} a_\alpha^{\ 2} \sum_{\alpha=x,y,z} b_\alpha^{\ 2}\right)+4\left(\sum_{\alpha=x,y,z} a_\alpha b_\alpha \sum_{\beta=x,y,z} a_\beta b_\beta\right)\sin^2\theta\cos^2\theta \\
&-4\left(\sum_{\alpha=x,y,z} b_\alpha^{\ 2} \sum_{\alpha=x,y,z} a_\alpha b_\alpha\right)\sin\theta\cos^3\theta \\
&+\left(\sum_{\alpha=x,y,z} b_\alpha^{\ 2}\right)^2 \cos^4\theta
\end{aligned} \tag{120}$$

Summarizing the second and fourth terms of the right side of Eq. (115), we obtain

$$\begin{aligned}
&\left[\sum_{\alpha=x,y,z}\left(\tilde{a}_\alpha\right)^2\right]^2+\left[\sum_{\alpha=x,y,z}\left(\tilde{b}_\alpha\right)^2\right]^2 \\
&=\left[\left(\sum_{\alpha=x,y,z} a_\alpha^{\ 2}\right)^2+\left(\sum_{\alpha=x,y,z} b_\alpha^{\ 2}\right)^2\right]\cos^4\theta \\
&+4\left[\left(\sum_{\alpha=x,y,z} a_\alpha^{\ 2}-\sum_{\alpha=x,y,z} b_\alpha^{\ 2}\right)\left(\sum_{\alpha=x,y,z} a_\alpha b_\alpha\right)\right]\cos^3\theta\sin\theta \\
&+4\left[\left(\sum_{\alpha=x,y,z} a_\alpha^{\ 2}\sum_{\alpha=x,y,z} b_\alpha^{\ 2}\right)+2\left(\sum_{\alpha=x,y,z} a_\alpha b_\alpha\right)^2\right]\cos^2\theta\sin^2\theta \\
&-4\left[\left(\sum_{\alpha=x,y,z} a_\alpha^{\ 2}-\sum_{\alpha=x,y,z} b_\alpha^{\ 2}\right)\sum_{\alpha=x,y,z} a_\alpha b_\alpha\right]\cos\theta\sin^3\theta \\
&+\left[\left(\sum_{\alpha=x,y,z} a_\alpha^{\ 2}\right)^2+\left(\sum_{\alpha=x,y,z} b_\alpha^{\ 2}\right)^2\right]\sin^4\theta
\end{aligned} \tag{121}$$

Summarizing all, we obtain

$$
\begin{aligned}
3V &= \sum_{\alpha=x,y,z}\left[(\tilde{a}_\alpha)^2\right]^2 - \frac{1}{3}\left[\sum_{\alpha=x,y,z}(\tilde{a}_\alpha)^2\right]^2 + \sum_{\alpha=x,y,z}\left[\left(\tilde{b}_\alpha\right)^2\right]^2 - \frac{1}{3}\left[\sum_{\alpha=x,y,z}\left(\tilde{b}_\alpha\right)^2\right]^2 \\
&= \left\{\sum_{\alpha=x,y,z}\left(a_\alpha^{\ 4}+b_\alpha^{\ 4}\right) - \frac{1}{3}\left[\left(\sum_{\alpha=x,y,z}a_\alpha^{\ 2}\right)^2 + \left(\sum_{\alpha=x,y,z}b_\alpha^{\ 2}\right)^2\right]\right\}\cos^4\theta \\
&+4\left\{\sum_{\alpha=x,y,z}\left(a_\alpha^{\ 3}b_\alpha - a_\alpha b_\alpha^{\ 3}\right) - \frac{1}{3}\left[\left(\sum_{\alpha=x,y,z}\left(a_\alpha^{\ 2}-b_\alpha^{\ 2}\right)\right)\left(\sum_{\alpha=x,y,z}a_\alpha b_\alpha\right)\right]\right\}\cos^3\theta\sin\theta \\
&+4\left\{3\sum_{\alpha=x,y,z}a_\alpha^{\ 2}b_\alpha^{\ 2} - \frac{1}{3}\left[\left(\sum_{\alpha=x,y,z}a_\alpha^{\ 2}\sum_{\alpha=x,y,z}b_\alpha^{\ 2}\right) + 2\left(\sum_{\alpha=x,y,z}a_\alpha b_\alpha\right)^2\right]\right\}\cos^2\theta\sin^2\theta \\
&-4\left\{\sum_{\alpha=x,y,z}\left(a_\alpha^{\ 3}b_\alpha - a_\alpha b_\alpha^{\ 3}\right) - \frac{1}{3}\left[\left(\sum_{\alpha=x,y,z}\left(a_\alpha^{\ 2}-b_\alpha^{\ 2}\right)\right)\sum_{\alpha=x,y,z}a_\alpha b_\alpha\right]\right\}\cos\theta\sin^3\theta \\
&+\left\{\sum_{\alpha=x,y,z}\left(a_\alpha^{\ 4}+b_\alpha^{\ 4}\right) - \frac{1}{3}\left[\left(\sum_{\alpha=x,y,z}a_\alpha^{\ 2}\right)^2 + \left(\sum_{\alpha=x,y,z}b_\alpha^{\ 2}\right)^2\right]\right\}\sin^4\theta \\
&= K_1\cos^4\theta + K_2\cos^3\theta\sin\theta + K_3\cos^2\theta\sin^2\theta - K_2\cos\theta\sin^3\theta + K_1\sin^4\theta
\end{aligned} \tag{122}
$$

where

$$
K_1 = \sum_{\alpha=x,y,z}\left(a_\alpha^{\ 4}+b_\alpha^{\ 4}\right) - \frac{1}{3}\left[\left(\sum_{\alpha=x,y,z}a_\alpha^{\ 2}\right)^2 + \left(\sum_{\alpha=x,y,z}b_\alpha^{\ 2}\right)^2\right] \tag{123}
$$

$$
K_2 = 4\left\{\sum_{\alpha=x,y,z}\left(a_\alpha^{\ 3}b_\alpha - a_\alpha b_\alpha^{\ 3}\right) - \frac{1}{3}\left[\left(\sum_{\alpha=x,y,z}\left(a_\alpha^{\ 2}-b_\alpha^{\ 2}\right)\right)\left(\sum_{\alpha=x,y,z}a_\alpha b_\alpha\right)\right]\right\} \tag{124}
$$

$$
K_3 = 4\left\{3\sum_{\alpha=x,y,z}a_\alpha^{\ 2}b_\alpha^{\ 2} - \frac{1}{3}\left[\left(\sum_{\alpha=x,y,z}a_\alpha^{\ 2}\sum_{\alpha=x,y,z}b_\alpha^{\ 2}\right) + 2\left(\sum_{\alpha=x,y,z}a_\alpha b_\alpha\right)^2\right]\right\} \tag{125}
$$

We obtain the condition where V has the local maximum as below.

$$\begin{aligned}
&\frac{\partial(3V)}{\partial\theta} \\
&= -4K_1\cos^3\theta\sin\theta \\
&\quad +K_2\left(-3\cos^2\theta\sin^2\theta+\cos^4\theta\right) \\
&\quad +K_3\left(-2\cos\theta\sin^3\theta+2\cos^3\theta\sin\theta\right) \\
&\quad -K_2\left(-\sin^4\theta+3\cos^2\theta\sin^2\theta\right) \\
&\quad +4K_1\sin^3\theta\cos\theta \\
&= K_2\cos^4\theta+\left(-4K_1+2K_3\right)\cos^3\theta\sin\theta \\
&\quad -6K_2\cos^2\theta\sin^2\theta+\left(4K_1-2K_3\right)\cos\theta\sin^3\theta+K_2\sin^4\theta \\
&= K_2\cos^4\theta-4K_4\cos^3\theta\sin\theta-6K_2\cos^2\theta\sin^2\theta+4K_4\cos\theta\sin^3\theta+K_2\sin^4\theta \\
&= 0
\end{aligned} \tag{126}$$

where

$$K_4 = K_1 - \frac{1}{2}K_3 \tag{127}$$

Dividing Eq. (126) by $\cos^4\theta$, we obtain

$$1-4\frac{K_4}{K_2}\tan\theta-6\tan^2\theta+4\frac{K_4}{K_2}\tan^3\theta+\tan^4\theta=0 \tag{128}$$

Arranging it, we obtain

$$\frac{K_2}{K_4}=\frac{4\left(\tan\theta-\tan^3\theta\right)}{1-6\tan^2\theta+\tan^4\theta}=\tan\left(4\theta\right) \tag{129}$$

where we utilize a lemma given by

$$\tan\left(4\theta\right)=\frac{4\left(\tan\theta-\tan^3\theta\right)}{1-6\tan^2\theta+\tan^4\theta} \tag{130}$$

We can then obtain

$$4\theta = \tan^{-1}\left(\frac{K_2}{K_4}\right) \tag{131}$$

We impose that the angle limitation is

$$-\frac{\pi}{2} < 4\theta < \frac{\pi}{2} \tag{132}$$

We assume that θ_0 is the one root of it. $\theta_0 - \pi$ can become the other root as shown in Figure 2.

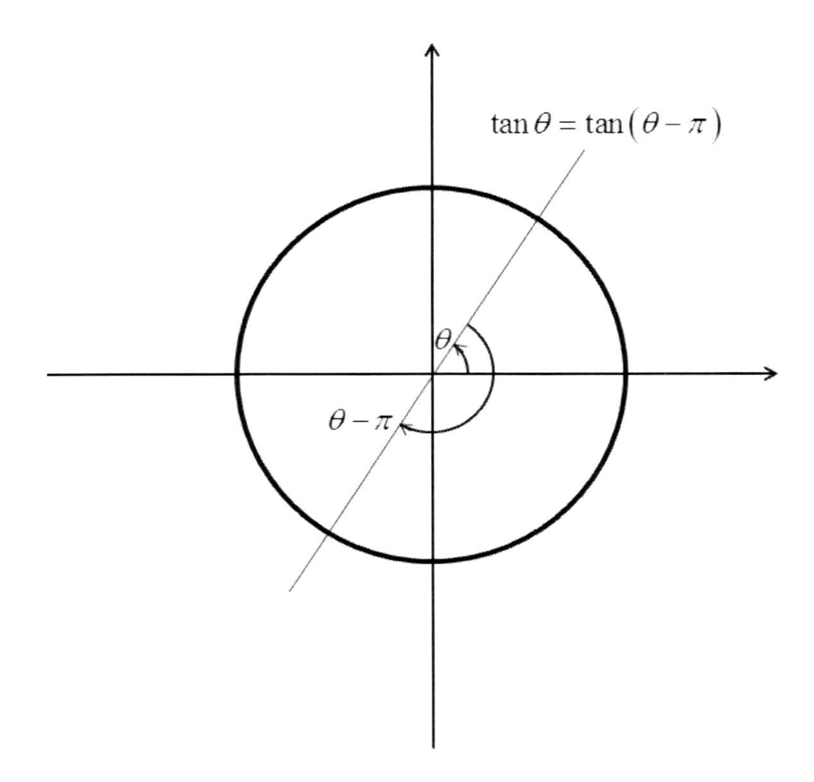

Figure 2. The two angles that provide the same value of tangent.

Therefore, we obtain two roots for that. One corresponds to the local maximum and the other local minimum. We then select the local maximum.

Further differentiate Eq. (126) with respect to θ, we obtain

$$\begin{aligned}
&\frac{\partial^2(3V)}{\partial\theta^2}\\
&= K_2\left(-4\cos^3\theta\sin\theta\right)\\
&-4K_4\left(-3\cos^2\theta\sin^2\theta+\cos^4\theta\right)\\
&-6K_2\left(-2\cos\theta\sin^3\theta+2\cos^3\theta\sin\theta\right)\\
&+4K_4\left(-\sin^4\theta+3\cos^2\theta\sin^2\theta\right)\\
&+K_2\left(4\cos\theta\sin^3\theta\right)\\
&= K_2\left(-4\cos^3\theta\sin\theta+12\cos\theta\sin^3\theta-12\cos^3\theta\sin\theta+4\cos\theta\sin^3\theta\right)\\
&-4K_4\left(-3\cos^2\theta\sin^2\theta+\cos^4\theta+\sin^4\theta-3\cos^2\theta\sin^2\theta\right)\\
&=16K_2\left(\cos\theta\sin^3\theta-\cos^3\theta\sin\theta\right)\\
&-4K_4\left(\cos^4\theta+\sin^4\theta-6\cos^2\theta\sin^2\theta\right)\\
&=16K_2\cos\theta\sin\theta\left(\sin^2\theta-\cos^2\theta\right)\\
&-4K_4\left[\left(\cos^2\theta-\sin^2\theta\right)^2-\left(2\cos\theta\sin\theta\right)^2\right]\\
&=-4\left(K_4\cos 4\theta+K_2\sin 4\theta\right)<0
\end{aligned} \tag{133}$$

Therefore, the condition for the local maximum is given by

$$K_4\cos 4\theta+K_2\sin 4\theta>0 \tag{134}$$

We have the relationship of

$$\frac{K_2}{K_4}=\tan\left(4\theta\right)=\frac{\sin 4\theta}{\cos 4\theta} \tag{135}$$

We then obtain

$$\cos 4\theta=\frac{K_4}{K_2}\sin 4\theta \tag{136}$$

Substituting Eq. (136) into Eq. (134), we obtain

$$\frac{K_2^{\ 2}+K_4^{\ 2}}{K_2}\sin 4\theta > 0 \tag{137}$$

Multiplying $K_2^{\ 2}/\left(K_2^{\ 2}+K_4^{\ 2}\right)>0$ to Eq. (137), we obtain

$$K_2 \sin 4\theta > 0 \tag{138}$$

This is the condition for the local maximum.

If the evaluated 4θ do not hold Eq. (138), we change the angle to

$$4\theta \rightarrow 4\theta \pm \pi \tag{139}$$

We select the sign of $\pm$ so that the angle satisfy

$$-\frac{\pi}{4} < \theta < \frac{\pi}{4} \tag{140}$$

We consider the various cases and establish the evaluation process.

When $K_2 \sin 4\theta > 0$ is held, this ensures the below.

$$-\frac{\pi}{8} < \theta < \frac{\pi}{8} \tag{141}$$

We adopt the angle as it is.

When $K_2 \sin 4\theta > 0$ is not held, we then consider the second condition.

When $\sin 4\theta > 0$ is held, 4θ exist in the first quadrant. When we adopt the negative sign, we obtain

$$-\pi < 4\theta < -\frac{\pi}{2} \tag{142}$$

Therefore, corresponding angle region is

$$-\frac{\pi}{4} < \theta < -\frac{\pi}{8} \tag{143}$$

This is in the acceptable region.

When we adopt positive singe, we obtain

$$\pi < 4\theta < \frac{3\pi}{2} \tag{144}$$

Therefore, the corresponding angle region is

$$\frac{\pi}{4} < \theta < \frac{3\pi}{8} \tag{145}$$

This is the outside of the acceptable region.

We then consider the case of $\sin 4\theta < 0$. This means that 4θ exist in the fourth quadrant region.

When we adopt negative sign, we obtain

$$-\frac{3\pi}{2} < 4\theta < -\pi \tag{146}$$

Therefore, corresponding angle region is

$$-\frac{3\pi}{8} < \theta < -\frac{\pi}{4} \tag{147}$$

This is the outside of the acceptable region.

When we adopt positive sign, we obtain

$$\frac{\pi}{2} < 4\theta < \pi \tag{148}$$

Therefore, corresponding angle region is

$$\frac{\pi}{8} < \theta < \frac{\pi}{4} \tag{149}$$

This is in the acceptable region.

Table 1 summarizes the above procedure.

Table 1. Conditions for determine angle

Condition 1	Condition 2	4θ conversion	Range of θ
$K_2 \sin 4\theta \geq 0$	None	None	$-\frac{\pi}{8} < \theta < \frac{\pi}{8}$
$K_2 \sin 4\theta < 0$	$\sin 4\theta \geq 0$	$4\theta \to 4\theta - \pi$	$-\frac{\pi}{4} < \theta < -\frac{\pi}{8}$
	$\sin 4\theta < 0$	$4\theta \to 4\theta + \pi$	$\frac{\pi}{8} < \theta < \frac{\pi}{4}$

A crude Valimax method, which we explain up to here, is inadequate when the factor $d_\alpha^2 = 1 - S_{e_\alpha}^{(2)}$ is large. When d_α^2 is large, the absolute value of factor loads are also apt to become large. Consequently, the rotation angle is dominantly determined by the factor load with large instead of whole data. Since we want to evaluate the angle related to the whole data, the result with the crude Valimax method may be unexpected one. Therefore, we modify V by changing the elements to $\left(a_\alpha / d_\alpha \right)^2$, $\left(b_\alpha / d_\alpha \right)^2$.

Therefore, we evaluate

$$
\begin{aligned}
S = & \frac{1}{3} \sum_{\alpha=x,y,z} \left[\left(\frac{\tilde{a}_\alpha}{d_\alpha} \right)^2 \right]^2 - \frac{1}{3^2} \left[\sum_{\alpha=x,y,z} \left(\frac{\tilde{a}_\alpha}{d_\alpha} \right)^2 \right]^2 \\
& + \frac{1}{3} \sum_{\alpha=x,y,z} \left[\left(\frac{\tilde{b}_\alpha}{d_\alpha} \right)^2 \right]^2 - \frac{1}{3^2} \left[\sum_{\alpha=x,y,z} \left(\frac{\tilde{b}_\alpha}{d_\alpha} \right)^2 \right]^2
\end{aligned} \tag{150}
$$

and decide the θ which gives the maximum S. This is called as a criteria Valimax method. The corresponding factors are given by

$$
\tan(4\theta) = \frac{G_2}{G_4} \tag{151}
$$

where

$$G_1 = \sum_{\alpha=x,y,z}\left[\left(\frac{a_\alpha}{d_\alpha}\right)^4 + \left(\frac{b_\alpha}{d_\alpha}\right)^4\right] - \frac{1}{3}\left\{\left[\sum_{\alpha=x,y,z}\left(\frac{a_\alpha}{d_\alpha}\right)^2\right]^2 + \left[\sum_{\alpha=x,y,z}\left(\frac{b_\alpha}{d_\alpha}\right)^2\right]^2\right\} \tag{152}$$

$$G_2 = 4\left\{\sum_{\alpha=x,y,z}\left[\left(\frac{a_\alpha}{d_\alpha}\right)^3\left(\frac{b_\alpha}{d_\alpha}\right) - \left(\frac{a_\alpha}{d_\alpha}\right)\left(\frac{b_\alpha}{d_\alpha}\right)^3\right] - \frac{1}{3}\left[\sum_{\alpha=x,y,z}\left[\left(\frac{a_\alpha}{d_\alpha}\right)^2 - \left(\frac{b_\alpha}{d_\alpha}\right)^2\right]\sum_{\alpha=x,y,z}\left(\frac{a_\alpha}{d_\alpha}\right)\left(\frac{b_\alpha}{d_\alpha}\right)\right]\right\} \tag{153}$$

$$G_3 = 4\left\{3\sum_{\alpha=x,y,z}\left(\frac{a_\alpha}{d_\alpha}\right)^2\left(\frac{b_\alpha}{d_\alpha}\right)^2 - \frac{1}{3}\left[\sum_{\alpha=x,y,z}\left(\frac{a_\alpha}{d_\alpha}\right)^2\sum_{\alpha=x,y,z}\left(\frac{b_\alpha}{d_\alpha}\right)^2 + 2\left[\sum_{\alpha=x,y,z}\left(\frac{a_\alpha}{d_\alpha}\right)\left(\frac{b_\alpha}{d_\alpha}\right)\right]^2\right]\right\} \tag{154}$$

$$G_4 = G_1 - \frac{1}{2}G_3 \tag{155}$$

$$G_2 \sin 4\theta > 0 \tag{156}$$

3.4. Factor Point and Independent Load with Two-Factors

We review the process up to here. The basic equation for two-variable factor analysis is given by

$$\begin{pmatrix} x_i \\ y_i \\ z_i \end{pmatrix} = f_i\begin{pmatrix} a_x \\ a_y \\ a_z \end{pmatrix} + g_i\begin{pmatrix} b_x \\ b_y \\ b_z \end{pmatrix} + \begin{pmatrix} e_{xi} \\ e_{yi} \\ e_{zi} \end{pmatrix} \tag{157}$$

We decide the factor load, and variances of e_x, e_y, e_z. We further need to evaluate f_i, g_i and e_x, e_y, e_z themselves.

We assume that x, y, z are commonly expressed with f and g. This means that the predictive value of $\hat{f}_i$ $\hat{g}_i$ and can be expressed with the linear sum of x, y, z given by

$$\hat{f}_i = h_{fx}x_i + h_{fy}y_i + h_{fz}z_i \tag{158}$$

$$\hat{g}_i = h_{gx}x_i + h_{gy}y_i + h_{gz}z_i \tag{159}$$

We focus on f_i first. This is the same analysis for one-factor analysis and we obtain

$$\begin{pmatrix} h_{fx} \\ h_{fy} \\ h_{fz} \end{pmatrix} = \begin{pmatrix} r_{xx} & r_{xy} & r_{zx} \\ r_{xy} & r_{yy} & r_{yz} \\ r_{zx} & r_{yz} & r_{zz} \end{pmatrix}^{-1} \begin{pmatrix} a_x \\ a_y \\ a_z \end{pmatrix} \tag{160}$$

Using the h_{fx}, h_{fy}, h_{fz}, the factor point is given by

$$\hat{f}_i = h_{fx}x_i + h_{fy}y_i + h_{fz}z_i \tag{161}$$

We can perform the same analysis for g, and obtain

$$\begin{pmatrix} h_{gx} \\ h_{gy} \\ h_{gz} \end{pmatrix} = \begin{pmatrix} r_{xx} & r_{xy} & r_{zx} \\ r_{xy} & r_{yy} & r_{yz} \\ r_{zx} & r_{yz} & r_{zz} \end{pmatrix}^{-1} \begin{pmatrix} b_x \\ b_y \\ b_z \end{pmatrix} \tag{162}$$

Using the h_{gx}, h_{gy}, h_{gz}, the factor point is given by

$$\hat{g}_i = h_{gx}x_i + h_{gy}y_i + h_{gz}z_i \tag{163}$$

Summarizing above, we can express as

$$\begin{pmatrix} h_{fx} & h_{gx} \\ h_{fy} & h_{gy} \\ h_{fz} & h_{gz} \end{pmatrix} = \begin{pmatrix} r_{xx} & r_{xy} & r_{zx} \\ r_{xy} & r_{yy} & r_{yz} \\ r_{zx} & r_{yz} & r_{zz} \end{pmatrix}^{-1} \begin{pmatrix} a_x & b_x \\ a_y & b_y \\ a_z & b_z \end{pmatrix} \tag{164}$$

Independent factor is expressed by

$$\begin{pmatrix} e_{xi} \\ e_{yi} \\ e_{zi} \end{pmatrix} = \begin{pmatrix} x_i \\ y_i \\ z_i \end{pmatrix} - \left[f_i \begin{pmatrix} a_x \\ a_y \\ a_z \end{pmatrix} + g_i \begin{pmatrix} b_x \\ b_y \\ b_z \end{pmatrix} \right] \tag{165}$$

This can be further expressed for any member i as

$$\begin{pmatrix} e_{x1} & e_{y1} & e_{z1} \\ e_{x2} & e_{y2} & e_{z2} \\ \vdots & \vdots & \vdots \\ e_{xn} & e_{yn} & e_{zn} \end{pmatrix} = \begin{pmatrix} x_1 & y_1 & z_1 \\ x_2 & y_2 & z_2 \\ \vdots & \vdots & \vdots \\ x_n & y_n & z_n \end{pmatrix} - \begin{pmatrix} f_1 & g_1 \\ f_2 & g_2 \\ \vdots & \vdots \\ f_n & g_n \end{pmatrix} \begin{pmatrix} a_x & a_y & a_z \\ b_x & b_y & b_z \end{pmatrix} \quad (166)$$

4. General Factor Number

We expand the analysis to any number of factors in this section. We treat p variables, q factors, and n data here. The restriction is

$$q \le p \quad (167)$$

The analysis is basically the same as the ones for two-factor analysis. We repeat the process again here without skipping to understand the derivation process clearly.

4.1. Description of the Variables

We treat various kinds of variables, and hence clarify the definition in the start point.

The kind number of variables is p, and each variable has data number of n. The probability variables are denoted as

$$X_1, X_2, \cdots, X_p \quad (168)$$

We express all of them symbolically as

$$X_\alpha \ for \ \alpha = 1, 2, \cdots, p \quad (169)$$

Each data is expressed by

$$X_1 = x_{11}, x_{21}, \cdots, x_{i1}, \cdots, x_{n1}$$
$$X_2 = x_{12}, x_{22}, \cdots, x_{i2}, \cdots, x_{n2}$$
$$\cdots$$
$$X_p = x_{1p}, x_{2p}, \cdots, x_{ip}, \cdots, x_{np} \quad (170)$$

We express all of them symbolically as

$$X_{i\alpha} \quad (171)$$

We sometimes express 2 variable symbolically as

$$X_\alpha, X_\beta \quad (172)$$

There are q kinds of factor point denoted by

$$f_1, f_2, \cdots, f_q \quad (173)$$

We express this symbolically as

$$f_k \ for \ k = 1, 2, \cdots, q \quad (174)$$

When we express that for each data, we then express the factor points as

$$f_1 : f_{11}, f_{21}, \cdots, f_{n1}$$
$$f_2 : f_{12}, f_{22}, \cdots, f_{n2}$$
$$\cdots$$
$$f_q : f_{1q}, f_{2q}, \cdots, f_{nq} \quad (175)$$

We express this symbolically as

$$f_{ik} \quad (176)$$

When we express two factor point symbolically as

$$f_k, f_l \tag{177}$$

We have q kinds of factor loads expressed by

$$\boldsymbol{a}_1, \boldsymbol{a}_2, \cdots, \boldsymbol{a}_q \tag{178}$$

Each factor load has p kinds of element and is expressed by

$$\boldsymbol{a}_1 = \begin{pmatrix} a_{11} \\ a_{21} \\ \vdots \\ a_{p1} \end{pmatrix}$$
$$\boldsymbol{a}_2 = \begin{pmatrix} a_{12} \\ a_{22} \\ \vdots \\ a_{p2} \end{pmatrix}$$
$$\cdots$$
$$\boldsymbol{a}_q = \begin{pmatrix} a_{1q} \\ a_{2q} \\ \vdots \\ a_{pq} \end{pmatrix} \tag{179}$$

We express the factor load symbolically as

$$\boldsymbol{a}_k \tag{180}$$

The corresponding elements are also expressed symbolically as

$$\boldsymbol{a}_k : a_{\alpha k} \tag{181}$$

Independent factor has p kinds and each has n data, which is denoted as

$$e_1, e_2, \cdots, e_p \tag{182}$$

It can be expressed symbolically as

$$e_\alpha \ for\ \alpha = 1, 2, \cdots, p \tag{183}$$

We can express each independent factor related to data as

$$\begin{aligned} &e_1 : e_{11}, e_{21}, \cdots, e_{n1} \\ &e_2 : e_{12}, e_{22}, \cdots, e_{n2} \\ &\cdots \\ &e_p : e_{1p}, e_{2p}, \cdots, e_{np} \end{aligned} \tag{184}$$

It can be expressed symbolically as

$$e_{i\alpha} \tag{185}$$

When we express two independent factors symbolically as

$$e_\alpha, e_\beta \tag{186}$$

We can obtain eigenvalues of p kinds given by

$$\lambda_1, \lambda_2, \cdots, \lambda_p \tag{187}$$

We utilize the factor number of them given by

$$\lambda_1, \lambda_2, \cdots, \lambda_q \tag{188}$$

4.2. Basic Equation

We assume normalized p variable of $x_1, x_2, \cdots, x_p$ and q kinds of factor of $f_1, f_2, \cdots, f_q$. The factor load associated with the factor point f_1 given by $\begin{pmatrix} a_{11} & a_{21} & \cdots & a_{p1} \end{pmatrix}$. In general, the factor load associated with the factor point f_k is given by

$$\boldsymbol{a}_k = \begin{pmatrix} a_{1k} \\ a_{2k} \\ \vdots \\ a_{pk} \end{pmatrix} \tag{189}$$

Therefore, the data are expressed by

$$\begin{pmatrix} x_1 \\ x_2 \\ \vdots \\ x_p \end{pmatrix} = f_1 \begin{pmatrix} a_{11} \\ a_{21} \\ \vdots \\ a_{p1} \end{pmatrix} + f_2 \begin{pmatrix} a_{12} \\ a_{22} \\ \vdots \\ a_{p2} \end{pmatrix} + \cdots + f_q \begin{pmatrix} a_{1q} \\ a_{2q} \\ \vdots \\ a_{pq} \end{pmatrix} + \begin{pmatrix} e_1 \\ e_2 \\ \vdots \\ e_p \end{pmatrix} \tag{190}$$

Focusing on data i, which correspond to the all data for member i, we express it as

$$\begin{pmatrix} x_{i1} \\ x_{i2} \\ \vdots \\ x_{ip} \end{pmatrix} = f_{i1} \begin{pmatrix} a_{11} \\ a_{21} \\ \vdots \\ a_{p1} \end{pmatrix} + f_{i2} \begin{pmatrix} a_{12} \\ a_{22} \\ \vdots \\ a_{p2} \end{pmatrix} + \cdots + f_{iq} \begin{pmatrix} a_{1q} \\ a_{2q} \\ \vdots \\ a_{pq} \end{pmatrix} + \begin{pmatrix} e_{i1} \\ e_{i2} \\ \vdots \\ e_{ip} \end{pmatrix} \quad for\ i = 1, 2, \cdots, n \tag{191}$$

Note that $a_{\alpha k}$ is independent of each member i, factor point is independent of x_α, but depend on i. $e_{i\alpha}$ depends both on i and x_α.

To solve Eq. (191), we impose some restrictions below.

We assume that the factor point f_k is normalized, that is,

$$E[f_k] = 0 \tag{192}$$

$$V[f_k] = 1 \tag{193}$$

The average of independent factor is zero, that is,

$$E\left[e_{x_\alpha}\right] = 0 \tag{194}$$

where $\alpha = 1, 2, \cdots, p$.

The covariance between f_k and e_α is zero, that is,

$$Cov\left[f_k, f_l\right] = \delta_{kl} \tag{195}$$

$$Cov\left[f_k, e_\alpha\right] = 0 \tag{196}$$

$$Cov\left[e_\alpha, e_\beta\right] = \delta_{\alpha\beta} \tag{197}$$

We define the vectors below.

$$\boldsymbol{f}_1 = \begin{pmatrix} f_{11} \\ f_{21} \\ \vdots \\ f_{n1} \end{pmatrix} \tag{198}$$

$$\boldsymbol{f}_2 = \begin{pmatrix} f_{12} \\ f_{22} \\ \vdots \\ f_{n2} \end{pmatrix} \tag{199}$$

. . .

$$\boldsymbol{f}_q = \begin{pmatrix} f_{1q} \\ f_{2q} \\ \vdots \\ f_{nq} \end{pmatrix} \tag{200}$$

$$\boldsymbol{e}_1 = \begin{pmatrix} e_{11} \\ e_{21} \\ \vdots \\ e_{n1} \end{pmatrix} \tag{201}$$

$$\boldsymbol{e}_2 = \begin{pmatrix} e_{12} \\ e_{22} \\ \vdots \\ e_{n2} \end{pmatrix} \tag{202}$$

$$\boldsymbol{e}_p = \begin{pmatrix} e_{1p} \\ e_{2p} \\ \vdots \\ e_{np} \end{pmatrix} \tag{203}$$

$$X = \begin{pmatrix} x_{11} & x_{21} & \cdots & x_{n1} \\ x_{12} & x_{22} & \cdots & x_{n2} \\ \vdots & \vdots & \ddots & \vdots \\ x_{1p} & x_{2p} & \cdots & x_{np} \end{pmatrix} \tag{204}$$

$$A = \begin{pmatrix} a_{11} & a_{12} & \cdots & a_{1q} \\ a_{21} & a_{22} & \cdots & a_{2q} \\ \vdots & \vdots & \ddots & \vdots \\ a_{p1} & a_{p2} & \cdots & a_{pq} \end{pmatrix} \tag{205}$$

$$F = \begin{pmatrix} f_{11} & f_{21} & \cdots & f_{n1} \\ f_{12} & f_{22} & \cdots & f_{n2} \\ \vdots & \vdots & \ddots & \vdots \\ f_{1q} & f_{2q} & \cdots & f_{nq} \end{pmatrix} \tag{206}$$

$$\Gamma = \begin{pmatrix} e_{11} & e_{21} & \cdots & e_{n1} \\ e_{12} & e_{22} & \cdots & e_{n2} \\ \vdots & \vdots & \ddots & \vdots \\ e_{1p} & e_{2p} & \cdots & e_{np} \end{pmatrix} \tag{207}$$

Then the data are expressed with a matrix form as

$$X = AF + \Gamma \tag{208}$$

Multiplying X^T to the Eq. (90) from the right side, and expanding it, we obtain

$$XX^T = \begin{pmatrix} x_{11} & x_{21} & \cdots & x_{n1} \\ x_{12} & x_{22} & \cdots & x_{n2} \\ \vdots & \vdots & \ddots & \vdots \\ x_{1p} & x_{2p} & \cdots & x_{np} \end{pmatrix} \begin{pmatrix} x_{11} & x_{12} & \cdots & x_{1p} \\ x_{21} & x_{22} & \cdots & x_{2p} \\ \vdots & \vdots & \ddots & \vdots \\ x_{n1} & x_{n2} & \cdots & x_{np} \end{pmatrix}$$
$$= \begin{pmatrix} nS_{11}^{(2)} & nS_{12}^{(2)} & \cdots & nS_{1p}^{(2)} \\ nS_{21}^{(2)} & nS_{22}^{(2)} & \cdots & nS_{2p}^{(2)} \\ \vdots & \vdots & \ddots & \vdots \\ nS_{p1}^{(2)} & nS_{p2}^{(2)} & \cdots & nS_{pp}^{(2)} \end{pmatrix}$$
$$= n \begin{pmatrix} r_{11} & r_{12} & \cdots & r_{1p} \\ r_{21} & r_{22} & \cdots & r_{2p} \\ \vdots & \vdots & \ddots & \vdots \\ r_{p1} & r_{p2} & \cdots & r_{pp} \end{pmatrix} \tag{209}$$

where X^T is the transverse matrix of X . Since the variables are normalized, the covariance is equal to the correlation factor and hence we obtain

$$r_{\alpha\beta} = S_{\alpha\beta}^{(2)} \tag{210}$$

$$r_{\alpha\alpha} = 1 \quad for\ \alpha = 1, 2, \cdots, p \tag{211}$$

Similarly, we multiply the transverse matrix to Eq. (16) from the right side, we obtain

$$\begin{aligned} &(AF + \Gamma)(AF + \Gamma)^T \\ &= (AF + \Gamma)(F^T A^T + \Gamma^T) \\ &= AFF^T A^T + AF\Gamma^T + \Gamma F^T A^T + \Gamma\Gamma^T \\ &= A(FF^T)A^T + A(F\Gamma^T) + (F\Gamma^T)^T A^T + \Gamma\Gamma^T \end{aligned} \tag{212}$$

We evaluate the terms in Eq. (81) as below.

$$FF^T = \begin{pmatrix} f_{11} & f_{21} & \cdots & f_{n1} \\ f_{12} & f_{22} & \cdots & f_{n2} \\ \vdots & \vdots & \ddots & \vdots \\ f_{1q} & f_{2q} & \cdots & f_{nq} \end{pmatrix} \begin{pmatrix} f_{11} & f_{12} & \cdots & f_{1q} \\ f_{21} & f_{22} & \cdots & f_{2q} \\ \vdots & \vdots & \ddots & \vdots \\ f_{n1} & f_{n2} & \cdots & f_{nq} \end{pmatrix}$$

$$= \begin{pmatrix} \sum_{i=1}^{n} f_{i1}^{2} & \sum_{i=1}^{n} f_{i1} f_{i2} & \cdots & \sum_{i=1}^{n} f_{i1} f_{iq} \\ \sum_{i=1}^{n} f_{i2} f_{i1} & \sum_{i=1}^{n} f_{i2}^{2} & \cdots & \sum_{i=1}^{n} f_{i2} f_{iq} \\ \vdots & \vdots & \ddots & \vdots \\ \sum_{i=1}^{n} f_{iq} f_{i1} & \sum_{i=1}^{n} f_{iq} f_{i2} & \cdots & \sum_{i=1}^{n} f_{iq}^{2} \end{pmatrix}$$

$$= n \begin{pmatrix} 1 & 0 & \cdots & 0 \\ 0 & 1 & \cdots & 0 \\ \vdots & \vdots & \ddots & \vdots \\ 0 & 0 & \cdots & 1 \end{pmatrix} \tag{213}$$

$$F\Gamma^T = \begin{pmatrix} f_{11} & f_{21} & \cdots & f_{n1} \\ f_{12} & f_{22} & \cdots & f_{n2} \\ \vdots & \vdots & \ddots & \vdots \\ f_{1q} & f_{2q} & \cdots & f_{nq} \end{pmatrix} \begin{pmatrix} e_{11} & e_{12} & \cdots & e_{1q} \\ e_{21} & e_{22} & \cdots & e_{2q} \\ \vdots & \vdots & \ddots & \vdots \\ e_{n1} & e_{n2} & \cdots & e_{nq} \end{pmatrix}$$

$$= \begin{pmatrix} \sum_{i=1}^{n} f_{i1} e_{i1} & \sum_{i=1}^{n} f_{i1} e_{i2} & \cdots & \sum_{i=1}^{n} f_{i1} e_{iq} \\ \sum_{i=1}^{n} f_{i2} e_{i1} & \sum_{i=1}^{n} f_{i2} e_{i2} & \cdots & \sum_{i=1}^{n} f_{i2} e_{iq} \\ \vdots & \vdots & \ddots & \vdots \\ \sum_{i=1}^{n} f_{iq} e_{i1} & \sum_{i=1}^{n} f_{iq} e_{i2} & \cdots & \sum_{i=1}^{n} f_{iq} e_{iq} \end{pmatrix}$$

$$= \begin{pmatrix} 0 & 0 & \cdots & 0 \\ 0 & 0 & \cdots & 0 \\ \vdots & \vdots & \ddots & \vdots \\ 0 & 0 & \cdots & 0 \end{pmatrix} \tag{214}$$

Therefore, we obtain

$$\frac{1}{n} XX^T = \begin{pmatrix} 1 & 0 & \cdots & 0 \\ 0 & 1 & \cdots & 0 \\ \vdots & \vdots & \ddots & \vdots \\ 0 & 0 & \cdots & 1 \end{pmatrix} = AA^T + \frac{1}{n} \Gamma\Gamma^T \tag{215}$$

We further expand the term AA^T and $(1/n)\Gamma\Gamma^T$ as

$$
\begin{aligned}
AA^T &= \begin{pmatrix} a_{11} & a_{12} & \cdots & a_{1q} \\ a_{21} & a_{22} & \cdots & a_{2q} \\ \vdots & \vdots & \ddots & \vdots \\ a_{p1} & a_{p2} & \cdots & a_{pq} \end{pmatrix} \begin{pmatrix} a_{11} & a_{21} & \cdots & a_{p1} \\ a_{12} & a_{22} & \cdots & a_{p2} \\ \vdots & \vdots & \ddots & \vdots \\ a_{1q} & a_{2q} & \cdots & a_{pq} \end{pmatrix} \\
&= \begin{pmatrix} a_{11}^2 + a_{12}^2 + \cdots + a_{1q}^2 & a_{11}a_{21} + a_{12}a_{22} + \cdots + a_{1q}a_{2q} & \cdots & a_{11}a_{p1} + a_{12}a_{p2} + \cdots + a_{1q}a_{pq} \\ a_{21}a_{11} + a_{22}a_{12} + \cdots + a_{2q}a_{1q} & a_{21}^2 + a_{22}^2 + \cdots + a_{2q}^2 & \cdots & a_{21}a_{p1} + a_{22}a_{p2} + \cdots + a_{2q}a_{pq} \\ \vdots & \vdots & \ddots & \vdots \\ a_{p1}a_{11} + a_{p2}a_{12} + \cdots + a_{pq}a_{1q} & a_{p1}a_{21} + a_{p2}a_{22} + \cdots + a_{pq}a_{2q} & \cdots & a_{p1}^2 + a_{p2}^2 + \cdots + a_{pq}^2 \end{pmatrix} \\
&= \begin{pmatrix} a_{11}^2 & a_{11}a_{21} & \cdots & a_{11}a_{p1} \\ a_{21}a_{11} & a_{21}^2 & \cdots & a_{21}a_{p1} \\ \vdots & \vdots & \ddots & \vdots \\ a_{p1}a_{11} & a_{p1}a_{21} & \cdots & a_{p1}^2 \end{pmatrix} \\
&+ \begin{pmatrix} a_{12}^2 & a_{12}a_{22} & \cdots & a_{12}a_{p2} \\ a_{22}a_{12} & a_{22}^2 & \cdots & a_{22}a_{p2} \\ \vdots & \vdots & \ddots & \vdots \\ a_{p2}a_{12} & a_{p2}a_{22} & \cdots & a_{p2}^2 \end{pmatrix} \\
&+ \cdots \\
&+ \begin{pmatrix} a_{1q}^2 & a_{1q}a_{2q} & \cdots & a_{1q}a_{pq} \\ a_{2q}a_{1q} & aa_{2q}^2 & \cdots & a_{2q}a_{pq} \\ \vdots & \vdots & \ddots & \vdots \\ a_{pq}a_{1q} & a_{pq}a_{2q} & \cdots & a_{pq}^2 \end{pmatrix} \\
&= \begin{pmatrix} a_{11} & a_{21} & \cdots & a_{p1} \end{pmatrix} \begin{pmatrix} a_{11} \\ a_{21} \\ \vdots \\ a_{p1} \end{pmatrix} \\
&+ \begin{pmatrix} a_{12} & a_{22} & \cdots & a_{p2} \end{pmatrix} \begin{pmatrix} a_{12} \\ a_{22} \\ \vdots \\ a_{p2} \end{pmatrix} \\
&+ \cdots \\
&+ \begin{pmatrix} a_{1q} & a_{2q} & \cdots & a_{pq} \end{pmatrix} \begin{pmatrix} a_{1q} \\ a_{2q} \\ \vdots \\ a_{pq} \end{pmatrix} \\
&= \sum_{k=1}^{q} \begin{pmatrix} a_{1k} & a_{2k} & \cdots & a_{pk} \end{pmatrix} \begin{pmatrix} a_{1k} \\ a_{2k} \\ \vdots \\ a_{pk} \end{pmatrix}
\end{aligned}
\tag{216}
$$

$$\frac{1}{n}\Gamma\Gamma^T = \frac{1}{n}\begin{pmatrix} e_{11} & e_{21} & \cdots & e_{n1} \\ e_{12} & e_{22} & \cdots & e_{n2} \\ \vdots & \vdots & \ddots & \vdots \\ e_{1p} & e_{2p} & \cdots & e_{np} \end{pmatrix}\begin{pmatrix} e_{11} & e_{12} & \cdots & e_{1p} \\ e_{21} & e_{22} & \cdots & e_{2p} \\ \vdots & \vdots & \ddots & \vdots \\ e_{n1} & e_{n2} & \cdots & e_{np} \end{pmatrix}$$

$$= \begin{pmatrix} \frac{1}{n}\sum_{i=1}^{n} e_{i1}^2 & \frac{1}{n}\sum_{i=1}^{n} e_{i1}e_{i2} & \cdots & \frac{1}{n}\sum_{i=1}^{n} e_{i1}e_{ip} \\ \frac{1}{n}\sum_{i=1}^{n} e_{i2}e_{i1} & \frac{1}{n}\sum_{i=1}^{n} e_{i2}^2 & \cdots & \frac{1}{n}\sum_{i=1}^{n} e_{i2}e_{ip} \\ \vdots & \vdots & \ddots & \vdots \\ \frac{1}{n}\sum_{i=1}^{n} e_{ip}e_{i1} & \frac{1}{n}\sum_{i=1}^{n} e_{ip}e_{i2} & \cdots & \frac{1}{n}\sum_{i=1}^{n} e_{ip}^2 \end{pmatrix}$$

$$= \begin{pmatrix} S_{e_1}^{(2)} & 0 & \cdots & 0 \\ 0 & S_{e_2}^{(2)} & \cdots & 0 \\ \vdots & \vdots & \ddots & \vdots \\ 0_{ix_1} & 0 & \cdots & S_{e_p}^{(2)} \end{pmatrix} \tag{217}$$

Finally, we obtain

$$\begin{pmatrix} 1 & r_{12} & \cdots & r_{1p} \\ r_{21} & 1 & \cdots & r_{2p} \\ \vdots & \vdots & \ddots & \vdots \\ r_{p1} & r_{p2} & \cdots & 1 \end{pmatrix} = \sum_{k=1}^{q}\begin{pmatrix} a_{1k} & a_{2k} & \cdots & a_{pk} \end{pmatrix}\begin{pmatrix} a_{1k} \\ a_{2k} \\ \vdots \\ a_{pk} \end{pmatrix} + \begin{pmatrix} S_{e_1}^{(2)} & 0 & \cdots & 0 \\ 0 & S_{e_2}^{(2)} & \cdots & 0 \\ \vdots & \vdots & \ddots & \vdots \\ 0_{ix_1} & 0 & \cdots & S_{e_p}^{(2)} \end{pmatrix} \tag{218}$$

This is the basic equation for the factor analysis.

4.3. Factor Load

We can expand the covariance matrix as below as shown in the one factor analysis given by

$$\begin{pmatrix} 1 & r_{12} & \cdots & r_{1p} \\ r_{21} & 1 & \cdots & r_{2p} \\ \vdots & \vdots & \ddots & \vdots \\ r_{p1} & r_{p2} & \cdots & 1 \end{pmatrix} = \lambda_1 \boldsymbol{v}_1 \boldsymbol{v}_1^T + \lambda_2 \boldsymbol{v}_2 \boldsymbol{v}_2^T + \cdots + \lambda_p \boldsymbol{v}_p \boldsymbol{v}_p^T \tag{219}$$

We regards the first term $\left(\sqrt{\lambda_1}\boldsymbol{v}_1\right)\left(\sqrt{\lambda_1}\boldsymbol{v}_1\right)^T + \left(\sqrt{\lambda_2}\boldsymbol{v}_2\right)\left(\sqrt{\lambda_2}\boldsymbol{v}_2\right)^T + \cdots + \left(\sqrt{\lambda_q}\boldsymbol{v}_q\right)\left(\sqrt{\lambda_q}\boldsymbol{v}_q\right)^T$ as factor load in the factor analysis, that is,

$$\begin{pmatrix} 1 & r_{12} & \cdots & r_{1p} \\ r_{21} & 1 & \cdots & r_{2p} \\ \vdots & \vdots & \ddots & \vdots \\ r_{p1} & r_{p2} & \cdots & 1 \end{pmatrix} = \left(\sqrt{\lambda_1}\boldsymbol{v}_1\right)\left(\sqrt{\lambda_1}\boldsymbol{v}_1\right)^T + \left(\sqrt{\lambda_2}\boldsymbol{v}_2\right)\left(\sqrt{\lambda_2}\boldsymbol{v}_2\right)^T + \cdots + \left(\sqrt{\lambda_q}\boldsymbol{v}_q\right)\left(\sqrt{\lambda_q}\boldsymbol{v}_q\right)^T + \begin{pmatrix} S_{e_1}^{(2)} & 0 & \cdots & 0 \\ 0 & S_{e_2}^{(2)} & \cdots & 0 \\ \vdots & \vdots & \ddots & \vdots \\ 0 & 0 & \cdots & S_{e_p}^{(2)} \end{pmatrix} \tag{220}$$

This is not the final form factor loadings, but is the first approximated one. The first approximated independent load can be evaluated as

$$\begin{aligned} \boldsymbol{a}_1 &= \sqrt{\lambda_1}\boldsymbol{v}_1 \\ \boldsymbol{a}_2 &= \sqrt{\lambda_2}\boldsymbol{v}_2 \\ &\cdots \\ \boldsymbol{a}_q &= \sqrt{\lambda_q}\boldsymbol{v}_q \end{aligned} \tag{221}$$

We regard these as first approximated ones, that is, we denote them as

$$\lambda_{1\_1}, \boldsymbol{v}_{1\_1}, \lambda_{2\_1}, \boldsymbol{v}_{2\_1}, \cdots, \lambda_{q\_1}, \boldsymbol{v}_{q\_1} \tag{222}$$

$$\begin{pmatrix} S_{e_1\_1}^{(2)} & 0 & \cdots & 0 \\ 0 & S_{e_2\_1}^{(2)} & \cdots & 0 \\ \vdots & \vdots & \ddots & \vdots \\ 0 & 0 & \cdots & S_{e_p\_1}^{(2)} \end{pmatrix} = \begin{pmatrix} 1 & r_{12} & \cdots & r_{1p} \\ r_{21} & 1 & \cdots & r_{2p} \\ \vdots & \vdots & \ddots & \vdots \\ r_{p1} & r_{p2} & \cdots & 1 \end{pmatrix} - \left(\sqrt{\lambda_{1\_1}}\boldsymbol{v}_{1\_1}\right)\left(\sqrt{\lambda_{1\_1}}\boldsymbol{v}_{1\_1}\right)^T + \left(\sqrt{\lambda_{2\_1}}\boldsymbol{v}_{2\_1}\right)\left(\sqrt{\lambda_{2\_1}}\boldsymbol{v}_{2\_1}\right)^T + \cdots + \left(\sqrt{\lambda_{q\_1}}\boldsymbol{v}_{q\_1}\right)\left(\sqrt{\lambda_{q\_1}}\boldsymbol{v}_{q\_1}\right)^T \tag{223}$$

The non-diagonal elements on the right side of Eq. (96) are not 0 in general. Therefore, Eq. (96) is invalid from the rigorous point of view. However, we ignore the non-diagonal elements and regard the diagonal element on both sides are the same.

We then form the new covariance matrix given by

$$R_1 = \begin{pmatrix} 1 - S_{e_1\_1}^{(2)} & r_{12} & \cdots & r_{1p} \\ r_{21} & 1 - S_{e_2\_1}^{(2)} & \cdots & r_{2p} \\ \vdots & \vdots & \ddots & \vdots \\ r_{p1} & r_{p2} & \cdots & 1 - S_{e_p\_1}^{(2)} \end{pmatrix} \tag{224}$$

We evaluate the eigenvalue and eigenvectors of the matrix, and evaluate below.

$$\lambda_{1\_2}, \boldsymbol{v}_{1\_2}, \lambda_{2\_2}, \boldsymbol{v}_{2\_2}, \cdots, \lambda_{q\_2}, \boldsymbol{v}_{q\_2} \tag{225}$$

The diagonal elements can then be evaluated as

$$\begin{pmatrix} S_{e_1\_2}^{(2)} & 0 & \cdots & 0 \\ 0 & S_{e_2\_2}^{(2)} & \cdots & 0 \\ \vdots & \vdots & \ddots & \vdots \\ 0 & 0 & \cdots & S_{e_p\_2}^{(2)} \end{pmatrix}$$
$$= \begin{pmatrix} 1 & r_{12} & \cdots & r_{1p} \\ r_{21} & 1 & \cdots & r_{2p} \\ \vdots & \vdots & \ddots & \vdots \\ r_{p1} & r_{p2} & \cdots & 1 \end{pmatrix}$$
$$-\left(\sqrt{\lambda_{1\_2}}\boldsymbol{v}_{1\_2}\right)\left(\sqrt{\lambda_{1\_2}}\boldsymbol{v}_{1\_2}\right)^T + \left(\sqrt{\lambda_{2\_2}}\boldsymbol{v}_{2\_2}\right)\left(\sqrt{\lambda_{2\_2}}\boldsymbol{v}_{2\_2}\right)^T + \cdots + \left(\sqrt{\lambda_{q\_2}}\boldsymbol{v}_{q\_2}\right)\left(\sqrt{\lambda_{q\_2}}\boldsymbol{v}_{q\_2}\right)^T \tag{226}$$

We evaluate the δ below.

$$\delta = \frac{1}{q}\left\{\left[S_{e_1\_2}^{(2)} - S_{e_1\_1}^{(2)}\right]^2 + \left[S_{e_2\_2}^{(2)} - S_{e_2\_1}^{(2)}\right]^2 + \cdots + \left[S_{e_q\_2}^{(2)} - S_{e_q\_1}^{(2)}\right]^2\right\} \tag{227}$$

If it is smaller than the setting value of δ_c, we regard the factor load as

$$\begin{aligned}
\boldsymbol{a}_1 &= \sqrt{\lambda_{1_2}}\boldsymbol{v}_{1_2} \\
\boldsymbol{a}_2 &= \sqrt{\lambda_{2_2}}\boldsymbol{v}_{2_2} \\
&\cdots \\
\boldsymbol{a}_q &= \sqrt{\lambda_{q_2}}\boldsymbol{v}_{q_2}
\end{aligned} \tag{228}$$

If $\delta > \delta_c$, we construct updated covariance matrix given by

$$R_2 = \begin{pmatrix} 1 - S_{e_1\_2}^{(2)} & r_{12} & \cdots & r_{1p} \\ r_{21} & 1 - S_{e_2\_2}^{(2)} & \cdots & r_{2p} \\ \vdots & \vdots & \ddots & \vdots \\ r_{p1} & r_{p2} & \cdots & 1 - S_{e_p\_2}^{(2)} \end{pmatrix} \tag{229}$$

and evaluate the eigenvalues and eigenvectors given by

$$\lambda_{1\_3}, \boldsymbol{v}_{1\_3}, \lambda_{2\_3}, \boldsymbol{v}_{2\_3}, \cdots, \lambda_{q\_3}, \boldsymbol{v}_{q\_3} \tag{230}$$

We evaluate the diagonal elements as

$$\begin{aligned}
&\begin{pmatrix} S_{e_1_3}^{(2)} & 0 & \cdots & 0 \\ 0 & S_{e_2_3}^{(2)} & \cdots & 0 \\ \vdots & \vdots & \ddots & \vdots \\ 0 & 0 & \cdots & S_{e_p_3}^{(2)} \end{pmatrix} \\
&= \begin{pmatrix} 1 & r_{12} & \cdots & r_{1p} \\ r_{21} & 1 & \cdots & r_{2p} \\ \vdots & \vdots & \ddots & \vdots \\ r_{p1} & r_{p2} & \cdots & 1 \end{pmatrix} \\
&-\left(\sqrt{\lambda_{1_3}}\boldsymbol{v}_{1_3}\right)\left(\sqrt{\lambda_{1_3}}\boldsymbol{v}_{1_3}\right)^T + \left(\sqrt{\lambda_{2_3}}\boldsymbol{v}_{2_3}\right)\left(\sqrt{\lambda_{2_3}}\boldsymbol{v}_{2_3}\right)^T + \cdots + \left(\sqrt{\lambda_{q_3}}\boldsymbol{v}_{q_3}\right)\left(\sqrt{\lambda_{q_3}}\boldsymbol{v}_{q_3}\right)^T
\end{aligned} \tag{231}$$

We then evaluate δ given by

$$\delta = \frac{1}{q}\left\{\left[S_{e_1\_3}^{(2)} - S_{e_1\_2}^{(2)}\right]^2 + \left[S_{e_2\_3}^{(2)} - S_{e_2\_2}^{(2)}\right]^2 + \cdots + \left[S_{e_q\_3}^{(2)} - S_{e_q\_2}^{(2)}\right]^2\right\} \tag{232}$$

If δ is smaller than δ_c, we decide the factor load as

$$\begin{aligned} \boldsymbol{a}_1 &= \sqrt{\lambda_{1\_3}}\boldsymbol{v}_{1\_3} \\ \boldsymbol{a}_2 &= \sqrt{\lambda_{2\_3}}\boldsymbol{v}_{2\_3} \\ &\cdots \\ \boldsymbol{a}_q &= \sqrt{\lambda_{q\_3}}\boldsymbol{v}_{q\_3} \end{aligned} \tag{233}$$

If δ is larger than δ_c we repeat this process until

$$\delta = \frac{1}{q}\left\{\left[S_{e_1\_m}^{(2)} - S_{e_1\_m-1}^{(2)}\right]^2 + \left[S_{e_2\_m}^{(2)} - S_{e_2\_m-1}^{(2)}\right]^2 + \cdots + \left[S_{e_q\_m}^{(2)} - S_{e_q\_m-1}^{(2)}\right]^2\right\} \tag{234}$$

$\delta \le \delta_c$ is held. We finally obtain the factor load given by

$$\begin{aligned} \boldsymbol{a}_1 &= \sqrt{\lambda_{1\_m}}\boldsymbol{v}_{1\_m} = \begin{pmatrix} a_{11} \\ a_{21} \\ \vdots \\ a_{p1} \end{pmatrix} \\ \boldsymbol{a}_2 &= \sqrt{\lambda_{2\_m}}\boldsymbol{v}_{2\_m} = \begin{pmatrix} a_{12} \\ a_{22} \\ \vdots \\ a_{p2} \end{pmatrix} \\ &\cdots \\ \boldsymbol{a}_q &= \sqrt{\lambda_{q\_m}}\boldsymbol{v}_{q\_m} = \begin{pmatrix} a_{1q} \\ a_{2q} \\ \vdots \\ a_{pq} \end{pmatrix} \end{aligned} \tag{235}$$

On the other hand, we obtain the final diagonal elements as

$$\begin{aligned} d_1^2 &= 1 - S_{e_1\_m}^{(2)} \\ d_2^2 &= 1 - S_{e_2\_m}^{(2)} \\ &\cdots \\ d_q^2 &= 1 - S_{e_q\_m}^{(2)} \end{aligned} \tag{236}$$

4.4. Uncertainty of Load Factor and Rotation

We will finally obtain X, A, F, Γ in the end. However, in the mathematical stand point of view, it is expressed by

$$\begin{aligned} X &= AF + \Gamma \\ &= ATT^{-1}F + \Gamma \end{aligned} \tag{237}$$

We assume that T is the rotation matrix, Eq. (90) can be modified as

$$X = AT(\theta)T(-\theta)F + \Gamma \tag{238}$$

where

$$\begin{cases} A' = AT(\theta) \\ F' = T(-\theta)F \end{cases} \tag{239}$$

Therefore, if A and F are the root of Eq. (89), A and F are also the roots of it. We will discuss this uncertainty later.

In the case of two-factor analysis, we can consider one angle. However, in the q-factor analysis, we have ${}_qC_2$ kinds of planes to be considered.

We consider two factors among them. The corresponding factor loads are

$$\boldsymbol{a}_k = \begin{pmatrix} a_{1k} \\ a_{2k} \\ \vdots \\ a_{pk} \end{pmatrix}, \boldsymbol{a}_l = \begin{pmatrix} a_{1l} \\ a_{2l} \\ \vdots \\ a_{pl} \end{pmatrix} \quad for\ k \neq l \tag{240}$$

The conversion is expressed by the matrix as

$$\begin{aligned} &\begin{pmatrix} a_{1k} & a_{1l} \\ a_{2k} & a_{2l} \\ \vdots & \vdots \\ a_{pk} & a_{pl} \end{pmatrix} \begin{pmatrix} \cos\theta & -\sin\theta \\ \sin\theta & \cos\theta \end{pmatrix} \\ &= \begin{pmatrix} a_{1k}\cos\theta + a_{1l}\sin\theta & -a_{1k}\sin\theta + a_{1l}\cos\theta \\ a_{2k}\cos\theta + a_{2l}\sin\theta & -a_{2k}\sin\theta + a_{2l}\cos\theta \\ \vdots & \vdots \\ a_{pk}\cos\theta + a_{pl}\sin\theta & -a_k\sin\theta + a_l\cos\theta \end{pmatrix} \end{aligned} \tag{241}$$

Therefore, the converted factor load is given by

$$\tilde{\boldsymbol{a}}_k = \begin{pmatrix} a_{1k}\cos\theta + a_{1l}\sin\theta \\ a_{2k}\cos\theta + a_{2l}\sin\theta \\ \vdots \\ a_{pk}\cos\theta + a_{pl}\sin\theta \end{pmatrix} = \begin{pmatrix} \tilde{a}_{1k} \\ \tilde{a}_{2k} \\ \vdots \\ \tilde{a}_{pk} \end{pmatrix} \tag{242}$$

$$\tilde{\boldsymbol{a}}_l = \begin{pmatrix} -a_{1k}\sin\theta + a_{1l}\cos\theta \\ -a_{2k}\sin\theta + a_{2l}\cos\theta \\ \vdots \\ -a_k\sin\theta + a_l\cos\theta \end{pmatrix} = \begin{pmatrix} \tilde{a}_{1l} \\ \tilde{a}_{2l} \\ \vdots \\ \tilde{a}_{pl} \end{pmatrix} \tag{243}$$

We decide the angle θ so that it maximizes the S below.

$$\begin{aligned} S &= \frac{1}{p}\sum_{\alpha=1}^{p}\left[\left(\frac{\tilde{a}_{\alpha k}}{d_\alpha}\right)^2\right]^2 - \frac{1}{p^2}\left[\sum_{\alpha=1}^{p}\left(\frac{\tilde{a}_{\alpha k}}{d_\alpha}\right)^2\right]^2 \\ &\quad + \frac{1}{p}\sum_{\alpha=1}^{p}\left[\left(\frac{\tilde{a}_{\alpha l}}{d_\alpha}\right)^2\right]^2 - \frac{1}{p^2}\left[\sum_{\alpha=1}^{p}\left(\frac{\tilde{a}_{\alpha l}}{d_\alpha}\right)^2\right]^2 \end{aligned} \tag{244}$$

The θ is solved as

$$\tan(4\theta) = \frac{G_2}{G_4} \tag{245}$$

The critical condition is given by

$$G_2 \sin 4\theta > 0 \tag{246}$$

Where each parameter is given by

$$G_1 = \sum_{\alpha=1}^{p}\left[\left(\frac{a_{\alpha k}}{d_\alpha}\right)^4 + \left(\frac{a_{\alpha l}}{d_\alpha}\right)^4\right] - \frac{1}{p}\left\{\left[\sum_{\alpha=1}^{p}\left(\frac{a_{\alpha k}}{d_\alpha}\right)^2\right]^2 + \left[\sum_{\alpha=1}^{p}\left(\frac{a_{\alpha l}}{d_\alpha}\right)^2\right]^2\right\} \tag{247}$$

$$G_2 = 4\left\{ \sum_{\alpha=x,y,z} \left[\left(\frac{a_{\alpha k}}{d_\alpha}\right)^3 \left(\frac{a_{\alpha l}}{d_\alpha}\right) - \left(\frac{a_{\alpha k}}{d_\alpha}\right)\left(\frac{a_{\alpha l}}{d_\alpha}\right)^3 \right] - \frac{1}{p}\left[\sum_{\alpha=1}^{p} \left[\left(\frac{a_{\alpha k}}{d_\alpha}\right)^2 - \left(\frac{a_{\alpha l}}{d_\alpha}\right)^2 \right] \sum_{\alpha=1}^{p} \left(\frac{a_{\alpha k}}{d_\alpha}\right)\left(\frac{a_{\alpha l}}{d_\alpha}\right) \right] \right\} \tag{248}$$

$$G_3 = 4\left\{ 3\sum_{\alpha=1}^{p} \left(\frac{a_{\alpha k}}{d_\alpha}\right)^2 \left(\frac{a_{\alpha l}}{d_\alpha}\right)^2 - \frac{1}{p}\left[\sum_{\alpha=1}^{p} \left(\frac{a_{\alpha k}}{d_\alpha}\right)^2 \sum_{\alpha=1}^{p} \left(\frac{a_{\alpha l}}{d_\alpha}\right)^2 + 2\left[\sum_{\alpha=1}^{p} \left(\frac{a_{\alpha k}}{d_\alpha}\right)\left(\frac{a_{\alpha l}}{d_\alpha}\right) \right]^2 \right] \right\} \tag{249}$$

$$G_4 = G_1 - \frac{1}{2} G_3 \tag{250}$$

$$G_2 \sin 4\theta > 0 \tag{251}$$

We select two factors. The factor k, l is expressed by the matrix below...

$$T(\theta) = \begin{pmatrix} 1 & & & & \cdots & & & & 0 \\ & \ddots & & & & & & & \\ & & 1 & & & & & & \\ & & & \overset{(k,k)}{\cos\theta} & & \overset{(k,l)}{-\sin\theta} & & & \\ & & & & 1 & & & & \\ \vdots & & & & \ddots & & & & \vdots \\ & & & & & 1 & & & \\ & & & \overset{(l,k)}{\sin\theta} & & \overset{(l,l)}{\cos\theta} & & & \\ & & & & & & 1 & & \\ & & & & & & & \ddots & \\ 0 & & & & \cdots & & & & 1 \end{pmatrix} \tag{252}$$

We then form

$$AT(\theta) \tag{253}$$

We perform whole combination of ${}_pC_2$.

When we perform the Valimax rotation, the two factor loads changes in general. Next, we select the same one and different one in the next selection, both factor load changes. Therefore, we need to many cycles so that the change becomes small.

4.5. Factor Point and Independent Load with General Number-Factors

We review the process up to here. The basic equation for general number p of variables factor analysis is given by

$$\begin{pmatrix} x_{i1} \\ x_{i2} \\ \vdots \\ x_{ip} \end{pmatrix} = f_{i1}\begin{pmatrix} a_{11} \\ a_{21} \\ \vdots \\ a_{p1} \end{pmatrix} + f_{i2}\begin{pmatrix} a_{12} \\ a_{22} \\ \vdots \\ a_{p2} \end{pmatrix} + \cdots + f_{iq}\begin{pmatrix} a_{1q} \\ a_{2q} \\ \vdots \\ a_{pq} \end{pmatrix} + \begin{pmatrix} e_{i1} \\ e_{i2} \\ \vdots \\ e_{ip} \end{pmatrix} \quad for\ i = 1, 2, \cdots, n \tag{254}$$

We decide factor load, and variance of $e_1, e_2, \cdots, e_p$. We further need to evaluate f_i, g_i and $e_{i1}, e_{i2}, \cdots, e_{ip}$ themselves.

We assume that $x_1, x_2, \cdots, x_p$ are commonly expressed with $f_{1k}, f_{2k}, \cdots, f_{pk}$ This means that the predictive value of $\hat{f}_{ik}$ can be expressed with the linear sum of $x_1, x_2, \cdots, x_p$ given by

$$\hat{f}_{ik} = h_{1k}x_{i1} + h_{2k}x_{i2} + \cdots + h_{pk}x_{ip} \tag{255}$$

Assuming f_{ik} as real data, the square of the deviation is given by

$$\begin{aligned} Q_k &= \sum_{i=1}^{n}\left(f_{ik} - \hat{f}_{ik}\right)^2 \\ &= \sum_{i=1}^{n}\left[f_{ik} - \left(h_{1k}x_{i1} + h_{k2}x_{i2} + \cdots + h_{pk}x_{ip}\right)\right]^2 \end{aligned} \tag{256}$$

We decide $h_{1k}, h_{2k}, \cdots, h_{pk}$ so that it minimize Eq. (256). Partial differentiating Eq. (256) with respect to h_{1k}, and set it as 0, we obtain

$$\begin{aligned} \frac{\partial Q_k}{\partial h_{1k}} &= \frac{\partial}{\partial h_{1k}}\sum_{i=1}^{n}\left[f_{ik} - \left(h_{1k}x_{i1} + h_{k2}x_{i2} + \cdots + h_{pk}x_{ip}\right)\right]^2 \\ &= -2\sum_{i=1}^{n}\left[f_{ik} - \left(h_{1k}x_{i1} + h_{k2}x_{i2} + \cdots + h_{pk}x_{ip}\right)\right]x_{i1} = 0 \end{aligned} \tag{257}$$

This leads to

$$\sum_{i=1}^{n}\left(h_{1k}x_{i1}x_{i1}+h_{k2}x_{i2}x_{i1}+\cdots+h_{pk}x_{ip}x_{i1}\right)=\sum_{i=1}^{n}f_{ik}x_{i1} \tag{258}$$

The left side of Eq. (258) is modified as

$$\begin{aligned}\sum_{i=1}^{n}\left(h_{1k}x_{i1}x_{i1}+h_{k2}x_{i2}x_{i1}+\cdots+h_{pk}x_{ip}x_{i1}\right)&=h_{1k}\sum_{i=1}^{n}x_{i1}x_{i1}+h_{2k}\sum_{i=1}^{n}x_{i2}x_{i1}+\cdots+h_{pk}\sum_{i=1}^{n}x_{ip}x_{i1}\\&=n\left(h_{1k}r_{11}+h_{2k}r_{21}+\cdots+h_{pk}r_{p1}\right)\end{aligned} \tag{259}$$

The right side of Eq. (258) is modified as

$$\begin{aligned}&\sum_{i=1}^{n}f_{ik}x_{i1}\\&=\sum_{i=1}^{n}f_{ik}\left(a_{11}f_{i1}+a_{12}f_{i2}+\cdots+a_{1q}f_{iq}\right)\\&=a_{11}\sum_{i=1}^{n}f_{ik}f_{i1}+a_{12}\sum_{i=1}^{n}f_{ik}f_{i2}+\cdots+a_{1q}\sum_{i=1}^{n}f_{ik}f_{iq}\\&=a_{1k}\sum_{i=1}^{n}f_{ik}f_{ik}\\&=na_{1k}\end{aligned} \tag{260}$$

Therefore, Eq. (258) is reduced to

$$h_{1k}r_{11}+h_{2k}r_{21}+\cdots+h_{pk}r_{p1}=a_{1k} \tag{261}$$

Performing the similar analysis with respect to $h_{\alpha k}$, we obtain

$$\begin{aligned}\frac{\partial Q_f}{\partial h_{\alpha k}}&=\frac{\partial}{\partial h_{\alpha k}}\sum_{i=1}^{n}\left[f_{ik}-\left(h_{1k}x_{i1}+h_{2k}x_{i2}+\cdots+h_{pk}x_{ip}\right)\right]^2\\&=-2\sum_{i=1}^{n}\left[f_{ik}-\left(h_{1k}x_{i1}+h_{2k}x_{i2}+\cdots+h_{pk}x_{ip}\right)\right]x_{i\alpha}=0\end{aligned} \tag{262}$$

We then obtain

$$\sum_{i=1}^{n}\left(h_{1k}x_{i1}x_{i\alpha}+h_{2k}x_{i2}x_{i\alpha}+\cdots+h_{pk}x_{ip}x_{i\alpha}\right)=\sum_{i=1}^{n}f_{ik}x_{i\alpha} \tag{263}$$

The left side of Eq. (263) is modified as

$$\begin{aligned}\sum_{i=1}^{n}\left(h_{1k}x_{i1}x_{i\alpha}+h_{2k}x_{i2}x_{i\alpha}+\cdots+h_{pk}x_{ip}x_{i\alpha}\right)&=h_{1k}\sum_{i=1}^{n}x_{i1}x_{i\alpha}+h_{2k}\sum_{i=1}^{n}x_{i2}x_{i\alpha}+\cdots+h_{pk}\sum_{i=1}^{n}x_{ip}x_{i\alpha}\\&=n\left(h_{1k}r_{1\alpha}+h_{2k}r_{2\alpha}+\cdots+h_{pk}r_{p\alpha}\right)\end{aligned} \tag{264}$$

The right side of Eq. (263) is modified as

$$\begin{aligned}&\sum_{i=1}^{n}f_{ik}x_{i\alpha}\\&=\sum_{i=1}^{n}f_{ik}\left(a_{\alpha1}f_{i\alpha}+a_{\alpha2}f_{i2}+\cdots+a_{\alpha q}f_{iq}\right)\\&=a_{11}\sum_{i=1}^{n}f_{ik}f_{i1}+a_{12}\sum_{i=1}^{n}f_{ik}f_{i2}+\cdots+a_{1q}\sum_{i=1}^{n}f_{ik}f_{iq}\\&=a_{1k}\sum_{i=1}^{n}f_{ik}f_{ik}\\&=na_{1k}\end{aligned} \tag{265}$$

Therefore, Eq. (263) is reduced to

$$h_{1k}r_{1\alpha}+h_{2k}r_{2\alpha}+\cdots+h_{pk}r_{p\alpha}=a_{\alpha k} \tag{266}$$

Summarizing above, we obtain

$$\begin{pmatrix}1 & r_{12} & \cdots & r_{1p}\\ r_{21} & 1 & \cdots & r_{2p}\\ \vdots & \vdots & \ddots & \vdots\\ r_{p1} & r_{p2} & \cdots & 1\end{pmatrix}\begin{pmatrix}h_{1k}\\ h_{2k}\\ \vdots\\ h_{2k}\end{pmatrix}=\begin{pmatrix}a_{1k}\\ a_{2k}\\ \vdots\\ a_{2k}\end{pmatrix} \tag{267}$$

We then obtain the factor point factors as

$$\begin{pmatrix} h_{1k} \\ h_{2k} \\ \vdots \\ h_{pk} \end{pmatrix} = \begin{pmatrix} 1 & r_{12} & \cdots & r_{1p} \\ r_{21} & 1 & \cdots & r_{2p} \\ \vdots & \vdots & \ddots & \vdots \\ r_{p1} & r_{p2} & \cdots & 1 \end{pmatrix}^{-1} \begin{pmatrix} a_{1k} \\ a_{2k} \\ \vdots \\ a_{2k} \end{pmatrix} \tag{268}$$

We can then obtain factor points as

$$\hat{f}_{ik} = h_{1k} x_{i1} + h_{2k} x_{i2} + \cdots + h_{pk} x_{ip} \tag{269}$$

The independent factor is given by

$$\begin{pmatrix} e_{i1} \\ e_{i2} \\ \vdots \\ e_{ip} \end{pmatrix} = \begin{pmatrix} x_{i1} \\ x_{i2} \\ \vdots \\ x_{ip} \end{pmatrix} - \left[f_{i1} \begin{pmatrix} a_{11} \\ a_{21} \\ \vdots \\ a_{p1} \end{pmatrix} + f_{i2} \begin{pmatrix} a_{12} \\ a_{22} \\ \vdots \\ a_{p2} \end{pmatrix} + \cdots + f_{iq} \begin{pmatrix} a_{1q} \\ a_{2q} \\ \vdots \\ a_{pq} \end{pmatrix} \right] \quad for\ i = 1, 2, \cdots, n \tag{270}$$

The above results are totally expressed by

$$\begin{pmatrix} e_{11} & e_{12} & \cdots & e_{1p} \\ e_{21} & e_{22} & \cdots & e_{2p} \\ \vdots & \vdots & \ddots & \vdots \\ e_{n1} & e_{n2} & \cdots & e_{np} \end{pmatrix} = \begin{pmatrix} x_{11} & x_{12} & \cdots & x_{1p} \\ x_{21} & x_{22} & \cdots & x_{2p} \\ \vdots & \vdots & \ddots & \vdots \\ x_{n1} & x_{n2} & \cdots & x_{np} \end{pmatrix} - \begin{pmatrix} f_{11} & f_{12} & \cdots & f_{1q} \\ f_{21} & f_{22} & \cdots & f_{2q} \\ \vdots & \vdots & \ddots & \vdots \\ f_{n1} & f_{n2} & \cdots & f_{nq} \end{pmatrix} \begin{pmatrix} a_{11} & a_{21} & \cdots & a_{p1} \\ a_{12} & a_{22} & \cdots & a_{p2} \\ \vdots & \vdots & \ddots & \vdots \\ a_{1q} & a_{2q} & \cdots & a_{pq} \end{pmatrix} \tag{271}$$

5. SUMMARY

To summarize the results in this chapter—

We treat various kinds of variables in this analysis, and hence summarize the treatments systematically in this section.

5.1. Notation of Variables

Number of kinds of variables is p, data number is n, and factor number is q.

Symbolic expressions for the variable are α, β.

Symbolic expressions for the data are i, j

Symbolic expressions for the factors are k, l

The order of the subscript is given by $\left[(i, j),(\alpha, \beta),(k, l)\right]$

Based on the above, each expression is as follows.

p kinds variable are expressed as

$$x_1, x_2, \cdots, x_p$$

This can be expressed with symbolically as

$$x_\alpha \ for\ \alpha = 1, 2, \cdots, p$$

The detailed expression that denotes the each data value is given by

$$x_1 : x_{11}, x_{21}, \cdots, x_{n1}$$
$$x_2 : x_{12}, x_{22}, \cdots, x_{n2}$$
$$\cdots$$
$$x_p : x_{1p}, x_{2p}, \cdots, x_{np}$$

The corresponding symbolic expression is given by

$$x_{i\alpha}$$

Two kinds of variables can be expressed symbolically as

$$x_\alpha, x_\beta$$

The normalize variables are expressed by

$$u_1, u_2, \cdots, u_p$$

This can be expressed symbolically as

$$u_\alpha \ for\ \alpha = 1, 2, \cdots, p$$

The detailed expression that denotes the each normalized data value is given by

$u_1 : u_{11}, u_{21}, \cdots, u_{n1}$
$u_2 : u_{12}, u_{22}, \cdots, u_{n2}$
$\cdots$
$u_p : u_{1p}, u_{2p}, \cdots, u_{np}$

This can be expressed symbolically as

$$u_{i\alpha}$$

Two kinds of normalized variables are symbolically expressed as

$$u_\alpha, u_\beta$$

We have q kinds of factor points and they are denoted as

$$f_1, f_2, \cdots, f_q$$

We express them symbolically as

$$f_k \ for\ k = 1, 2, \cdots, q$$

The detailed expression to denote the each data value is given by

$f_1 : f_{11}, f_{21}, \cdots, f_{n1}$
$f_2 : f_{12}, f_{22}, \cdots, f_{n2}$
$\cdots$
$f_q : f_{1q}, f_{2q}, \cdots, f_{nq}$

This can be expressed symbolically as

$$f_{ik}$$

Two kinds of factor points are symbolically expressed as

$$f_k, f_l$$

There are q kinds of factor loadings as

$$\boldsymbol{a}_1, \boldsymbol{a}_2, \cdots, \boldsymbol{a}_q$$

Each factor loading has p elements and are expressed as

$$\boldsymbol{a}_1 = \begin{pmatrix} a_{11} \\ a_{21} \\ \vdots \\ a_{p1} \end{pmatrix}$$

$$\boldsymbol{a}_2 = \begin{pmatrix} a_{12} \\ a_{22} \\ \vdots \\ a_{p2} \end{pmatrix}$$

$\cdots$

$$\boldsymbol{a}_q = \begin{pmatrix} a_{1q} \\ a_{2q} \\ \vdots \\ a_{pq} \end{pmatrix}$$

The factor loading is symbolically expressed as

$$\boldsymbol{a}_k$$

The elements are expressed as

$$\boldsymbol{a}_k : a_{\alpha k}$$

Independent load has the same structure as the data x . It has p kinds of variables and each variable has n data. We express them as

$$e_1, e_2, \cdots, e_p$$

This can be symbolically expressed as

$$e_\alpha \ for \ \alpha = 1, 2, \cdots, p$$

The detailed expression to denote the each data value is given by

$$e_1 : e_{11}, e_{21}, \cdots, e_{n1}$$
$$e_2 : e_{12}, e_{22}, \cdots, e_{n2}$$
$$\cdots$$
$$e_p : e_{1p}, e_{2p}, \cdots, e_{np}$$

This can be expressed symbolically as

$$e_{i\alpha}$$

Two kinds of independent loadings are symbolically expressed as

$$e_\alpha, e_\beta$$

We have k kinds of eigenvalues. However, we utilize up to the q number of eigenvalue in the factor analysis given by

$$\lambda_1, \lambda_2, \cdots, \lambda_q$$

5.2. Basic Equation

The basic equation for the factor analysis is given by

$$\begin{pmatrix} u_{i1} \\ u_{i2} \\ \vdots \\ u_{ip} \end{pmatrix} = f_{i1} \begin{pmatrix} a_{11} \\ a_{21} \\ \vdots \\ a_{p1} \end{pmatrix} + f_{i2} \begin{pmatrix} a_{12} \\ a_{22} \\ \vdots \\ a_{p2} \end{pmatrix} + \cdots + f_{iq} \begin{pmatrix} a_{1q} \\ a_{2q} \\ \vdots \\ a_{pq} \end{pmatrix} + \begin{pmatrix} e_{i1} \\ e_{i2} \\ \vdots \\ e_{ip} \end{pmatrix} \quad for\ i = 1, 2, \cdots, n$$

This equation should hold the following equation given by

$$\begin{pmatrix} 1 & r_{12} & \cdots & r_{1p} \\ r_{21} & 1 & \cdots & r_{2p} \\ \vdots & \vdots & \ddots & \vdots \\ r_{p1} & r_{p2} & \cdots & 1 \end{pmatrix} = \sum_{k=1}^{q} \begin{pmatrix} a_{1k} & a_{2k} & \cdots & a_{pk} \end{pmatrix} \begin{pmatrix} a_{1k} \\ a_{2k} \\ \vdots \\ a_{pk} \end{pmatrix} + \begin{pmatrix} S_{e_1}^{(2)} & 0 & \cdots & 0 \\ 0 & S_{e_2}^{(2)} & \cdots & 0 \\ \vdots & \vdots & \ddots & \vdots \\ 0_{ix_1} & 0 & \cdots & S_{e_p}^{(2)} \end{pmatrix}$$

5.3. Normalization of Variables

The average μ_α and variance σ_α are given by

$$\mu_\alpha = \frac{1}{n}\sum_{i=1}^{n} x_{i\alpha}$$

$$\sigma_\alpha = \sqrt{\frac{1}{n-1}\sum_{i=1}^{n}\left(x_{i\alpha} - \mu_\alpha\right)^2}$$

When the data are regarded as the one in population set, the variance is given by

$$\sigma_\alpha = \sqrt{\frac{1}{n}\sum_{i=1}^{n}\left(x_{i\alpha} - \mu_\alpha\right)^2}$$

The normalized variable is given by

$$u_{i\alpha} = \frac{x_{i\alpha} - \mu_\alpha}{\sigma_\alpha}$$

5.4. Correlation Factor Matrix

The correlation factors are evaluated as

$$r_{\alpha\beta} = \frac{1}{n-1}\sum_{i=1}^{n} u_{i\alpha} u_{i\beta}$$

When data are ones in population, the corresponding correlation factor is given by

$$r_{\alpha\beta} = \frac{1}{n}\sum_{i=1}^{n} u_{i\alpha} u_{i\beta}$$

We can then obtain the correlation factor matrix given by

$$R = \begin{pmatrix} 1 & r_{12} & \cdots & r_{1p} \\ r_{21} & 1 & \cdots & r_{2p} \\ \vdots & \vdots & \ddots & \vdots \\ r_{p1} & r_{p2} & \cdots & 1 \end{pmatrix}$$

5.5. Factor Loading

Once the correlation matrix is obtained, it can be expanded using related eigenvalues and eigenvectors as

$$\begin{pmatrix} 1 & r_{12} & \cdots & r_{1p} \\ r_{21} & 1 & \cdots & r_{2p} \\ \vdots & \vdots & \ddots & \vdots \\ r_{p1} & r_{p2} & \cdots & 1 \end{pmatrix} = \lambda_{1\_1}\boldsymbol{v}_{1\_1}\boldsymbol{v}_{1\_1}{}^{T} + \lambda_{2\_1}\boldsymbol{v}_{2\_1}\boldsymbol{v}_{2\_1}{}^{T} + \cdots + \lambda_{p\_1}\boldsymbol{v}_{p\_1}\boldsymbol{v}_{p\_1}{}^{T}$$

We then approximate as follows as a first step.

$$\begin{pmatrix} S^{(2)}_{e_1\_1} & 0 & \cdots & 0 \\ 0 & S^{(2)}_{e_2\_1} & \cdots & 0 \\ \vdots & \vdots & \ddots & \vdots \\ 0_{ix_1} & 0 & \cdots & S^{(2)}_{e_p\_1} \end{pmatrix} = \begin{pmatrix} 1 & r_{12} & \cdots & r_{1p} \\ r_{21} & 1 & \cdots & r_{2p} \\ \vdots & \vdots & \ddots & \vdots \\ r_{p1} & r_{p2} & \cdots & 1 \end{pmatrix} - \left(\lambda_{1\_1}\boldsymbol{v}_{1\_1}\boldsymbol{v}_{1\_1}{}^{T} + \lambda_{2\_1}\boldsymbol{v}_{2\_1}\boldsymbol{v}_{2\_1}{}^{T} + \cdots + \lambda_{q\_1}\boldsymbol{v}_{q\_1}\boldsymbol{v}_{q\_1}{}^{T}\right)$$

We then approximate the second approximated correlation matrix as below, and obtain related eigenvalues and eigenvectors and expand the matrix as below.

$$\begin{pmatrix} 1-S^{(2)}_{e_1\_1} & r_{12} & \cdots & r_{1p} \\ r_{21} & 1-S^{(2)}_{e_2\_1} & \cdots & r_{2p} \\ \vdots & \vdots & \ddots & \vdots \\ r_{p1} & r_{p2} & \cdots & 1-S^{(2)}_{e_p\_1} \end{pmatrix} = \lambda_{1\_2}\boldsymbol{v}_{1\_2}\boldsymbol{v}_{1\_2}{}^{T} + \lambda_{2\_2}\boldsymbol{v}_{2\_2}\boldsymbol{v}_{2\_2}{}^{T} + \cdots + \lambda_{p\_2}\boldsymbol{v}_{p\_2}\boldsymbol{v}_{p\_2}{}^{T}$$

The second order independent factor is then given by

$$\begin{pmatrix} S^{(2)}_{e_1\_2} & 0 & \cdots & 0 \\ 0 & S^{(2)}_{e_2\_2} & \cdots & 0 \\ \vdots & \vdots & \ddots & \vdots \\ 0_{ix_1} & 0 & \cdots & S^{(2)}_{e_p\_2} \end{pmatrix} = \begin{pmatrix} 1-S^{(2)}_{e_1\_1} & r_{12} & \cdots & r_{1p} \\ r_{21} & 1-S^{(2)}_{e_2\_1} & \cdots & r_{2p} \\ \vdots & \vdots & \ddots & \vdots \\ r_{p1} & r_{p2} & \cdots & 1-S^{(2)}_{e_p\_1} \end{pmatrix} - \left(\lambda_{1\_2}\boldsymbol{v}_{1\_2}\boldsymbol{v}_{1\_2}{}^{T} + \lambda_{2\_2}\boldsymbol{v}_{2\_2}\boldsymbol{v}_{2\_2}{}^{T} + \cdots + \lambda_{q\_2}\boldsymbol{v}_{q\_2}\boldsymbol{v}_{q\_2}{}^{T}\right)$$

We evaluate the deviation of independent factor $DS_e^{(2)}$ as

$$DS_e^{(2)} = \begin{pmatrix} S_{e_1\_2}^{(2)} - S_{e_1\_1}^{(2)} \\ S_{e_2\_2}^{(2)} - S_{e_2\_1}^{(2)} \\ \vdots \\ S_{e_p\_2}^{(2)} - S_{e_p\_1}^{(2)} \end{pmatrix}$$

We then evaluate

$$\delta = \frac{1}{p}\sum_{\alpha=1}^{p}\left(S_{e_\alpha\_2}^{(2)} - S_{e_\alpha\_1}^{(2)}\right)^2$$

We repeat the above step until we obtain $\delta < \delta_c$, that is,

$$\delta = \frac{1}{p}\sum_{\alpha=1}^{p}\left(S_{e_\alpha\_m}^{(2)} - S_{e_\alpha\_m-1}^{(2)}\right)^2 < \delta_c$$

We then obtain factor loading as

$$\boldsymbol{a}_1 = \sqrt{\lambda_{1\_m}}\boldsymbol{v}_{1\_m} = \begin{pmatrix} a_{11} \\ a_{21} \\ \vdots \\ a_{p1} \end{pmatrix}$$

$$\boldsymbol{a}_2 = \sqrt{\lambda_{2\_m}}\boldsymbol{v}_{2\_m} = \begin{pmatrix} a_{12} \\ a_{22} \\ \vdots \\ a_{p2} \end{pmatrix}$$

…

$$\boldsymbol{a}_q = \sqrt{\lambda_{q\_m}}\boldsymbol{v}_{q\_m} = \begin{pmatrix} a_{1q} \\ a_{2q} \\ \vdots \\ a_{pq} \end{pmatrix}$$

5.6. Valimax Rotation

Factor loading and factor points have uncertainty with respect to rotation. We decide the rotation angle as follows.

The element rotated with an angle of θ is given by

$$\tilde{\boldsymbol{a}}_k = \begin{pmatrix} a_{1k}\cos\theta + a_{1l}\sin\theta \\ a_{2k}\cos\theta + a_{2l}\sin\theta \\ \vdots \\ a_{pk}\cos\theta + a_{pl}\sin\theta \end{pmatrix} = \begin{pmatrix} \tilde{a}_{1k} \\ \tilde{a}_{2k} \\ \vdots \\ \tilde{a}_{pk} \end{pmatrix}$$

$$\tilde{\boldsymbol{a}}_l = \begin{pmatrix} -a_{1k}\sin\theta + a_{1l}\cos\theta \\ -a_{2k}\sin\theta + a_{2l}\cos\theta \\ \vdots \\ -a_{k}\sin\theta + a_{l}\cos\theta \end{pmatrix} = \begin{pmatrix} \tilde{a}_{1l} \\ \tilde{a}_{2l} \\ \vdots \\ \tilde{a}_{pl} \end{pmatrix}$$

We then evaluate as below.

$$S = \frac{1}{p}\sum_{\alpha=1}^{p}\left[\left(\frac{\tilde{a}_{\alpha k}}{d_\alpha}\right)^2\right]^2 - \frac{1}{p^2}\left[\sum_{\alpha=1}^{p}\left(\frac{\tilde{a}_{\alpha k}}{d_\alpha}\right)^2\right]^2 + \frac{1}{p}\sum_{\alpha=1}^{p}\left[\left(\frac{\tilde{a}_{\alpha l}}{d_\alpha}\right)^2\right]^2 - \frac{1}{p^2}\left[\sum_{\alpha=1}^{p}\left(\frac{\tilde{a}_{\alpha l}}{d_\alpha}\right)^2\right]^2$$

The angle θ that maximize S is given by

$$\tan(4\theta) = \frac{G_2}{G_4}$$

$$G_2 \sin 4\theta > 0$$

where

$$G_1 = \sum_{\alpha=1}^{p}\left[\left(\frac{a_{\alpha k}}{d_\alpha}\right)^4 + \left(\frac{a_{\alpha l}}{d_\alpha}\right)^4\right] - \frac{1}{p}\left\{\left[\sum_{\alpha=1}^{p}\left(\frac{a_{\alpha k}}{d_\alpha}\right)^2\right]^2 + \left[\sum_{\alpha=1}^{p}\left(\frac{a_{\alpha l}}{d_\alpha}\right)^2\right]^2\right\}$$

$$G_2 = 4\left\{\sum_{\alpha=1}^{p}\left[\left(\frac{a_{\alpha k}}{d_\alpha}\right)^3\left(\frac{a_{\alpha l}}{d_\alpha}\right) - \left(\frac{a_{\alpha k}}{d_\alpha}\right)\left(\frac{a_{\alpha l}}{d_\alpha}\right)^3\right] - \frac{1}{p}\left[\sum_{\alpha=1}^{p}\left[\left(\frac{a_{\alpha k}}{d_\alpha}\right)^2 - \left(\frac{a_{\alpha l}}{d_\alpha}\right)^2\right]\sum_{\alpha=1}^{p}\left(\frac{a_{\alpha k}}{d_\alpha}\right)\left(\frac{a_{\alpha l}}{d_\alpha}\right)\right]\right\}$$

$$G_3 = 4\left\{3\sum_{\alpha=1}^{p}\left(\frac{a_{\alpha k}}{d_\alpha}\right)^2\left(\frac{a_{\alpha l}}{d_\alpha}\right)^2 - \frac{1}{p}\left[\sum_{\alpha=1}^{p}\left(\frac{a_{\alpha k}}{d_\alpha}\right)^2\sum_{\alpha=1}^{p}\left(\frac{a_{\alpha l}}{d_\alpha}\right)^2 + 2\left[\sum_{\alpha=1}^{p}\left(\frac{a_{\alpha k}}{d_\alpha}\right)\left(\frac{a_{\alpha l}}{d_\alpha}\right)\right]^2\right]\right\}$$

$$G_4 = G_1 - \frac{1}{2} G_3$$

We perform the above process to any combination of k, l.

The factor loading is denoted as $\boldsymbol{a}_k$, where we should be careful about the angle conversion as

Condition 1	Condition 2	4θ conversion	θ range
$K_2 \sin 4\theta \geq 0$	none	none	$-\frac{\pi}{8} < \theta < \frac{\pi}{8}$
$K_2 \sin 4\theta < 0$	$\sin 4\theta \geq 0$	$4\theta \to 4\theta - \pi$	$-\frac{\pi}{4} < \theta < -\frac{\pi}{8}$
	$\sin 4\theta < 0$	$4\theta \to 4\theta + \pi$	$\frac{\pi}{8} < \theta < \frac{\pi}{4}$

5.7. Factor Point

The coefficient for the factor can be obtained from

$$\begin{pmatrix} h_{1k} \\ h_{2k} \\ \vdots \\ h_{pk} \end{pmatrix} = \begin{pmatrix} 1 & r_{12} & \cdots & r_{1p} \\ r_{21} & 1 & \cdots & r_{2p} \\ \vdots & \vdots & \ddots & \vdots \\ r_{p1} & r_{p2} & \cdots & 1 \end{pmatrix}^{-1} \begin{pmatrix} a_{1k} \\ a_{2k} \\ \vdots \\ a_{2k} \end{pmatrix}$$

We then obtain factor point associated with a factor k is given by

$$\hat{f}_{ik} = h_{1k} x_{i1} + h_{2k} x_{i2} + \cdots + h_{pk} x_{ip}$$

5.8. Independent Factor

The independent factor is obtained as

$$\begin{pmatrix} e_{i1} \\ e_{i2} \\ \vdots \\ e_{ip} \end{pmatrix} = \begin{pmatrix} x_{i1} \\ x_{i2} \\ \vdots \\ x_{ip} \end{pmatrix} - \left[f_{i1} \begin{pmatrix} a_{11} \\ a_{21} \\ \vdots \\ a_{p1} \end{pmatrix} + f_{i2} \begin{pmatrix} a_{12} \\ a_{22} \\ \vdots \\ a_{p2} \end{pmatrix} + \cdots + f_{iq} \begin{pmatrix} a_{1q} \\ a_{2q} \\ \vdots \\ a_{pq} \end{pmatrix} \right] \quad for\ i = 1, 2, \cdots, n$$

This can be totally expressed as

$$\begin{pmatrix} e_{11} & e_{12} & \cdots & e_{1p} \\ e_{21} & e_{22} & \cdots & e_{2p} \\ \vdots & \vdots & \ddots & \vdots \\ e_{n1} & e_{n2} & \cdots & e_{np} \end{pmatrix} = \begin{pmatrix} x_{11} & x_{12} & \cdots & x_{1p} \\ x_{21} & x_{22} & \cdots & x_{2p} \\ \vdots & \vdots & \ddots & \vdots \\ x_{n1} & x_{n2} & \cdots & x_{np} \end{pmatrix} - \begin{pmatrix} f_{11} & f_{12} & \cdots & f_{1q} \\ f_{21} & f_{22} & \cdots & f_{2q} \\ \vdots & \vdots & \ddots & \vdots \\ f_{n1} & f_{n2} & \cdots & f_{nq} \end{pmatrix} \begin{pmatrix} a_{11} & a_{21} & \cdots & a_{p1} \\ a_{12} & a_{22} & \cdots & a_{p2} \\ \vdots & \vdots & \ddots & \vdots \\ a_{1q} & a_{2q} & \cdots & a_{pq} \end{pmatrix}$$

5.9. Appreciation of the Results

The score of each item is appreciated by the factor of number q.

The relationship between the factor and each item is expressed by $a_{\alpha k}$. For example, the elements for factor 1 are expressed by $a_{11}, a_{12}, \cdots, a_{1q}$. The most important one is the maximum one among $a_{11}, a_{12}, \cdots, a_{1q}$.

The point of each member for the factor is f_{ik}.

Therefore, we should set the target where large a with low f.

Chapter 12

CLUSTER ANALYSIS

ABSTRACT

One subject (a person for a special case) has various kinds of items. We define distance between the items using the data, and evaluate the similarity. Based on the similarity, we form clusters of the subject. We show various methods to evaluate the similarity.

Keywords: cluster analysis, arithmetic distance, tree diagram, chain effect, Ward method

1. INTRODUCTION

Let us consider scores of an examination of various subjects. The scores are scattered in general. We can deduce that some members are resembled. We want to make groups based on the results. On the other case, we can have data where each customer buys some products. It is convenient if we make some groups and perform proper advertisements depending on the characteristics of the groups. Therefore, we want to have some procedures to obtain the groups, which we study in this chapter.

2. SIMPLE EXAMPLE FOR CLUSTERING

Cluster analysis defines a distance between two categories and evaluates the similarity by the distance.

We have score data for English and mathematics as shown in Table 1. The distance between members A and B is denoted as $d(A,B)$, and it is defined as

$$d(A,B) = \sqrt{(x_{A1} - x_{B1})^2 + (x_{A2} - x_{B2})^2}$$
$$= \sqrt{(67-43)^2 + (64-76)^2}$$
$$= 26.8 \tag{1}$$

Table 1. Score of English and Mathematics

Member ID	English x_1	Mathematics x_2
A	67	64
B	43	76
C	45	72
D	28	92
E	77	64
F	59	40
G	28	76
H	28	60
I	45	12
J	47	80

We can evaluate the distance for the other combinations of students and obtain the data as shown in Table 2. The minimum distance among these data is the distance between members B and C. Therefore, we form a cluster of B and C first.

After forming a cluster, we modify the data for the cluster. The simplest modification is to use its average one given by

$$x_{BC1} = \frac{x_{B1} + x_{C1}}{2}, x_{BC2} = \frac{x_{B2} + x_{C2}}{2} \tag{2}$$

Using this data, we evaluate the distance between the members which is shown in Table 3.

Table 2. Distance of members

	A	B	C	D	E	F	G	H	I	J
A		26.8	23.4	48.0	10.0	25.3	40.8	39.2	56.5	25.6
B			4.5	21.9	36.1	39.4	15.0	21.9	64.0	5.7
C				26.2	33.0	34.9	17.5	20.8	60.0	8.2
D					56.4	60.5	16.0	32.0	81.8	22.5
E						30.0	50.4	49.2	61.1	34.0
F							47.5	36.9	31.3	41.8
G								16.0	66.2	19.4
H									50.9	27.6
I										68.0
J										

Table 3. The distance between members for second step

Member ID	English x_1	Mathematics x_2
A	67	64
B,C	44	74
D	28	92
E	77	64
F	59	40
G	28	76
H	28	60
I	45	12
J	47	80

	A	B,C	D	E	F	G	H	I	J
A		25.1	48.0	10.0	25.3	40.8	39.2	56.5	25.6
B,C			24.1	34.5	37.2	16.1	21.3	62.0	6.7
D				56.4	60.5	16.0	32.0	81.8	22.5
E					30.0	50.4	49.2	61.1	34.0
F						47.5	36.9	31.3	41.8
G							16.0	66.2	19.4
H								50.9	27.6
I									68.0
J									

In this case, the distance between the cluster BC and J is the minimum. Therefore, we form the cluster BCJ, and modify the data for the cluster as

$$x_{BCJ1} = \frac{x_{B1} + x_{C1} + x_{J1}}{3}, x_{BC2} = \frac{x_{B2} + x_{C2} + x_{J2}}{3} \tag{3}$$

Using this data, we evaluate the distance between the members which is shown in Table 4.

Table 4. The distance between members for third step

Member ID	English x_1	Mathematics x_2
A	67	64
B,C,J	45	76
D	28	92
E	77	64
F	59	40
G	28	76
H	28	60
I	45	12

	A	B,C,J	D	E	F	G	H	I
A		25.1	48.0	10.0	25.3	40.8	39.2	56.5
B,C,J			23.3	34.2	38.6	17.0	23.3	64.0
D				56.4	60.5	16.0	32.0	81.8
E					30.0	50.4	49.2	61.1
F						47.5	36.9	31.3
G							16.0	66.2
H								50.9
I								

In this case the distance between members A and E is the minimum. Therefore, we form a cluster AE, and modify the data as

$$x_{AE1} = \frac{x_{A1} + x_{E1}}{2}, x_{AE2} = \frac{x_{A2} + x_{E2}}{2} \tag{4}$$

Using this data, we evaluate the distance between the members which is shown in Table 5.

Table 5. The distance between members for fourth step

Member ID	English x_1	Mathematics x_2
A,E	72	64
B,C,J	45	76
D	28	92
F	59	40
G	28	76
H	28	60
I	45	12

	A,E	B,C,J	D	F	G	H	I
A,E		29.5	52.2	27.3	45.6	44.2	58.6
B,C,J			23.3	38.6	17.0	23.3	64.0
D				60.5	16.0	32.0	81.8
F					47.5	36.9	31.3
G						16.0	66.2
H							50.9
I							

In this case, the distance between D, G, and H is the minimum, and we form a cluster DGH, and modify the related data as

$$x_{DGH1} = \frac{x_{D1} + x_{G1} + x_{H1}}{3}, x_{DGH2} = \frac{x_{D2} + x_{G2} + x_{H2}}{3} \tag{5}$$

Using this data, we evaluate the distance between the members which is shown in Table 6.

Table 6. The distance between members for fifth step

Member ID	English x_1	Mathematics x_2
A,E	72	64
B,C,J	45	76
D,G,H	28	76
F	59	40
I	45	12

	A,E	B,C,J	D,GH	F	I
A,E		29.55	45.61	27.29	58.59
B,C,J			17.00	38.63	64.00
D,GH				47.51	66.22
F					31.30
I					

In this case, the distance between cluster BCJ and cluster DGH is the minimum, and we form a cluster BCJDGH, and modify the data as

$$x_{BCJDGH1} = \frac{x_{B1} + x_{C1} + x_{J1} + x_{D1} + x_{G1} + x_{H1}}{6}, x_{DGH2} = \frac{x_{B2} + x_{C2} + x_{J2} + x_{D2} + x_{G2} + x_{H2}}{6} \tag{6}$$

Using this data, we evaluate the distance between the members which is shown in Table 7.

Table 7. The distance between members for sixth step

Member ID	English x_1	Mathematics x_2
A,E	72	64
B,C,J,D,G,H	36.5	76
F	59	40
I	45	12

	A,E	B,C,J,D,G,H	F	I
A,E		37.47	27.29	58.59
B,C,J,D,G,H			42.45	64.56
F				31.30
I				

In this case the distance between cluster AE and F is the minimum. We then form a cluster AEF and modify the data as

$$x_{AEF1} = \frac{x_{A1} + x_{E1} + x_{F1}}{3}, x_{AEF2} = \frac{x_{A2} + x_{E2} + x_{F2}}{3} \tag{7}$$

Table 8. The distance between members for sixth step

Member ID	English x_1	Mathematics x_2
A,E,F	67.67	56.00
B,C,J,D,G,H	36.50	76.00
I	45.00	12.00

	A,E,F	B,C,J,D,G,H	I
A,E,F		37.03	49.50
B,C,J,D,G,H			64.56
I			

Using this data, we evaluate the distance between the members which are shown in Table 8. In this case, the distance between cluster AEF and cluster BCJDGH is the minimum. We hence form a cluster AEF BCJDGH and evaluate the data for the cluster.

Finally, we form the last cluster as shown in Table 9.

Table 9. The distance between members for eighth step

Member ID	English x_1	Mathematics x_2
A,E,F,B,C,J,D,G,H	46.89	69.33
I	45.00	12.00

	A,E,F,B,C,J,D,G,H	I
A,E,F,B,C,J,D,G,H		57.36
I		

After the above process, we can form a tree structure as shown in Figure 1. We can obtain some groups if we set the critical distance. For example, if we obtain 4 groups, we set the critical distance of 25 and obtain the groups of below.

[B,C,J,D,G,H], [A,E], F,I

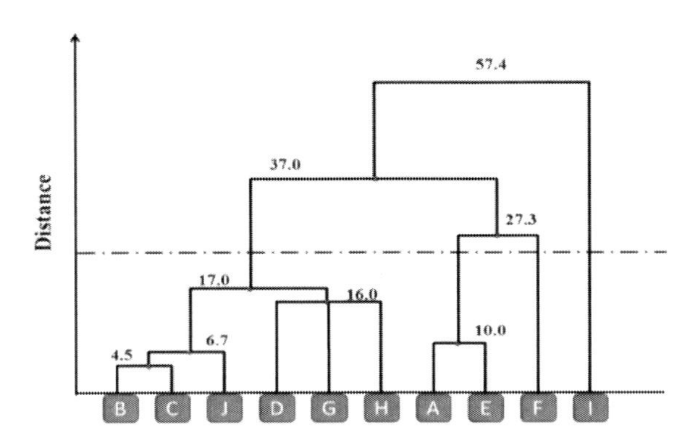

Figure 1. Tree diagram.

There are various definitions of distances. The process above is the simplest one among them.

We should mention one point in the above process.

The above process is valid when the two variables x_1 and x_2 are similar ones, which means the magnitude of the two variables are in the almost same order. However, they are quite different variables and hence the magnitude is significantly different, the distance is determined by the one which is larger. Therefore, we consider only one variable in this case.

In such case, we need to use normalized variables below.

$$z_{i1} = \frac{x_{i1} - \mu_{x_1}}{\sigma_{x_1}}, z_{i2} = \frac{x_{i2} - \mu_{x_2}}{\sigma_{x2}} \tag{8}$$

where

$$\mu_{x_1} = \frac{1}{n}\sum_i x_{i1}, \sigma_{x_1} = \frac{1}{n}\sum_i \left(x_{i1} - \mu_{x_1}\right)^2 \tag{9}$$

$$\mu_{x_2} = \frac{1}{n}\sum_i x_{i2}, \sigma_{x_2} = \frac{1}{n}\sum_i \left(x_{i2} - \mu_{x_2}\right)^2 \tag{10}$$

and n is the data number.

3. The Relationship between Arithmetic Distance and Variance

We evaluate the relationship between arithmetic distance and the variance.

The square of the distance is modified as

$$\begin{aligned}
d^2 &= \left(x_{ax_1} - x_{bx_1}\right)^2 + \left(x_{ax_2} - x_{bx_2}\right)^2 \\
&= \left[\left(x_{ax_1} - \bar{x}_{x_1}\right) - \left(x_{bx_1} - \bar{x}_{x_1}\right)\right]^2 + \left[\left(x_{ax_2} - \bar{x}_{x_2}\right) - \left(x_{bx_2} - \bar{x}_{x_2}\right)\right]^2 \\
&= \left(x_{ax_1} - \bar{x}_{x_1}\right)^2 + \left(x_{bx_1} - \bar{x}_{x_1}\right)^2 - 2\left(x_{ax_1} - \bar{x}_{x_1}\right)\left(x_{bx_1} - \bar{x}_{x_1}\right) \\
&\quad + \left(x_{ax_2} - \bar{x}_{x_2}\right)^2 + \left(x_{bx_2} - \bar{x}_{x_2}\right)^2 - 2\left(x_{ax_2} - \bar{x}_{x_2}\right)\left(x_{bx_2} - \bar{x}_{x_2}\right) \\
&= S_{ab}^{(2)} - 2\left[\left(x_{ax_1} - \bar{x}_{x_1}\right)\left(x_{bx_1} - \bar{x}_{x_1}\right) + \left(x_{ax_2} - \bar{x}_{x_2}\right)\left(x_{bx_2} - \bar{x}_{x_2}\right)\right]
\end{aligned} \tag{11}$$

where

$$\left(x_{ax_1}-\bar{x}_{x_1}\right)\left(x_{bx_1}-\bar{x}_{x_1}\right)=\left(x_{ax_1}-\frac{x_{ax_1}+x_{bx_1}}{2}\right)\left(x_{bx_1}-\frac{x_{ax_1}+x_{bx_1}}{2}\right)$$
$$=\frac{x_{ax_1}-x_{bx_1}}{2}\frac{x_{bx_1}-x_{ax_1}}{2}$$
$$=-\frac{1}{4}\left(x_{ax_1}-x_{bx_1}\right)^2 \quad (12)$$

Similarly, we obtain

$$\left(x_{ax_2}-\bar{x}_{x_2}\right)\left(x_{bx_2}-\bar{x}_{x_2}\right)=-\frac{1}{4}\left(x_{ax_2}-x_{bx_2}\right)^2 \quad (13)$$

Therefore, we obtain

$$d^2=S_{ab}^{(2)}+\frac{1}{2}\left[\left(x_{ax_1}-x_{bx_1}\right)^2+\left(x_{ax_2}-x_{bx_2}\right)^2\right]$$
$$=S_{ab}^{(2)}+\frac{1}{2}d^2 \quad (14)$$

This can be modified as

$$d^2=2S_{ab}^{(2)} \quad (15)$$

The arithmetic distance is double of the variance.

4. Various Distance

The distance is basically defined by

$$d_{ab}=\sqrt{\frac{1}{n}\sum_{i=1}^{n}\left(x_{ia}-x_{ib}\right)^2} \quad (16)$$

However, there are various definitions for the distance. We briefly mention the distance.

When each item is not identical, we multiply a weighted factor w_i, and the corresponding distance is given by

$$d_{ab} = \sqrt{\frac{1}{n}\sum_{i=1}^{n} w_i \left(x_{ia} - x_{ib}\right)^2} \tag{17}$$

This is called as weighted distance.

The generalized distance is given by

$$d_{ab} = \left[\frac{1}{n}\sum_{i=1}^{n} \left|x_{ia} - x_{ib}\right|^{\alpha}\right]^{\frac{1}{\alpha}} \tag{18}$$

The maximum deviation is apt to dominate the distance with increasing α.

5. Chain Effect

Chain effect means that one cluster absorbs one factor in order. In that special case, the group is always two for any stage of distance. This situation is not appropriate for our purpose. The above simple procedure is apt to suffer this chain effect, although the reason is not clear.

6. Ward Method for Two Variables

Ward method form a cluster in the standpoint that the square sum is the minimum. This method is regarded as the one that overcomes the chain effects, and hence is frequently used.

Table 10. Score of Japanese and English for members

Member ID	Japanese x_1	English x_2
1	5	1
2	4	2
3	1	5
4	5	4
5	5	5

Let us consider the distance between member 1 and 2.

We evaluate the means of Japanese and English scores for members 1 and 2 as shown in Table 11, and evaluate the sum of square given by

$$K_{12} = \sum_{i=1}^{2}\sum_{k=1}^{2}(x_{ik} - \bar{x}_k)^2$$
$$= (5-4.5)^2 + (4-4.5)^2 + (1-1.5)^2 + (2-1.5)^2$$
$$= 1.00 \tag{19}$$

We can evaluate the sum of square for the other combination of members as shown in Table 12. The minimum square value is that for the combination of member 4 and 5, and we form a cluster of 4 and 5 denoted as C_1.

Table 11. Score of Japanese and English for member 1 and 2

Member ID	Japanese x_1	English x_2
1	5	1
2	4	2
Mean	4.5	1.5

Table 12. Sum of square for member combinations

ID	1	2	3	4	5
1					
2	1				
3	16	9			
4	4.5	2.5	8.5		
5	8	5	8	0.5	

After forming a cluster we evaluate the distance between cluster C_1 and the member 1.

The corresponding data is shown in Table 13.

Table 13. Score of cluster C_1 and member 1

Cluster ID	Member ID	Japanese x_1	English x_2
1	1	5	1
C1	4	5	4
	5	5	5
Average		5.00	3.33

The mean of Japanese for this group of the cluster and member is 5 and that of English is 3.33. The square sum $K_{C_1 1}$ is evaluated as

$$\begin{aligned} K_{C_1 1} &= \sum_{i=1}^{3}\sum_{k=1}^{2}\left(x_{ik}-\bar{x}_k\right)^2 \\ &= \left(5-5.00\right)^2+\left(5-5.00\right)^2+\left(5-5.00\right)^2 \\ &\quad +\left(1-3.33\right)^2+\left(4-3.33\right)^2+\left(5-3.33\right)^2 \\ &= 8.67 \end{aligned} \tag{20}$$

We do not use this sum directly, but the deviation of the sum $\Delta K_{C_1 1}$ given by

$$\begin{aligned} \Delta K_{C_1 1} &= K_{C_1 1}-\left(K_{C_1}+K_1\right) \\ &= 8.67-\left(0.5+0\right) \\ &= 8.17 \end{aligned} \tag{21}$$

where K_1 is 0 since there is only one data.

The variance deviations for any combination are shown in Table 14. The minimum variance deviation is between members 1 and 2. Therefore, we form a cluster of members 1 and 2, and denote it as C2.

Table 14. Variance deviation for second step

ID	1	2	3	C1
1				
2	1			
3	16	9		
C1	8.17	4.83	10.83	

We then evaluate the variance deviation between C1, C2, and member 3, which is shown in Table 15. The minimum variance deviation is between cluster C1 and C2, and we form the cluster of C1 and C2. The corresponding tree diagram is shown in Figure 2.

Table 15. Variance deviation for step 2

ID	3	C1	C2
3			
C1	10.83		
C2	16.33	9.25	

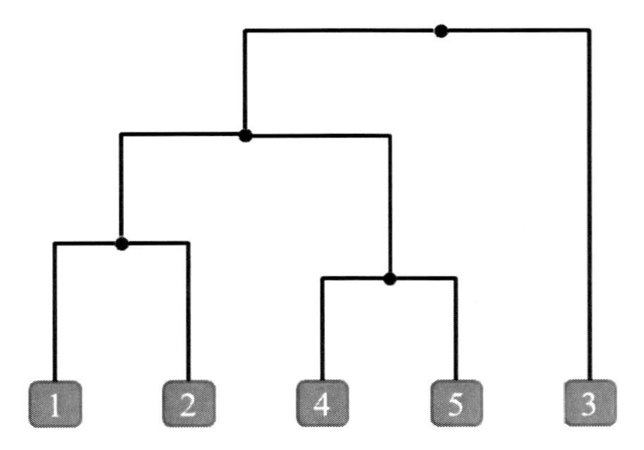

Figure 2. Tree diagram with Ward method.

7. Ward Method with Many Variables

We can easily extend the procedure for the two variables to that for many variables of more than two.

Let us consider that we form a cluster lm by merging cluster l and cluster m. Cluster l and cluster m has the data denoted as x_{ilk}, x_{imk}, and the merged number is denoted as n_l and n_m. We then obtain the variables below

$$K_l = \sum_{i=1}^{n} \sum_{k=1}^{n_l} \left(x_{ilk} - \bar{x}_{il} \right)^2 \tag{22}$$

$$K_m = \sum_{i=1}^{n} \sum_{k=1}^{n_m} \left(x_{imk} - \bar{x}_{im} \right)^2 \tag{23}$$

The merged variance is expressed by

$$K_{lm} = K_l + K_m + \Delta K_{lm} \tag{24}$$

The deviation of the variance is given by

$$\Delta K_{lm} = \frac{n_l n_m}{n_l + n_m} \sum_{i=1}^{n} \left(\bar{x}_{il} - \bar{x}_{im} \right)^2 \tag{25}$$

which is very important rule and is proved in the following

Square sum of K_{lm} is given by

$$K_{lm} = \sum_{i=1}^{n} \left[\sum_{k=1}^{n_l} \left(x_{ilk} - \overline{x}_{ilm} \right)^2 + \sum_{k=1}^{n_m} \left(x_{imk} - \overline{x}_{ilm} \right)^2 \right] \tag{26}$$

where

$$\overline{x}_{ilm} = \frac{\sum_{k=1}^{n_l} x_{ilk} + \sum_{k=1}^{n_m} x_{imk}}{n_l + n_m} \tag{27}$$

We modify this equation to relate it to K_l and K_m as

$$\begin{aligned}
K_{lm} &= \sum_{i=1}^{n} \left[\sum_{k=1}^{n_l} \left(x_{ilk} - \overline{x}_{il} + \overline{x}_{il} - \overline{x}_{ilm} \right)^2 + \sum_{k=1}^{n_m} \left(x_{imk} - \overline{x}_{im} + \overline{x}_{im} - \overline{x}_{ilm} \right)^2 \right] \\
&= \sum_{i=1}^{n} \left[\begin{array}{l} \sum_{k=1}^{n_l} \left(x_{ilk} - \overline{x}_{il} \right)^2 + 2\sum_{k=1}^{n_l} \left(x_{ilk} - \overline{x}_{il} \right)\left(\overline{x}_{il} - \overline{x}_{ilm} \right) + n_l \left(\overline{x}_{il} - \overline{x}_{ilm} \right)^2 \\ + \sum_{k=1}^{n_m} \left(x_{imk} - \overline{x}_{im} \right)^2 + 2\sum_{k=1}^{n_m} \left(x_{imk} - \overline{x}_{im} \right)\left(\overline{x}_{im} - \overline{x}_{ilm} \right) + n_m \left(\overline{x}_{im} - \overline{x}_{ilm} \right)^2 \end{array} \right] \\
&= \sum_{i=1}^{n} \left[\sum_{k=1}^{n_l} \left(x_{ilk} - \overline{x}_{il} \right)^2 + \sum_{k=1}^{n_m} \left(x_{imk} - \overline{x}_{im} \right)^2 \right] \\
&\quad + \sum_{i=1}^{n} \left[n_l \left(\overline{x}_{il} - \overline{x}_{ilm} \right)^2 + n_m \left(\overline{x}_{im} - \overline{x}_{ilm} \right)^2 \right] \\
&= K_l + K_m + \sum_{i=1}^{n} \left[n_l \left(\overline{x}_{il} - \overline{x}_{ilm} \right)^2 + n_m \left(\overline{x}_{im} - \overline{x}_{ilm} \right)^2 \right]
\end{aligned} \tag{28}$$

where we utilize

$$\sum_{k=1}^{n_l} \left(x_{ilk} - \overline{x}_{il} \right)\left(\overline{x}_{il} - \overline{x}_{ilm} \right) = \left(\overline{x}_{il} - \overline{x}_{ilm} \right) \sum_{k=1}^{n_l} \left(x_{ilk} - \overline{x}_{il} \right) = 0 \tag{29}$$

$$\sum_{k=1}^{n_m} \left(x_{imk} - \overline{x}_{im} \right)\left(\overline{x}_{im} - \overline{x}_{ilm} \right) = \left(\overline{x}_{im} - \overline{x}_{ilm} \right) \sum_{k=1}^{n_l} \left(x_{imk} - \overline{x}_{im} \right) = 0 \tag{30}$$

$$\overline{x}_{il} - \overline{x}_{ilm} = \frac{\sum_{k=1}^{n_l} x_{ilk}}{n_l} - \frac{\sum_{k=1}^{n_l} x_{ilk} + \sum_{k=1}^{n_m} x_{imk}}{n_l + n_m}$$

$$= \frac{\left(n_l + n_m\right)\sum_{k=1}^{n_l} x_{ilk} - n_l\left[\sum_{k=1}^{n_l} x_{ilk} + \sum_{k=1}^{n_m} x_{imk}\right]}{n_l\left(n_l + n_m\right)}$$

$$= \frac{n_m \sum_{k=1}^{n_l} x_{ilk} - n_l \sum_{k=1}^{n_m} x_{imk}}{n_l\left(n_l + n_m\right)}$$

$$= \frac{n_m n_l \frac{\sum_{k=1}^{n_l} x_{ilk}}{n_l} - n_l n_m \frac{\sum_{k=1}^{n_m} x_{imk}}{n_m}}{n_l\left(n_l + n_m\right)}$$

$$= \frac{n_m n_l\left(\overline{x}_{il} - \overline{x}_{im}\right)}{n_l\left(n_l + n_m\right)}$$

$$= \frac{n_m\left(\overline{x}_{il} - \overline{x}_{im}\right)}{n_l + n_m} \tag{31}$$

Similarly, we obtain

$$\overline{x}_{im} - \overline{x}_{ilm} = \frac{n_l\left(\overline{x}_{im} - \overline{x}_{in}\right)}{n_l + n_m} \tag{32}$$

Therefore, we obtain

$$n_l\left(\overline{x}_{il} - \overline{x}_{ilm}\right)^2 + n_m\left(\overline{x}_{im} - \overline{x}_{ilm}\right)^2 = n_l\left[\frac{n_m\left(\overline{x}_{il} - \overline{x}_{im}\right)}{n_l + n_m}\right]^2 + n_m\left[\frac{n_l\left(\overline{x}_{im} - \overline{x}_{in}\right)}{n_l + n_m}\right]^2$$

$$= \frac{\left(n_l n_m^2 + n_m n_n^2\right)\left(\overline{x}_{il} - \overline{x}_{im}\right)^2}{\left(n_l + n_m\right)^2}$$

$$= \frac{n_l n_m}{n_l + n_m}\left(\overline{x}_{il} - \overline{x}_{im}\right)^2 \tag{33}$$

Finally, Eq. (25) is proved.
The distance is normalized by the data number n and is given by

$$d_{lm} = \sqrt{\frac{\Delta K_{lm}}{n}} \tag{34}$$

The average in the cluster is given by

$$\bar{x}_{ilm} = \frac{\sum_{k=1}^{n_l} x_{ilk} + \sum_{k=1}^{n_m} x_{imk}}{n_l + n_m} \tag{35}$$

8. Example of Ward Method with N Variables

We apply the Ward method and form clusters.

Table 1 shows the CS data of members for various items. The score is 5 point perfect. The member number n is 20. In this case, the following five items influence the customer satisfaction.

a. Understanding customer
b. Prompt response
c. Flexible response
d. Producing ability
e. Providing useful information
f. Active proposal

We evaluate clustering of items using 20 member scores.

The dummy variable for items a-f is k, and that for members 1-20 is i.

Step 1: Sum of Square

The sum of square is evaluated as

$$K_l = \sum_{i=1}^{n} \sum_{k=1}^{n_l} \left(x_{ilk} - \bar{x}_{il} \right)^2 \tag{36}$$

We consider an item a, and evaluate the corresponding square sum.

Since we consider the item a, l is a. Since we do not form cluster, $k = a$. The element is only one, and the mean is the same as the data given by

$$\bar{x}_{ia} = x_{ia} \tag{37}$$

Table 16. Satisfaction score of members for various items

Member ID	a: Understanding customer	b:Prompt response	c:Flexible response	d: Producing ability	e:Providing uselul Informatio	f:Active proposal
1	4	4	3	4	5	4
2	4	3	4	3	3	2
3	3	1	2	3	2	1
4	5	5	5	5	5	5
5	4	4	4	4	4	4
6	2	2	2	2	2	2
7	5	5	5	5	5	5
8	4	3	3	3	3	3
9	2	1	1	1	1	2
10	2	1	1	1	1	3
11	4	5	5	5	5	5
12	3	3	3	2	2	2
13	3	3	3	3	3	3
14	3	4	4	3	3	3
15	4	3	3	3	3	3
16	1	1	1	1	1	1
17	5	5	5	5	5	5
18	4	1	1	1	3	3
19	4	4	3	3	4	3
20	4	4	4	3	3	3

Therefore, the variance is given by

$$K_a = 0 \quad (38)$$

The variances for the other items are the same and is given by

$$K_a = K_b = \cdots = K_f = 0 \quad (39)$$

This corresponds the initialization of K_l .

Step 2: Sum of the Variance for Whole Combinations

The whole combination is as below.

a-b, a-b, a-d, a-e, a-f,a-e,a-f
b-c,b-d,b-e,b-f

c-d,c-e,c-f
d-e,d-f
e-f

Table 17. Variance deviation of items a and b

Member ID	a:Understanding customer	b:Prompt response	Square of deviation
1	4	4	0
2	4	3	1
3	3	1	4
4	5	5	0
5	4	4	0
6	2	2	0
7	5	5	0
8	4	3	1
9	2	1	1
10	2	1	1
11	4	5	1
12	3	3	0
13	3	3	0
14	3	4	1
15	4	3	1
16	1	1	0
17	5	5	0
18	4	1	9
19	4	4	0
20	4	4	0
		Sum	20
		ΔK	10

We show the variance deviation evaluation of the combination between a and b. The corresponding variance is given by

$$\Delta K_{ab} = \frac{n_a n_b}{n_a + n_b} \sum_{i=1}^{20} \left(\bar{x}_{ia} - \bar{x}_{ib} \right)^2 \tag{40}$$

where a and b are single item, and hence $n_a = n_b = 1$. This also means the means equal to the data themselves, that is, $\bar{x}_{ia} = x_{ia}$ and $\bar{x}_{ib} = x_{ib}$. The corresponding sum of square is 20.

$$\sum_{i=1}^{20} \left(\bar{x}_{ia} - \bar{x}_{ib} \right)^2 = 20 \tag{41}$$

Therefore, the variance deviation is

$$\begin{aligned}\Delta K_{ab} &= \frac{n_a n_b}{n_a + n_b} \sum_{i=1}^{20} \left(\bar{x}_{ia} - \bar{x}_{ib} \right)^2 \\ &= \frac{1 \times 1}{1+1} \times 20 \\ &= 10\end{aligned} \tag{42}$$

We can evaluate the other combinations as shown in Table 18.

Table 18. Variance deviation for first step

Item	a	b	c	d	e	f
a		10	9	9	5.5	8
b			2	4	4.5	7
c				3	6.5	9
d					3.5	7
e						4.5
f						

Step 3: First Cluster Formation

The minimum variance deviation is that for ΔK_{bc}. Therefore, we form a cluster of b and c.

The corresponding distance d_{bc} is given by

$$\begin{aligned}d_{bc} &= \sqrt{\frac{\Delta K_{bc}}{n}} \\ &= \sqrt{\frac{2}{20}} \\ &= 0.316\end{aligned} \tag{43}$$

The variance associated with the cluster bc is given by

$$K_{bc} = K_b + K_c + \Delta K_{bc} \\ = \Delta K_{bc} \\ = 2 \quad (44)$$

The corresponding item number is given by

$$n_{bc} = 2 \quad (45)$$

Calculating means for the cluster, we obtain the updated table and obtain Table 19.

Table 19. Satisfaction score of members for various items and the first cluster

Cluster 1	a	[bc]			d	e	f
Member ID	a: Understanding customer	b:Prompt response	c:Flexible response	[bc] mean	d: Producing ability	e:Providing uselul Informati	f:Active proposal
1	4	4	3	3.5	4	5	4
2	4	3	4	3.5	3	3	2
3	3	1	2	1.5	3	2	1
4	5	5	5	5	5	5	5
5	4	4	4	4	4	4	4
6	2	2	2	2	2	2	2
7	5	5	5	5	5	5	5
8	4	3	3	3	3	3	3
9	2	1	1	1	1	1	2
10	2	1	1	1	1	1	3
11	4	5	5	5	5	5	5
12	3	3	3	3	2	2	2
13	3	3	3	3	3	3	3
14	3	4	4	4	3	3	3
15	4	3	3	3	3	3	3
16	1	1	1	1	1	1	1
17	5	5	5	5	5	5	5
18	4	1	1	1	1	3	3
19	4	4	3	3.5	3	4	3
20	4	4	4	4	3	3	3

Step 4: Sum of Square for Second Time

We then evaluate the variance deviation of the combinations below.

a-[bc], a-d, a-e, a-f,a-e,a-f
[bc]-d,[bc]-e,[bc]-f
d-e,d-f
e-f

We show the evaluation of variance between a and [bc] below

$$\Delta K_{a[bc]} = \frac{n_a n_{bc}}{n_a + n_{bc}} \sum_{i=1}^{20} \left(\bar{x}_{ia} - \bar{x}_{ibc} \right)^2$$
$$= \frac{1 \times 2}{1+2} \times 18$$
$$= 12 \tag{46}$$

The corresponding table is given by Table 20.

Table 20. Variance deviation of items a and a cluster [bc]

Member ID	a: Understanding customer	[bc] mean	Square of deviation
1	4	3.5	0.25
2	4	3.5	0.25
3	3	1.5	2.25
4	5	5	0
5	4	4	0
6	2	2	0
7	5	5	0
8	4	3	1
9	2	1	1
10	2	1	1
11	4	5	1
12	3	3	0
13	3	3	0
14	3	4	1
15	4	3	1
16	1	1	0
17	5	5	0
18	4	1	9
19	4	3.5	0.25
20	4	4	0
		Sum	18
		ΔK	12

We perform similar analysis for the other combinations and obtain Table 21.

Table 21. Variance deviation for second step

ID	a	bc	d	e	f
a		12	9	5.5	8
bc			4	6.67	10
d				3.5	7
e					4.5
f					

Step 5: Second Cluster Formation

The minimum variance deviation is that for ΔK_{de}. Therefore, we form a cluster of d and e.

The corresponding distance d_{bc} is given by

$$\begin{aligned} d_{de} &= \sqrt{\frac{\Delta K_{de}}{n}} \\ &= \sqrt{\frac{3.5}{20}} \\ &= 0.418 \end{aligned} \tag{47}$$

The variance associated with the cluster bc is given by

$$\begin{aligned} K_{de} &= K_d + K_e + \Delta K_{de} \\ &= \Delta K_{de} \\ &= 3.5 \end{aligned} \tag{48}$$

The corresponding item number is given by

$$n_{de} = 2 \tag{49}$$

We repeat the above process and obtain the tree diagram

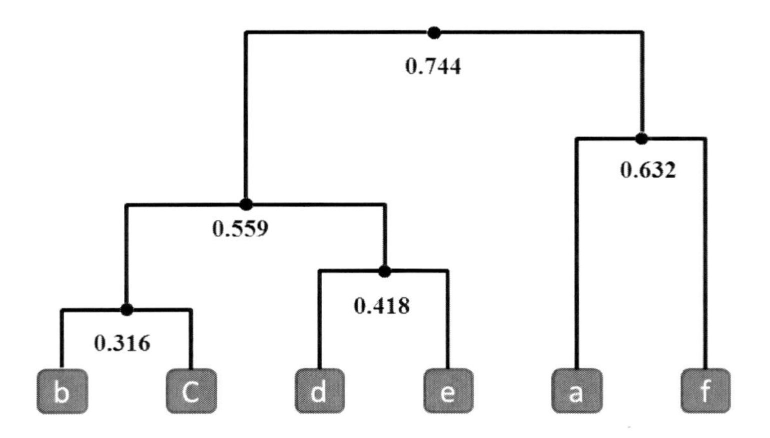

Figure 3. Tree diagram for the data.

SUMMARY

To summarize the results in this chapter—
The distance between two variables are given by

$$d(A,B) = \sqrt{(x_{A1} - x_{B1})^2 + (x_{A2} - x_{B2})^2}$$

We evaluate all combination of the variables, and select the minimum one, and form a cluster. After forming a cluster, we modify the data for the cluster. The simplest modification is to use its average one given by

$$x_{AB1} = \frac{x_{A1} + x_{B1}}{2}, x_{AB2} = \frac{x_{A2} + x_{B2}}{2}$$

Using this data, we evaluate the distance between the members.
We can use a modified distance as

$$d = \left[\frac{1}{n} \sum_{i=1}^{n} |x_{iA} - x_{iB}|^{\alpha} \right]^{\frac{1}{\alpha}}$$

There is the other method for forming clusters called as a Ward method. The Ward method forms a cluster in the standpoint that the square sum is the minimum.

Let us consider that we form cluster lm by merging cluster l and cluster m. Cluster l and cluster m has the data denoted as x_{ilk}, x_{imk}, and the merged number is denoted as n_l and n_m. We then obtain the variables below.

$$K_l = \sum_{i=1}^{n}\sum_{k=1}^{n_l}\left(x_{ilk} - \bar{x}_{il}\right)^2$$

$$K_m = \sum_{i=1}^{n}\sum_{k=1}^{n_m}\left(x_{imk} - \bar{x}_{im}\right)^2$$

The merged variance is expressed by

$$K_{lm} = K_l + K_m + \Delta K_{lm}$$

The deviation of the variance is given by

$$\Delta K_{lm} = \frac{n_l n_m}{n_l + n_m}\sum_{i=1}^{n}\left(\bar{x}_{il} - \bar{x}_{im}\right)^2$$

We search the minimum combination for this deviation of variance, and repeat the process.

Chapter 13

DISCRIMINANT ANALYSIS

ABSTRACT

We assume that we have two groups and have reference data where we clearly know to which group the data belong to. When we obtain one set data for a certain member or subject, the discriminant analysis give us the procedure to decide the group for the member or the subjects. The decision is done by evaluating the distance between the data and ones for each group.

Keywords: distance, determinant, Maharanobis' distance

1. INTRODUCTION

When we have a bad condition in health, we go to a hospital, and take some kinds of medical checks. The doctor has to decide whether the person is sick or not. This problem is generalized as follows. Two groups are set, and the corresponding data with the same items are also provided. We consider the situation where we obtain one set data for the items. Discriminant analysis helps us to decide to which group it belongs.

We perform matrix operation in this chapter. The basic matrix operation is shown in Chapter 15 of volume 3.

2. DISCRIMINANT ANALYSIS WITH ONE VARIABLE

We want to judge whether a person is healthy or disease using one inspection value x_1. We define a distance between a data and two groups and select the shorter one and decide the person who belongs to the group of shorter distance.

We show the inspection data for healthy and diseased people as shown in Table 1. We inspect a person and judge whether he is healthy or diseased.

First of all, we make two groups from the data in Table 1: healthy and disease. We then evaluate the averages and variances for the groups as shown in Table 2.

Table 1. Inspection data for healthy and diseased people

No.	Kind	Inspection x1
1	Healthy	50
2	Healthy	69
3	Healthy	93
4	Healthy	76
5	Healthy	88
6	Disease	43
7	Disease	56
8	Disease	38
9	Disease	21
10	Disease	25

Table 2. Average and variance for healthy and diseased people

	Healthy	Disease
	50	43
	69	56
	93	38
	76	21
	88	25
Average	75.2	36.6
Variance	230.96	159.44

The average and variance of healthy people are denoted as μ_{H1}, and $\sigma_{H1}^{(2)}$, and the average and variance of disease people are denoted as μ_{D1}, and $\sigma_{D1}^{(2)}$. These are evaluated as below.

$$\mu_{H1} = \frac{1}{5} \times (50 + 69 + 93 + 76 + 88) \\ = 75.2 \tag{1}$$

$$\sigma_{H1}^{(2)} = \frac{1}{5} \times \left[(50-75.2)^2 + (69-75.2)^2 + (93-75.2)^2 + (76-75.2)^2 + (88-75.2)^2 \right] \\ = 230.96 \tag{2}$$

$$\mu_{D1} = \frac{1}{5} \times (43+56+38+21+25) \\ = 36.6 \tag{3}$$

$$\sigma_{D1}^{(2)} = \frac{1}{5} \times \left[(43-36.6)^2 + (56-36.6)^2 + (38-36.6)^2 + (21-36.6)^2 + (25-36.6)^2 \right] \\ = 159.44 \tag{4}$$

We assume that the healthy and diseased people data follow normal distributions. The distribution for healthy people is given by

$$f(x_1) = \frac{1}{\sqrt{2\pi\sigma_{H1}^{(2)}}} \exp\left[-\frac{(x_1-\mu_{H1})^2}{2\sigma_{H1}^{(2)}} \right] \tag{5}$$

The distribution for disease people is given by

$$f(x_1) = \frac{1}{\sqrt{2\pi\sigma_{D1}^{(2)}}} \exp\left[-\frac{(x_1-\mu_{D1})^2}{2\sigma_{D1}^{(2)}} \right] \tag{6}$$

Figure 1 shows the above explanation schematically.

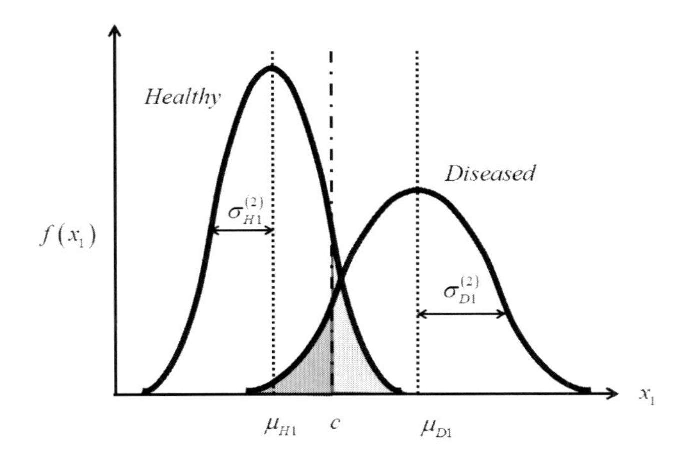

Figure 1. The probability distribution for healthy and diseased people.

When we have a data value of x_1, we want to judge whether the person is healthy or diseased. We define the distance as below.

$$D_H^2 = \frac{(x_1 - \mu_{H1})^2}{\sigma_{H1}^{(2)}} \tag{7}$$

$$D_D^2 = \frac{(x_1 - \mu_{D1})^2}{\sigma_D^{(2)}} \tag{8}$$

We can judge whether the person is healthy or diseased as

$$\begin{cases} D_H^2 < D_D^2 & \rightarrow \quad Healthy \\ D_H^2 > D_D^2 & \rightarrow \quad Disease \end{cases} \tag{9}$$

We evaluate the value of x_1 where the both distance is the same and denote it as c. We then have the relationship of

$$\frac{\mu_{H1} - c}{\sqrt{\sigma_{H1}^{(2)}}} = \frac{c - \mu_{D1}}{\sqrt{\sigma_{D1}^{(2)}}} \tag{10}$$

We then obtain

$$c = \frac{\mu_{H1}\sqrt{\sigma_{H1}^{(2)}} + \mu_{D1}\sqrt{\sigma_{D1}^{(2)}}}{\sqrt{\sigma_{H1}^{(2)}} + \sqrt{\sigma_{D1}^{(2)}}} \tag{11}$$

In this case, we obtain

$$c = \frac{75.2 \times \sqrt{230.96} + 36.6 \times \sqrt{159.44}}{\sqrt{230.96} + \sqrt{159.44}} = 57.68 \tag{12}$$

Summarizing above, we obtain general expressions for the decision as

$$\begin{cases} D_D^2 - D_H^2 > 0 \Rightarrow x_1 < c \rightarrow Helathy \\ D_D^2 - D_H^2 < 0 \Rightarrow x_1 > c \rightarrow Disease \end{cases} \tag{13}$$

3. Discriminant Analysis with Two Variables

We consider two inspections 1 and 2, and groups A and B. We here introduce vectors to extend the analysis for many variables.

Two inspection items are expressed by a vector

$$\mathbf{x} = \begin{pmatrix} x_1 \\ x_2 \end{pmatrix} \tag{14}$$

The vector associated with averages are given by

$$\boldsymbol{\mu}_A = \begin{pmatrix} \mu_{1A} \\ \mu_{2A} \end{pmatrix}, \boldsymbol{\mu}_B = \begin{pmatrix} \mu_{1B} \\ \mu_{2B} \end{pmatrix} \tag{15}$$

$$\Sigma_A = \begin{pmatrix} \sigma_{11A}^{(2)} & \sigma_{12A}^{(2)} \\ \sigma_{21A}^{(2)} & \sigma_{22A}^{(2)} \end{pmatrix}, \Sigma_B = \begin{pmatrix} \sigma_{11B}^{(2)} & \sigma_{12B}^{(2)} \\ \sigma_{21B}^{(2)} & \sigma_{22B}^{(2)} \end{pmatrix} \tag{16}$$

The determinant for Σ_A and Σ_B are given by

$$|\Sigma_A| = \begin{vmatrix} \sigma_{11A}^{(2)} & \sigma_{12A}^{(2)} \\ \sigma_{21A}^{(2)} & \sigma_{22A}^{(2)} \end{vmatrix} = \sigma_{11A}^{(2)}\sigma_{22A}^{(2)} - \sigma_{12A}^{(2)}\sigma_{21A}^{(2)} \tag{17}$$

$$|\Sigma_B| = \begin{vmatrix} \sigma_{11B}^{(2)} & \sigma_{12B}^{(2)} \\ \sigma_{21B}^{(2)} & \sigma_{22B}^{(2)} \end{vmatrix} = \sigma_{11B}^{(2)}\sigma_{22B}^{(2)} - \sigma_{12B}^{(2)}\sigma_{21B}^{(2)} \tag{18}$$

The distribution for two variables for the group A is given by

$$f_A(x_1, x_2) = \frac{1}{(2\pi)^{\frac{2}{2}}\sqrt{|\Sigma_A|}} \exp\left(-\frac{D_A^2}{2}\right) \tag{19}$$

$$D_A^2 = (x_1 - \mu_{1A}, x_2 - \mu_{2A}) \begin{pmatrix} \sigma_A^{11(2)} & \sigma_A^{12(2)} \\ \sigma_A^{21(2)} & \sigma_A^{22(2)} \end{pmatrix} \begin{pmatrix} x_1 - \mu_{1A} \\ x_2 - \mu_{2A} \end{pmatrix} \tag{20}$$

where

$$\begin{pmatrix} \sigma_A^{11(2)} & \sigma_A^{12(2)} \\ \sigma_A^{21(2)} & \sigma_A^{22(2)} \end{pmatrix} = \begin{pmatrix} \sigma_{11A}^{(2)} & \sigma_{12A}^{(2)} \\ \sigma_{21A}^{(2)} & \sigma_{22A}^{(2)} \end{pmatrix} = \Sigma_A^{-1} \tag{21}$$

The Maharanobis' distance associated with the group A is denoted as d_A and is given by

$$d_A = \sqrt{D_A^2} \tag{22}$$

The Maharanobis' distance associated with the group B is denoted as d_B and is given by

$$d_B = \sqrt{D_B^2} \tag{23}$$

where

$$D_B^2 = \left(x_1 - \mu_{1B}, x_2 - \mu_{2B}\right) \begin{pmatrix} \sigma_B^{11(2)} & \sigma_B^{12(2)} \\ \sigma_B^{21(2)} & \sigma_B^{22(2)} \end{pmatrix} \begin{pmatrix} x_1 - \mu_{1B} \\ x_2 - \mu_{2B} \end{pmatrix} \tag{24}$$

We evaluate d_A and d_B, and decide that the data belong to which group.

4. Discriminant Analysis for Multi Variables

We extend the analysis in the previous section to a multi variables case, where we assume p variables.

The data are expressed by a vector having p elements given by

$$\mathbf{x} = \begin{pmatrix} x_1 \\ x_2 \\ \vdots \\ x_p \end{pmatrix} \tag{25}$$

It follows the normal distribution with $N\left(\mu^{[k]}, |\Sigma|\right)$, and is expressed by

$$f\left(x_1, x_2, \cdots, x_p\right) = \frac{1}{(2\pi)^{\frac{p}{2}}\sqrt{|\Sigma|}}\exp\left(-\frac{D^2}{2}\right) \tag{26}$$

where

$$\Sigma = \begin{pmatrix} \sigma_{11}^{(2)} & \sigma_{12}^{(2)} & \cdots & \sigma_{1p}^{(2)} \\ \sigma_{21}^{(2)} & \sigma_{22}^{(2)} & \cdots & \sigma_{2p}^{(2)} \\ \vdots & \vdots & \ddots & \vdots \\ \sigma_{p1}^{(2)} & \sigma_{p2}^{(2)} & \cdots & \sigma_{pp}^{(2)} \end{pmatrix} \tag{27}$$

$$\begin{aligned} D^2 &= \left(x_1 - \mu_1, x_2 - \mu_2, \cdots, x_p - \mu_p\right) \begin{pmatrix} \sigma^{11(2)} & \sigma^{12(2)} & \cdots & \sigma^{1p(2)} \\ \sigma^{21(2)} & \sigma^{22(2)} & \cdots & \sigma^{2p(2)} \\ \vdots & \vdots & \ddots & \vdots \\ \sigma^{p1(2)} & \sigma^{p2(2)} & \cdots & \sigma^{pp(2)} \end{pmatrix} \begin{pmatrix} x_1 - \mu_1 \\ x_2 - \mu_2 \\ \vdots \\ x_p - \mu_p \end{pmatrix} \\ &= \sum_{i=1}^{p}\sum_{j=1}^{p}\left(x_i - \mu_i\right)\left(x_j - \mu_j\right)\sigma^{ij(2)} \end{aligned} \tag{28}$$

We also define a vector of $\boldsymbol{\mu}$ as

$$\boldsymbol{\mu} = \begin{pmatrix} \mu_1 \\ \mu_2 \\ \vdots \\ \mu_p \end{pmatrix} \tag{29}$$

The distance d is given by

$$d = \sqrt{D^2} \tag{30}$$

We can select the group for shorter distance.

SUMMARY

To summarize the results in this chapter—

The distance of the member and a group can be evaluated as

$$d = \sqrt{D^2}$$

where

$$D^2 = \left(x_1 - \mu_1, x_2 - \mu_2, \cdots, x_p - \mu_p\right) \begin{pmatrix} \sigma^{11(2)} & \sigma^{12(2)} & \cdots & \sigma^{1p(2)} \\ \sigma^{21(2)} & \sigma^{22(2)} & \cdots & \sigma^{2p(2)} \\ \vdots & \vdots & \ddots & \vdots \\ \sigma^{p1(2)} & \sigma^{p2(2)} & \cdots & \sigma^{pp(2)} \end{pmatrix} \begin{pmatrix} x_1 - \mu_1 \\ x_2 - \mu_2 \\ \vdots \\ x_p - \mu_p \end{pmatrix}$$

We evaluate the distance with respect to each group, and select the group with smaller distance.

REFERENCES

[1] W. L. Carlson and B. Thorne, *Applied Statistical Methods*, 1997, Presence-Hall, Inc., New Jersey, U.S.A.

[2] R. A. Barnett, M. R. Ziegler, and K. E. Byleen, *College Mathematics for Business, Economics, Life science, and Social sciences 12th edition*, 2011, Pearson Education, Inc., 2011, U. S. A.

[3] G.Maruyama, *Probability and statistics*, 1956, Kyoritu, Japan, in Japanese. 丸山儀四郎、“確率および統計入門”、共立出版、日本、1956.

[4] A. Kobari, *Introduction to probability and statistics*, 1973, Iwanami Shoten, Japan. 永田靖、小針　宏、“確率・統計入門”、岩波書店、東京、1973.

[5] Y. Tanaka and K. Wakimoto, *Multivariate statistical analysis*, 1994, Gendai Sugakusha, Japan, in Japanese. 田中豊、脇本和昌、“多変量統計解析法”、現代数学社、京都、1994.

[6] Y. Nagata and M. Munechika, *Multivariate analysis*, 2007, Science Company, Japan, in Japanese. 永田靖、棟近雅彦、“多変量解析法入門”、サイエンス社、東京、2007.

[7] Y. Wakui and S. Wakui, *Covariance Structural Analysis*, Nihon Jitsugyo Publisher, 2003, Japan, in Japanese. 涌井良幸、涌井貞美、“共分散構造分析”、日本実業出版社、東京、2003.

[8] Y. Wakui, *Bayes Statistics as a Tool*, Nihon Jitsugyo Publisher, 2009, Japan, in Japanese. 涌井良幸、“道具としてのベイズ統計”、日本実業出版社、東京、2009.

[9] H. Cramer, *Mathematical Methods of Statistics*, 1999, Princeton University Press, U. S. A.

[10] D. M. Levine,T. C. Krehbiel, and M. L. Berenson, *Business Statistics*, 2013, Pearson Education Inc., U. S. A.

ABOUT THE AUTHOR

Kunihiro Suzuki, PhD
Fujitsu Limited, Tokyo, Japan
Email: Suzuki.kunihiro@jp.fujitsu.com

Kunihiro Suzuki was born in Aomori, Japan in 1959. He received his BS, MS, and PhD degrees in electronic engineering from Tokyo Institute of Technology, Tokyo, Japan, in 1981, 1983, and 1996, respectively.

He joined Fujitsu Laboratories Ltd., Atsugi, Japan in 1983 and was engaged in design and modeling of high-speed bipolar and MOS transistors. He studied process modeling as a visiting researcher at the Swiss Federal Institute of Technology, Zurich, Switzerland in 1996 and 1997. He moved to Fujitsu Limited, Tokyo, Japan in 2010, where he was engaged in a division that is responsible for supporting sales division. His current interests are statistics and queuing theory for business.

His research covers theory and technology in both semiconductor device and process. To analyze and fabricate high-speed devices, he also organizes a group that includes physicists, mathematicians, process engineers, system engineers, and members for analysis such as SIMS and TEM. The combination of theory and experiment and the aid from various members make his group special to do various original works. His models and experimental data are systematic and valid for wide range conditions and can contribute to academic and practical product fields.

He is the author and co-author of more than 100 refereed papers in journals, more than 50 papers in international technical conference proceedings, and more than 90 papers in domestic technical conference proceedings.

INDEX

A

analysis of variance, v, 277, 285, 290, 305
arithmetic distance, 431, 436, 437
average, ix, 3, 4, 5, 6, 46, 55, 56, 64, 65, 66, 75, 77, 84, 86, 92, 93, 126, 148, 227, 228, 231, 232, 233, 234, 235, 236, 239, 240, 241, 242, 244, 248, 249, 250, 253, 254, 255, 257, 261, 269, 270, 271, 274, 277, 279, 281, 282, 285, 287, 290, 291, 292, 293, 295, 299, 301, 302, 311, 321, 322, 328, 331, 341, 356, 357, 364, 375, 402, 423, 432, 444, 451, 454

B

Bayesian network, 179, 180, 210, 211, 212, 217, 219, 220, 221
Bayesian updating, 179, 187, 191, 197, 208, 223, 224, 227, 228
beta distribution, 223, 229, 230, 269
beta function, 223
binomial distribution, 223, 232

C

chain effect, 431, 438
child node, 179, 210, 212, 213, 217
cluster analysis, v, 431
correlation factor, v, ix, 1, 6, 7, 8, 9, 10, 12, 13, 15, 16, 17, 20, 21, 22, 23, 24, 27, 43, 44, 45, 50, 51, 52, 53, 54, 56, 58, 59, 62, 63, 64, 65, 66, 67, 69, 71, 74, 76, 77, 80, 81, 93, 117, 122, 123, 143, 144, 152, 153, 154, 155, 157, 164, 341, 356, 357, 405, 424
covariance, 1, 2, 4, 5, 6, 10, 15, 17, 75, 133, 150, 169, 173, 176, 364, 367, 368, 369, 370, 375, 379, 380, 381, 382, 402, 405, 408, 410, 411, 461

D

degree of freedom adjusted coefficient of determination, 82, 124
determinant, 18, 20, 79, 82, 93, 121, 122, 123, 124, 138, 142, 144, 157, 162, 453, 457
determination coefficient, 79, 80, 122, 143
Dirichlet distribution, 223, 229, 232, 233, 270
discriminant analysis, v, ix, 453, 457, 458
distance, 93, 126, 147, 163, 351, 355, 358, 359, 431, 432, 433, 434, 435, 436, 437, 438, 439, 443, 447, 450, 451, 453, 456, 458, 459, 460

E

eigenvalue, 339, 344, 346, 347, 351, 353, 369, 370, 381, 382, 410, 423
eigenvector, 339, 346, 347, 349, 350, 370, 382

F

factor analysis, v, ix, 306, 314, 327, 361, 362, 363, 367, 368, 369, 371, 374, 377, 379, 380, 396, 397, 398, 408, 409, 413, 416, 423
factor loading, 354, 361, 363, 369, 375, 381, 409, 421, 422, 424, 426, 428

factor point, 361, 363, 371, 374, 375, 383, 384, 385, 396, 397, 399, 400, 401, 402, 416, 418, 421, 427, 428
forced regression, 77

G

Gamma distribution, 233, 234, 235, 271
Gamma function, 223
Gibbs sampling, 223, 244, 249, 250, 259, 275

H

hierarchical Bayesian theory, ix, 223, 261

I

independent load, 361, 363, 364, 369, 371, 374, 375, 381, 396, 409, 416, 423
inverse Gamma distribution, 223, 239, 273

J

Jacobian, 19, 21, 24, 25, 36, 37

L

latent variable, 169, 170, 171, 172, 175
leverage ratio, 69, 83, 84, 85, 114, 124, 125, 126, 127, 145, 146, 147, 148
likelihood function, 223, 224, 228, 229, 230, 232, 233, 234, 236, 237, 240, 263, 264, 269, 270, 271, 272, 275
logarithmic regression, 69, 95
logistic regression, 100

M

Metropolis-Hastings (MH) algorithm, 223, 250, 251, 252, 256, 257, 258, 259, 261, 275
Monty Hall problem, 185, 187
multicollinearity, 117, 157
multinomial distribution, 223, 232
multiple regression, v, viii, 117, 124, 125, 140, 146, 160, 167, 340, 355, 357, 359

N

normal distribution, 18, 21, 22, 27, 30, 32, 33, 44, 49, 50, 55, 56, 65, 71, 83, 87, 89, 90, 125, 128, 135, 146, 149, 150, 236, 239, 255, 262, 265, 272, 299, 455, 458

O

observed variable, 169, 170, 176, 177
orthogonal table, v, ix, 305, 317, 328

P

parent node, 179, 210
partial correlation, 117, 155, 157, 164
partial correlation factor, 117, 155, 157, 164
path diagram, 70, 169, 170, 171, 176, 177, 340, 361, 362
Poisson distribution, 233
population, v, ix, 4, 8, 23, 24, 53, 62, 66, 70, 80, 81, 82, 86, 117, 118, 128, 141, 149, 181, 223, 224, 229, 233, 268, 274, 424
population ratio, 229, 233
posterior probability, 179, 188, 189, 191, 223, 228, 235, 242, 244, 247, 250, 252, 253, 258, 259, 261, 268
prediction, vii, 17, 50, 51, 52, 54, 55, 64, 86, 117, 127, 149, 179, 227, 282, 284
principal component analysis, v, ix, 339, 340, 347, 350, 351, 355, 359, 361, 362, 367, 368, 379
priori probability, 179, 223, 228, 230, 233, 235, 240, 268, 269, 270
pseudo correlation, 1, 12

R

rank correlation, 8
regression, v, viii, 12, 23, 69, 70, 71, 72, 74, 76, 77, 78, 79, 80, 81, 83, 84, 86, 90, 92, 93, 94, 95, 97, 100, 108, 109, 110, 111, 112, 113, 114, 117, 118, 121, 122, 123, 124, 125, 127, 135, 137, 138, 139, 140, 141, 142, 144, 145, 146, 149, 150, 151, 155, 157, 158, 159, 160, 161, 162, 163, 164, 165, 167, 339, 340, 355, 356, 357, 359

regression line, 69, 72, 74, 76, 77, 79, 80, 81, 83, 86, 90, 93, 100, 111, 118, 122, 127, 135, 141, 149, 150, 163
rotation matrix, 361, 380, 384, 413

S

sample correlation factor, 6, 8, 23, 24, 27, 43, 44, 45, 51, 52, 53, 62, 63, 66, 81, 93
standard deviation, 56, 231, 235, 240, 261, 266, 351
standard normal distribution, 18, 27, 30, 32, 33, 49, 56, 83, 89, 125, 146
structural equation modeling, v, viii, 169, 170
studentized range distribution, 277, 283, 299

T

three prisoners, 184

V

Valimax method, 361, 385, 395

W

Ward method, 431, 438, 441, 444, 452
weighted distance, 438

Quaternion Matrix Computations

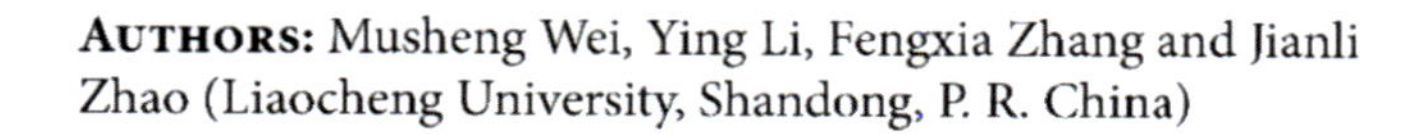

Authors: Musheng Wei, Ying Li, Fengxia Zhang and Jianli Zhao (Liaocheng University, Shandong, P. R. China)

Series: Mathematics Research Developments

Book Description: In this monograph, the authors describe state-of-the-art real structure-preserving algorithms for quaternion matrix computations, especially the LU, the Cholesky, the QR and the singular value decomposition of quaternion matrices, direct and iterative methods for solving quaternion linear systems, generalized least squares problems, and quaternion right eigenvalue problems.

Hardcover ISBN: 978-1-53614-121-4
Retail Price: $160

Simulated Annealing: Introduction, Applications and Theory

Editors: Alex Scollen and Thomas Hargraves

Series: Mathematics Research Developments

Book Description: The opening chapter of this book aims to present and analyze the application of the simulated annealing algorithm in solving parameter optimization problems of various manufacturing processes.

Hardcover ISBN: 978-1-53613-674-6
Retail Price: $195